Climate Change and the Kyoto Protocol

Climate Change and the Kyoto Protocol

The Role of Institutions and Instruments to Control Global Change

Edited by

Michael Faure
Maastricht University, The Netherlands

Joyeeta Gupta
Free University Amsterdam, The Netherlands

Andries Nentjes
University of Groningen, The Netherlands

Edward Elgar
Cheltenham, UK • Northampton MA, USA

© Michael Faure, Joyeeta Gupta, Andries Nentjes 2003

All rights reserved. No part of this publication may be reproduced, stored in a retrieval system or transmitted in any form or by any means, electronic, mechanical or photocopying, recording, or otherwise without the prior permission of the publisher.

Published by
Edward Elgar Publishing Limited
Glensanda House
Montpellier Parade
Cheltenham
Glos GL50 1UA
UK

Edward Elgar Publishing, Inc.
136 West Street
Suite 202
Northampton
Massachusetts 01060
USA

A catalogue record for this book
is available from the British Library

Climate change and the Kyoto protocol : the role of institutions and instruments to control global change / edited by Michael Faure, Joyeeta Gupta, Andries Nentjes.
    p. cm
  Includes bibliographical references and index.
  1. Climatic changes—Government policy. 2. Global environmental change. 3. Emissions trading. I. Faure, Michael (Michael G.) II. Gupta, Joyeeta, 1964– III. Nentjes, A.

QC981.8.C5C5134 2003
363.738'74526–dc21
                                    2002041380

ISBN 1 84376 245 5

Printed and bound in Great Britain by MPG Books Ltd, Bodmin, Cornwall

Contents

List of figures	vii
List of tables	viii
List of contributors	x

PART I KYOTO

1. Key instrumental and institutional design issues in climate change policy 3
   Michael Faure, Joyeeta Gupta and Andries Nentjes
2. The Kyoto mechanisms and the economics of their design 25
   Luke Brander
3. Alternative design options for emissions trading: a survey and assessment of the literature 45
   Jan-Tjeerd Boom and Andries Nentjes
4. To design and implement climate change measures and the need to strike a balance between environmental protection and international trade law 68
   David Grimeaud
5. Developing carbon trading in Europe: does grandfathering distort competition and lead to state aid? 108
   Edwin Woerdman
6. Legal aspects of the Dutch approach to CO_2 reduction 128
   Chris Backes and Reinske Teuben
7. Legal feasibility of emissions trading: learning points from emissions trading for ozone-depleting substances 147
   Marjan Peeters
8. CDM in climate policies in the Netherlands: a promising tool? 171
   Rianne de Leeuw and Ekko C. van Ierland
9. Optimal institutional arrangements and instruments for the promotion of energy from renewable sources 195
   Jan C. Bongaerts and George Dogbe
10. Domestic capacity, regional institution and global negotiations: lessons from the Netherlands–EU Kyoto Protocol negotiation 230
    Norichika Kanie

11 Global environmental change regimes: impact assessment on
    the basis of an extended GTAP model   248
    Shunli Wang, Peter Nijkamp and Onno Kuik

PART II AFTER KYOTO

12 The multi-sector convergence approach to global burden
    sharing of greenhouse gas reductions   279
    Jos J.C. Bruggink
13 The Dutch energy transition and its institutional problems:
    report from a stakeholder assessment   292
    Matthijs Hisschemöller
14 Modulating dynamics in transport for climate protection   312
    René Kemp and Ellen Moors
15 Institutional change in Europe and the implications for climate
    control measures   340
    Graham Bennett

Index   351

Figures

1.1	Per capita emission allowances	15
2.1	Supply and demand for emission reductions from the Kyoto mechanisms	27
6.1	Tradable reductions	137
6.2	Tradable reductions and BAT	143
8.1	Exclusion, selection and contract-award criteria in the ERUPT tender procedure	174
8.2	The former position of CDM under the Minister of Development Cooperation	175
8.3	Project emissions and four baselines	184
9.1	EU15 energy imports, 1980–1999	197
9.2	Share of renewables in the total energy balance, EU15, 1980–1999	198
9.3	EU renewable energy, production by member states	200
9.4	Share of renewable energy production by technology, EU15, 1997	202
9.5	Electricity consumption as a share of total energy consumption, EU15	203
9.6	Share of hydro in renewable electricity production, EU15	205
10.1	Three modes of leadership in international multilateral negotiation	231
11.1	First level of production structure for output of industry j in region r in GTAP-E	253
11.2	The value-added nest and composite intermediates nest in GTAP-E	254
11.3	The production of energy in GTAP-E	255
11.4	Impact of carbon tax as a result of CDM	265
11.5	Decomposition of carbon leakage for ROW by regimes USA_P and USA_P+CDM	266
11.6	Relative magnitude of CDM effect with regard to US participation regimes	269
11.7	Relative magnitude of CDM effect with regard to US non-participation case	270
12.1	Per capita total emission allowances	288
14.1	Types of policies involved in managing the transition to sustainable transport	335

Tables

1.1	Conflicting views on transition towards a decarbonized economy	17
2.1	Use of the Kyoto mechanisms and domestic action in meeting the estimated Annex I reduction requirement, 2010	29
2.2	Kyoto mechanisms and domestic action as a percentage of the Annex I reduction requirement, 2010	29
2.3	US and rest of Annex I reduction requirements, 2010	40
8.1	Results of calculations of baselines and emission reductions for the Philippines case study	183
8.2	Results of calculations of baselines and emission reductions for the Egypt case study	186
8.3	Total costs and emission reductions for scenarios 2 and 3 compared to baseline scenario 1	189
9.1	EU15 primary energy consumption and production, 1980–1999	196
9.2	Share of renewable energy in inland energy consumption	199
9.3	Capacities of renewable energy installed in gross MW, 1997	201
9.4	Share of hydro in inland renewable electricity generation	204
9.5	Total renewable inland electricity generation	206
9.6	Comparative cost of renewable and other electricity, selected IEA countries	207
9.7	Schemes for electricity from RES in the EU member states	222
10.1	Greenhouse gas reduction target in 2010 relative to 1990	238
11.1	Regime possibilities under analysis	257
11.2	CO_2 emission content in the GTAP-E model	258
11.3	Carbon emission target and carbon tax for USA_P and USA_NP	259
11.4	Indicators for international competitiveness for USA_P and USA_NP	261
11.5	Emission reductions and carbon tax for CDM regimes	264
11.6	Macroeconomic indicators for CDM regimes	268
11.7	Emission reductions, carbon tax and macroeconomic indicators for different assumptions on baseline calculations under USA_P+CDM	272
12.1	Input data for MSC burden-sharing rule	288
13.1	Findings for the four sectors in 2050	302

13.2 Views on institutions needed for realizing the transition towards −80 per cent greenhouse gas emissions by 2050   304

Contributors

Chris Backes Professor of environmental law, Centre of Environmental Law and Policy (CELP)/NILOS, Utrecht University, The Netherlands.

Graham Bennett Director of Syzygy in Nijmegen, The Netherlands.

Jan C. Bongaerts Professor at the Technische Universität-Bergakademie Freiberg, Germany.

Jan-Tjeerd Boom Unit of Economics, The Royal Veterinary and Agricultural University, Frederiksberg, Denmark.

Luke Brander Researcher, Institute for Environmental Studies (IVM), Free University of Amsterdam, The Netherlands.

Jos J.C. Bruggink Policy Unit, Energy Research Centre, The Netherlands.

Rianne (G.J.) de Leeuw CSTM, Cartesius Institute, University of Twente, The Netherlands.

George Dogbe Research Fellow, Centre for Advanced Mineral and Energy Research, University of Alberta, Canada.

Michael Faure Professor of comparative and environmental law, Academic Director of METRO, Maastricht University, The Netherlands.

David Grimeaud Research Fellow, METRO, Maastricht University, The Netherlands.

Joyeeta Gupta Institute for Environmental Studies (IVM), Free University of Amsterdam, The Netherlands.

Matthijs Hisschemöller Senior Researcher, Institute for Environmental Studies (IVM), Free University of Amsterdam, The Netherlands.

Norichika Kanie Associate Professor, Graduate School of Decision Science and Technology, Tokyo Institute of Technology, Japan.

Contributors

René Kemp Senior Research Fellow, MERIT, Maastricht University, The Netherlands.

Onno Kuik Institute for Environmental Studies (IVM), Free University of Amsterdam, The Netherlands.

Ellen Moors Assistant Professor, Department of Innovation Studies, University of Utrecht, The Netherlands.

Andries Nentjes Professor of economics, Department of Economics and Public Finance, Faculty of Law, University of Groningen, The Netherlands.

Peter Nijkamp Professor of economics, Department of Spatial Economics, Free University of Amsterdam, The Netherlands.

Marjan Peeters Senior Researcher, METRO, Maastricht University, The Netherlands.

Reinske Teuben Centre of Environmental Law and Policy (CELP)/NILOS, Utrecht University, The Netherlands.

Ekko C. van Ierland Environmental Economics and Natural Resources Group, Wageningen University, The Netherlands.

Shunli Wang Department of Spatial Economics, Free University of Amsterdam, The Netherlands.

Edwin Woerdman Post Doctoral Research Fellow, Faculty of Law, University of Groningen, The Netherlands.

PART I

Kyoto

1. Key instrumental and institutional design issues in climate change policy

Michael Faure, Joyeeta Gupta and Andries Nentjes

1 GOAL OF THIS BOOK

Institutions and instruments to control global environmental change have evolved over the past decade; in particular we see the beginnings of an innovative governance structure to mitigate climate change. The United Nations Framework Convention on Climate Change (FCCC) adopted in 1992 provided a framework for policy making to deal with climate change. In 1997 the Kyoto Protocol (KP) to the Climate Convention was adopted which highlighted commitments for all countries, including quantitative commitments for the developed countries. It also provided for a range of so-called flexible instruments to help promote the implementation of the quantitative commitments of the developed countries and the more qualitative commitments of the developing countries. Although the adoption of the Kyoto Protocol was seen as a major legal achievement, the enthusiasm of the participating countries waned in the following period. There were clear signals that the US would be unwilling to ratify and this was finally confirmed with the withdrawal of the US from the negotiations in 2001. Other developed countries were increasingly reluctant to ratify in the absence of the US, but are trying to rally forces to ensure that the regime does not completely break down. Difficulties encountered in the aftermath of the Kyoto Protocol are partly of a political nature, but can also be attributed to the incomplete understanding of what exactly has been agreed in the Protocol. Although the Marrakesh Accords of 2001 provide more detail, there remain a large number of challenges regarding the implications of the Protocol for individual countries. We believe that the social sciences can help to better understand such problems and perhaps even suggest more effective approaches to solve them.

Usually researchers communicate their scientific work only with colleagues from within their own discipline. A common language and a shared discipline in formulating research questions and methods to answer these research questions has yet to be developed because of the constraints in the scientific world. The price the scientific community pays for these habits is fragmentation of the knowledge that is needed to tackle complex policy questions.

Research on climate change mitigation is hardly an exception to this observation, although the Intergovernmental Panel on Climate Change (IPCC) has done useful work in assessing and collating the state of the art in various disciplines. Another relevant initiative is the International Human Dimensions Programme of Global Environmental Change (IHDP) which offers a framework for research on economic, social and cultural processes and their relation with the physical environment. A conference under the auspices of the national HDP project, Institutional Dimensions of Global Environmental Change on 21-22 June 2001, brought together researchers from the various social sciences: law, political science, sociology and economics to present and discuss their scientific work that although diversified in perspective is united by one and the same subject: the role of institutions and instruments to control global change. To get more focus and coherence the contributions concentrate on climate change. They give a kaleidoscopic picture of the state of the art in social science research on institutions and instruments to mitigate climate change, ranging from the global and regional levels, in particular the European Union (EU), through to the national level, with institutions and instruments in the Netherlands as case studies.

The Kyoto Protocol of 1997 is and will stay a milestone in the process of ensuring that climate change remains on the political agenda and promoting internationally coordinated action. The first part of this book presents discussions and analyses of the Kyoto Protocol, mainly focusing on the design and functioning of the flexibility mechanisms. The second part of the book looks behind Kyoto: what type of institutional change is necessary to realize the transition to a carbonless society in the twenty-first century and what might the future have in store? This chapter presents the key issues that are discussed in this book and an integrated analysis of the conclusions of different authors.

2 THE KYOTO PROTOCOL

The Relative Advantage of Different Instruments

The Kyoto Protocol defines three instruments by which the developed countries listed in Annex I of the Convention (Annex I countries) can obtain part of their greenhouse gas (GHG) emission reductions from non-domestic sources: emissions trading, joint implementation (JI) and the clean development mechanism (CDM). Brander (Chapter 2, this book) discusses their most important, as yet unresolved design aspects, such as supplementarity, hot air, adaptation tax, the inclusion of sinks, crediting periods, compliance rules and liability provisions. The chapter discusses the potential effects of current proposals on the relative use of the Kyoto mechanisms in achieving the Kyoto commitments. A comparison of quantitative estimates of four recent studies shows that the Annex I emissions reduction requirement might be realized by domestic action amounting to 30 per cent, emissions trading and JI 30 per cent and CDM 40 per cent at least, assuming that there are no ceilings on using the flexibility mechanisms. Non-ratification of the Protocol by the US will reduce the price of tradable CO_2 equivalents, thus reducing the compliance cost of ratifying parties.

Alternative Design Options for Emissions Trading

International emissions trading and joint implementation are the instruments that allow the transfer of GHG emissions between private parties in countries of the Annex I category. Boom and Nentjes (Chapter 3, this book) argue that the adequate functioning of the two international flexibility mechanisms depends on how well the national instruments are designed and implemented. Two blueprints for national flexibility in controlling CO_2 emissions are discussed: the cap-and-trade scheme and the credit-trading scheme. The blueprints are schemes that include all sectors of the economy from energy-intensive industry through to heating, electricity consumption and car use by households. In the cap-and-trade scheme, emission allowances are grandfathered in specified quantities to participants (that is, are allocated to specific participants on the basis of current emission levels). In the credit-trading scheme, emissions are allocated on the basis of performance standards. Both schemes combine downstream distribution of allowances to energy users with upstream monitoring of compliance as distinctive features. International emissions trading between private parties can develop by allowing exchange of emission allowances between emission sources in different, but similarly organized national cap-and-trade schemes.

The chapter proposes that in order to foster environmental effectiveness only Annex I parties with national cap-and-trade schemes that meet well-defined criteria in terms of monitoring emissions, allowance registration and enforcement should be allowed to participate in the select club of Annex I parties engaged in international permit trading between private parties.

In terms of effectiveness and efficiency, the cap-and-trade variety is a first-best solution; nationally as well as internationally. Credit trading requires frequent adjustment of the performance standards and is administratively a bit more complicated than cap and trade; it also gives weak incentives to reduce fossil fuel intensity per unit of output. Therefore national and international credit trading are a second-best option, but they will meet less resistance from interest groups.

International Trade Law and Emissions Trading

Although economists may devise their blueprints of (effective and efficient) flexibility mechanisms, it is the actual discussions on shaping the instrument in successive meetings of the Conference of the Parties (COPs) to the FCCC that determine how these instruments take shape, as Grimeaud (Chapter 4, this book) makes clear. The Marrakesh Accords confirm that legal entities (private parties) may participate in international emissions trading; still leaving it to national authorities to determine under which conditions nationally regulated entities will participate in such a regime. Emissions trade between Annex I parties and transfer of parts of assigned amounts (allocated emissions to Annex I countries) will be allowed only in so far as certain eligibility criteria have been fulfilled, such as adequate monitoring and reporting.

The Kyoto Protocol specifies that achievement of the objectives on climate change shall not result in violation of international trade law. Grimeaud identifies potential international trade disputes that may arise if the US does not ratify the KP and the EU does. Here climate policies and measures of the latter may restrict the ability of the former to sell energy-related and GHG-emitting products to the latter. Even trade between ratifying Annex I parties might be restricted by differences in national technical, fiscal and emission trading rules that do not meet the World Trade Organization (WTO) requirement of non-discrimination and where it cannot be shown that no alternative measures, less restrictive to international trade, exist.

Since emission allowances cannot be considered as goods or services under the WTO, Annex I parties will be able to adopt rules on national emissions trading that limit the export or import by domestic regulated companies of allowances and that prohibit the import or export of allowances coming from specific Annex I parties who would not comply with their eligibility requirements. Yet, on the other hand, it may well be that such national rules on emissions trading would indirectly affect the international trade in energy and energy-related products.

Emissions Trading, Competition Distortion and State Aid

The question whether the introduction of an emission-trading scheme could create trade restrictions also haunts the EU where, in particular, it relates to trade between member states. A potential source of conflicts is differences between the domestic permit allocation procedures of the member states where one of them might hand out permits for free (grandfathering) and the others auction the allowances. Could such differences distort competition and be conceived as state aid in a European emission trading market? Woerdman

(Chapter 5, this book) argues that that depends on whether one takes an efficiency or an equity perspective. He argues that there is no problem from an efficiency perspective. However, the competitive distortion and state aid issues are relevant from an equity perspective.

From an efficiency perspective, firms receiving permits for free do not have a cost advantage over firms in other member states that auction the permits, because the former firms have to include the opportunity cost of using the permits to cover the firm's emission in the product price. This means that grandfathered firms are not advantaged, so that there is no state aid. In this view, there is also no need to harmonize permit allocation procedures.

However, grandfathered permits are a capital gift to the firm, inducing a windfall profit, so that an identical firm abroad which has to buy its permits has a higher cash outflow and hence fewer financial resources. Therefore, from an equity perspective, competition (or: the level playing field) is distorted and state aid occurs because the mere allocation of permits leads to unequal changes of the financial positions and competitive relations between firms across the EU. *Ceteris paribus*, a grandfathered firm in one member state is then advantaged because it has more financial resources than its auctioned competitor in another one. In this view it is also desirable, or even necessary, to harmonize permit allocation procedures.

The European Community guidelines on state aid for environmental protection, revised in 2001, place a stronger emphasis on cost internationalization than the previous guidelines of 1994. This seems to support grandfathering because of its opportunity costs. However, the provisions for investment and operating aid suggest that firms may receive no more than a certain percentage (for instance 50 per cent) of their permits for free during a limited period of time (for instance five years) as a transition phase towards an auctioned scheme.

In its decision of April 2000 on carbon trading in Denmark, the European Commission considered grandfathering to be state aid, but nevertheless exempted it by using both economic, legal and political arguments. Although it mentioned neither the opportunity cost argument nor the desire for a level playing field, grandfathering was interpreted as a wealth transfer that could affect the equal treatment of firms. This sets a political (albeit not legal) precedent in the EU to interpret grandfathering in terms of fairness.

In the context of economic instruments for environmental policy, the Commission has to decide whether permit allocation differences among member states are compatible with the rules (and exemptions) on state aid. If equity considerations play a role in this decision, the issues of competitive distortion and state aid become relevant in developing a European carbon trading market.

Trading of Ozone-Depleting Chemicals

The US has been the pioneer in establishing emission-trading schemes. The only European Union programme in place is the transferable production quota of ozone-depleting substances (ODSs). As a party to the Vienna Convention on the Ozone Layer (1985) and the Montreal Protocol on Substances that Deplete the Ozone Layer (1987) the European Union cooperates with the other parties in a scheme of phasing out the production of ODSs. Production quotas have been grandfathered to established producers in proportion to ODS output in a reference year. The quota can be transferred among producers. Peeters (Chapter 7, this book) discusses and assesses the legal aspects of the transfer provisions in the European ODS regulations including the discretion for administrative intervention with trade in products containing ozone-depleting substances and the compliance provisions.

Implementing Kyoto in the Netherlands

The flexibility provisions in the Kyoto Protocol and in particular Article 17 on emissions trading, which was inserted in the Protocol despite strong resistance from the European Union, has subsequently stimulated several EU member states to consider options for increasing flexibility in national implementation. The European Commission played its role in the process by publishing a Green Paper on emissions trading for energy-intensive industries in April 2000. In the Netherlands the efforts to implement the recommendations of the Rio de Janeiro Conference of 1998 have been rather disastrous. Instead of stabilizing the GHG emissions in the year 2000 at 1990 levels, actual emissions were considerably higher. More radical efforts and instruments might be needed to avoid a repetition of the same failure to achieve the Kyoto commitments in the first decade of the twenty-first century. Transferability of emissions, or emission reductions, will reduce the total cost of emission control. Such options which help to keep costs low are politically relevant since they mitigate against resistance which might arise in the future when more stringent emission targets for a next commitment period have to be agreed upon. Backes and Teuben (Chapter 6, this book) discuss the recent Dutch proposal for domestic NO_x emission trading for energy-intensive industry. It is basically a scheme of uniform performance standards with the flexibility to sell emission credits to a firm, which emits less than the standard allows. The potential purchasers are firms facing difficulties in meeting the new stringent standards. There are signs that the NO_x credit-trading scheme may set the precedent that will be followed in the design of CO_2 emissions trading for Dutch energy-intensive industry.

An observer without legal training might think that adding credit trading to a scheme of performance standards that is already in place is a minor adjustment, fitting within the existing framework of environmental legislation. However, the legal interpretation tells another story, as Backes and Teuben show. The European IPPC directive, elaborated in the (Dutch) ALARA (as low as reasonably achievable) and European BAT (best available techniques) principles are incompatible with emissions trading.

ALARA has been interpreted as representing a minimum as well as maximum standard. Requirements based on the ALARA principle cannot be complied with by buying emissions. Moving pollution from one location to another is in conflict with the present location-based environmental licence system in which the individual installation is the object of environmental regulation.

Unlike ALARA, the BAT principle embedded in EU legislation only represents a minimum requirement: more stringent (national) requirements are allowed. However, unconstrained emissions trading would allow a single enterprise to emit more than the BAT-based standards of its environmental licence permit by buying up emission allowances in the market.

CDM in the Netherlands

Another promising option for promoting the implementation of climate policies is the clean development mechanism. De Leeuw and van Ierland (Chapter 8, this book) point out that to realize CDM's potential, many complications related to the practical application of the instrument have to be overcome. They assess the experience in the Netherlands, where the government has broken new ground as a donor of CDM projects by auctioning its CDM subsidies. In order to identify the practical bottlenecks, four CDM project cases are discussed in depth. The case studies reveal that the main problems are related to the criteria of additionality (that is, the project leads to additional emission reductions that could not otherwise have been achieved) and sustainable development. Another important topic is the monitoring of the projects and the danger of carbon leakage if the monitoring takes place only at project level. Since developing countries do not have national quantitative assigned amounts, nor are they required to present a complete industry/project-wise breakdown of GHG emissions in their national inventory, it will be extremely difficult to assess the net impacts of a CDM project.

Thus, many difficulties will have to be overcome before the government of the Netherlands can implement CDM while adhering to its own criteria, which require that emission reductions achieved abroad through CDM must be more efficient than those achieved domestically, of high quality and that the emission reduction technologies contribute to sustainable development. The combination of efficiency and high-quality emission reductions might prove to be conflicting. It will require either a political choice to favour one criterion over the other, or a very specific and restrictive set of conditions. However, measures to enhance the quality of emission reductions generated through CDM will almost inevitably increase transaction costs, raise the price of emission reductions generated through CDM and diminish its significance as an instrument of climate policy.

Promoting the Use of Renewable Energy Sources

Among the measures to mitigate climate change a more intensive exploitation of renewable energy sources (RES) is an option. Bongaerts and Dogbe (Chapter 9, this book) analyse the position of electricity generation from RES in the European Union. Although the EU can be seen as an important 'player' in international policy making on climate change, one has to take into account that with respect to promotion of electricity from RES, it has little or no powers to achieve its ambitious target of doubling the share of RES in total energy production from 6 to 12 per cent in 2010. Any real contribution will have to come from the actions of the member states, which is why studying their policies is of interest.

The instruments that have been applied aim at demand and supply management. Demand management creates a market for electricity from RES by a statutory obligation for certain groups of customers to purchase minimal quantities of 'green' electricity. A number of countries have introduced flexibility or have planned to do so by allowing transfer of (trade in) such purchase quotas. Feed-in tariffs, obliging grid operators to buy electricity from RES at a regulated minimum price are also widely used. Supply management instruments are mainly of the fiscal type: lower taxes, tax rebates and subsidies. The authors argue that demand management instruments are probably more effective than supply management instruments. Examples such as 'green electricity' certificates – sometimes in combination with renewable obligations – or Ireland's Alternative Energy Requirement (AER) illustrate the point. While supply-side instruments do not have predictable and specific effects on supply, demand management instruments can meet the objectives almost by definition. Moreover, supply-side instruments, requiring fiscal measures, typically depend on public budgets, which are always limited.

Leadership in Multilateral Negotiations

There are different theories about how countries can show leadership in international regime formation. For countries that do not have structural power, that is, are middle powers, leadership can be shown through unilateral action (directional leadership) or instrumental leadership. In order to maximize the potential impact of such leadership, middle powers can use regional organizations. Kanie (Chapter 10, this book) tackles this hypothesis by examining how the Netherlands, a small country, has been able to play a leading role in the negotiations leading to the Kyoto Protocol, through the European Union acting as an intermediating regional institution.

In the 1990s it became clear that in order to take its aspired leadership role in the FCCC process leading up to the adoption of the Kyoto Protocol the EU would first have to develop a coherent position. To be credible this needed an internal agreement about burden sharing (that is, sharing of the collective EU commitment) within the EU. Since earlier discussions ended in disagreement when the presidency of the European Council was taken over by the Netherlands in 1997, it was clear that the key to a leadership policy at EU level called for a convincing approach to burden sharing backed up by scientific rigour. The Triptych approach was thus devised to calculate intermediate emission reduction targets for countries, taking into account the sectoral differences in emission reduction potential between countries and served as a basis for identifying country-specific commitments.

In January 1997 the Dutch presidency of the EU sent a proposal based on the Triptych approach to the member states aiming at 15 per cent reduction for the EU relative to 1990. Since the MS representatives could agree on no more than internal burden sharing of a 10 per cent EU emission reduction it was decided that in order not to undermine the EU's ability to take a lead in the global negotiation the EU would propose a 15 per cent reduction target. If the Kyoto agreement were to exceed a reduction of 10 per cent, then the remaining 5 per cent would have to be renegotiated after the Kyoto Conference. Thus, the Member States agreed that 10 per cent would be enough until Kyoto and so avoided acrimonious discussion about the division of the remaining 5 per cent.

In the summer and autumn of 1997 it became apparent that the EU proposal contained the most ambitious target for the reduction of GHGs. The G77 (the group of developing countries) expressed support for the EU proposal later on followed by a group of influential environmental non-governmental organizations (ENGOs). In Kyoto the EU Troika, consisting of the Netherlands, Luxembourg and the UK, negotiated on behalf of the EU. The chief negotiator of the Netherlands was Dr Bert Metz, the main architect of the Triptych approach and the main Dutch negotiator since COP1. After considerable negotiation, 38 industrialized countries agreed to reduce their combined GHG emissions by 5 per cent relative to 1990 levels between 2008 and 2012. This included differentiated reduction targets of 6 per cent for Japan, 7 per cent for the USA and 8 per cent for the EU.

This case shows how a regional organization can be an important device when a middle-power country wants to exert greater influence in global negotiations. Three factors made the Dutch case possible. First, the intellectual as well as instrumental (diplomatic) capacity at the national (domestic) level made it possible for the Netherlands to make use of the EU as a device to extend its potential. Without domestic capacity, a regional organization cannot be used as a device that makes its diplomatic influence higher. Second, the coherence of the EU as a coalition made it possible to be a recognizable important power bloc in the global negotiation process. (Of course, the economic scale was another important factor that made the EU one of the most important parties in the final stage of the negotiation.) A third important factor is the institutional design of the EU. Because the EU presidency can exercise a fairly large degree of political power both internally and externally and because the EU Troika can negotiate externally on behalf of the EU, the climate change political/diplomatic potential of the Netherlands could be readily transferred to the EU negotiations.

The Influence of Climate Change on Trade

The general economic idea behind the call for reducing climate change is that the voluntarily agreed commitment to reduce the emission of greenhouse gases by Annex I countries will lead to a Pareto-optimal level of welfare. The starting point is therefore the basic economic insight that internalizing externalities leads to efficiency. However, political practice shows that there are many difficulties, for example, in ratifying and implementing the Kyoto Protocol, because although there is much to gain for everyone, there are inevitably winners and losers.

One issue is how the various policy mechanisms contained in the Kyoto Protocol (emissions trading, joint implementation and clean development mechanism) may affect international trade. That is the topic of the chapter by Wang, Nijkamp and Kuik (Chapter 11, this book). They have selected from the set of available general equilibrium models the so-called 'GTAP model', which allows the measurement of the impact of possible future regimes to achieve emission reductions, in comparison with a business-as-usual scenario.

They come to various interesting results concerning, for example, the implementation of emission reductions with or without CDM and they look at the impact of various scenarios including the US or not. The result is, not surprisingly, that at the global level, the business-as-usual scenario will not be optimal. The implementation of, for example, CDM activities actually improves welfare and is thus not costly. However, the problem remains that it may be more advantageous for one single country not to participate. That is obviously the well-known free-rider problem. It will be the major challenge of future (re)negotiations of the Kyoto Protocol to see how this can be remedied. The authors apply this model to a situation where they assume that the US would not go along with the emission reduction measures and conclude that it may indeed be beneficial for a single country like the US not to engage in emission reduction measures. However, the actual emission reductions that will then be achieved at world level, will be lower in this US stand-alone case. Thus, applying this GTAP model to climate change issues, illustrates some of the trade effects of the implementation of various instruments and the effects of various political scenarios (like the US joining or not).

3 AFTER KYOTO

The Multi-sector Convergence Approach

The UN Framework Convention on Climate Change recognizes the problems of burden sharing by noting that countries have common but differentiated responsibilities and respective capabilities in dealing with the problem. Accordingly, the Kyoto Protocol provides emission-related targets for the developed countries (Annex I signatories). These include targets which allow some countries actually to increase their emissions. The second round involving further quantitative commitments from developed countries and for certain categories of developing countries is likely to be much more complex. The vast differences between the average developing and developed country with respect to historical cumulative emissions, present income levels and future adverse impacts lead to difficult ethical questions. In Chapter 12, Bruggink outlines the role of principles of distributive justice in establishing global burden-sharing rules. He summarizes lessons learned from international negotiations among Annex I countries so far and presents a new, pragmatic approach to global burden sharing. Global burden sharing can illustrate the consequences of normative choices quantitatively and thus provide a road map towards common ground in the dialogue on global participation.

Climate change actions involve costs and benefits that are distributed across nations in different and uncertain ways. Presumably, international negotiations on the distribution of costs and benefits are based not only on pure self-interest, but also on principles of distributive justice. There are, however, many principles of distributive justice that can be considered for deriving burden-sharing formulas in the case of climate change actions. Four commonly used criteria are:

- Historical contribution to the problem: countries that are the cause of the problem should pay to solve it; often referred to as the guilt principle or the polluter-pays principle.
- Ability to pay for the solution: countries that can afford the economic

burden should shoulder it; often called the capacity-to-pay principle.
- Equality of rights: everybody has the right to an equal share of allowable greenhouse gas emissions. The rich should pay the poor if they wish to use more than their fair share of the global commons (equality principle). This principle is often used as a distant target to strive for and is prominent in burden-sharing rules based on the contraction-and-convergence approach.
- Historical claims, or the principle of grandfathering, that present levels of emissions constitute a fair initial distribution of emission rights. This is often invoked as a convenient starting point for assigning emission rights, but is not very appealing from an ethical perspective.

That a formal burden-sharing rule can form an excellent vehicle for persuasion and communication has been clearly demonstrated in the case of differentiation of the European Kyoto target among member states. The differentiated emission reduction targets for EU member states have been calculated using the so-called 'Triptych model'. The model is based on a sectoral approach to target setting, which allows the application of different principles for each sector. The domestic sector is subject to the equality principle of distributive justice (per capita convergence). The principle for the energy-intensive industry is based on a norm for efficiency improvement (historical rights). The principle for power generation is based on the national generation mix (not actually an equity argument, but based on considerations of acceptable marginal costs of abatement). Reduction targets were based on baseline reference scenarios that allowed for higher growth in lower-income countries. The model judiciously combined elements of equity principles with realistic considerations regarding cost effectiveness and carbon leakage.

The multi-sector convergence approach has been developed in a joint study by CICERO (Centre for International Climate and Environmental Research, Oslo) and ECN (Energy Research Centre of the Netherlands, Petten) specifically for global burden sharing where the adoption of quantitative commitments by non-Annex I countries forms the key issue. Its main features are based on a mix of general principles of fairness, sectoral target levels and country-specific allowances. It combines the ethical elements of the contraction-and-convergence approach with the sectoral detailing of the Triptych approach. Initial short-term commitments are based on expected greenhouse gas emissions by sector in 2010. Ultimate long-term commitments are based on equal per capita emission norms per sector in 2100. This amounts to a gradual transition from grandfathering to equality on the basis of sectoral emission rights. These rules are similar to the well-known set-up of the contraction-and-convergence approach; but with reduction rates not determined on the national level, but on the sectoral level similar to the Triptych approach.

A numerical case study illustrates the basic features of the multi-sector convergence approach. The distribution of emission trajectories it generates for major emitters are shown in Figure 1.1. At each point in time countries are divided into two groups – those with national per capita emissions above the global average and those with national per capita emissions below this average. For countries with national per capita emissions above the global average for the corresponding year a geometric convergence from their 2010 sectoral per capita emissions to the final per capita sectoral allowance in 2100 is assumed. For the second group, a specific annual growth rate of emissions (3 per cent) is assumed until they have reached the global average. Then they are allowed an adjustment period of 15 years after which they are subject to the convergence conditions of the first group of countries. Figure 1.1 presents the calculated emission trajectories for four major global emitters: the US, the EU15, India and China.

The evolution from grandfathering to equalization over the greater part of a century in the multi-sectoral convergence approach is intended to provide a road map with a persuasive function in horizontal negotiations between nations at very different levels of economic development. On the other hand the sectoral orientation will be important for vertical negotiations within nations between governments and sectoral interest groups. Combining these two features in a hybrid approach will provide international negotiators with

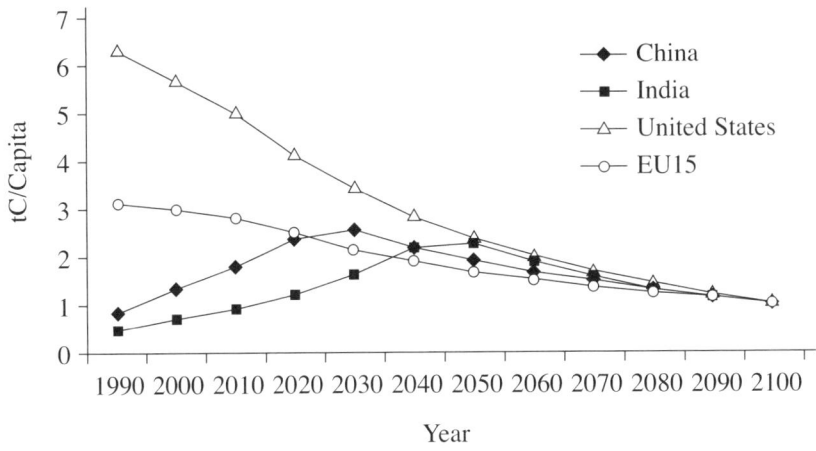

Source: See Chapter 12, this volume (Figure 12.1).

Figure 1.1   Per capita emission allowances

an effective communication tool on both the international and national levels.

The COOL Project

Hisschemöller (Chapter 13, this book) reports on findings from the project Climate OptiOns for the Long term (COOL), which aims at generating strategic recommendations for long-term climate policy in the Netherlands. The COOL project addressed the question how can greenhouse gas emission reductions of up to 80 per cent for the Netherlands and Western Europe, be realized by 2050?

The project has taken a participatory approach. Dialogue groups at the global, European and national levels have assessed the strengths and weaknesses of policy options for reducing greenhouse gas emissions. The dialogue groups included participants from government agencies, environmental and consumer NGOs, business and agriculture.

Climate policies for the decades to come face a serious dilemma. On the one hand, government intervention in most countries has decreased in the last few decades and is expected to decrease further as a consequence of the liberalization of energy markets. On the other hand, the transition towards a carbonless economy calls for huge investment and government involvement. These observations appear quite irreconcilable. In order to structure the dialogue and to articulate conflicting assumptions with respect to technological feasibility and institutional form, three trajectories or regimes for managing the transition towards a carbonless economy have been spelled out and labelled regulation, emissions trading and shared responsibility.

The regulation regime achieves the 80 per cent emission reduction goal by 2050 by means of progressive standard setting and long-term standards for specific sectors. In the COOL National Dialogue, arguments in support of this trajectory have been put forward for the housing and transport sectors. The trajectory fits in nicely with Dutch policy to improve the energy efficiency of buildings by adjusting the standards with reference to the latest technologies. The dialogue groups were confident that significant emission reductions in the future might be realized, providing that the regime is technology forcing in its emission requirements.

An emission-trading regime can be set up by entitling private parties to specified quantities of emissions and allowing transfer of the entitlements. The scheme requires high-quality monitoring of emissions and registration of allowance ownership, hand in hand with strict enforcement.

The shared responsibility trajectory realizes the 80 per cent emission reduction target by private parties who cooperate to realize self-established emission targets on a voluntary basis. In contrast to regulation and trading, which both reflect conceptions of steering, shared responsibility assumes that parties themselves take the initiative and develop climate policies. The government comes in only after societal demand for policy has emerged. The underlying assumption here is that if there is broad social agreement that significant emission reductions are needed, then parties are committed to contribute as much as they feel is possible. The three regimes do not necessarily exclude one another. They may even be complementary over time.

The National Dialogue Integration Workshop concluded the dialogue. A majority of participants indicated that, given the analysis of options in their dialogue groups, 80 per cent emission reduction in the Netherlands by 2050 could be feasible from a technological point of view, provided that the institutional and social barriers are removed during the coming era. However, many within the majority were sceptical with respect to the acceptability of two major reduction options: CO_2 removal and storage and biomass. Large-scale removal of CO_2 is considered a non-sustainable option. In relation to biomass it is doubted whether institutional arrangements can be created to safeguard sustainable production and use of biomass.

A key question is how the relation between regime and technical feasibility is seen. Table 1.1 articulates, in an ideal typical sense, the 'extreme positions' with respect to this question. Cell A articulates the view that competitive technologies, including CO_2 removal and storage and mixing natural gas with hydrogen, are available. The major problem addressed in cell A is that these must get a fair chance on the market. Allowing the market to set a price for CO_2 facilitates the acceptance of CO_2-neutral technologies by companies and

Table 1.1 Conflicting views on transition towards a decarbonized economy

Technology available? Emission-trading regime	Dominant in the medium to long term	Dominant as soon as possible
Not yet	R&D through cross-sectoral cooperation. Major role of government in financing and R&D infrastructure D	Long-term standard setting for specific sectors or technologies, combine with ecotax to avoid externalities B
Yes	C Government support to force innovations that are still weak into the market (create lead customers)	A Acceptance by market parties and consumers

Source: See Chapter 13, this volume (Table 13.2).

consumers. Cell B articulates the view that emissions trading must, for specific sectors such as the car industry, be preceded by long-term standards in order to realize CO_2-neutral technologies that are not yet available. Once these technologies are available the policy approach shifts to cell A. Cell C articulates the problems that exist when technologies are available but the market is not able to deal with them due to market deficiencies or lack of early movers. Progressive standard setting is a regime capable of dealing with these problems. Cell D articulates a strategy that is most in contrast with an emission-trading regime. Technology development is forced by public–private partnership mega investments. Precedents are the Apollo and Manhattan projects in the US or the European Space programme. The major argument in support of this strategy is that it has proved to work, at least in the last century. The government is the only party capable of organizing and implementing mega investments directed at developing and implementing innovations.

The National Dialogue shows that diverging expectations as regards the technological innovations needed are linked to diverging views on preferable regimes for supporting the transition. Competing views emerge. They may have a bearing on problems that compete but also may emerge one after another in the process of transition. The most challenging question for long-term climate policy at this stage would be to find a way to deal in a tailor-made fashion with these problems over time.

Transport and Climate Change

The problems of managing transition, brought out in the COOL Dialogue, are also the subject of the chapter contributed by Kemp and Moors (Chapter 14, this book). They discuss transport and how climate protection goals can be achieved in the sector through what they call 'modulation policies'. Although the focus is on Dutch policy the diagnosis and remedies have a broader application to other European countries.

Within the existing policy belief system in the Netherlands, changes in technology and behaviour are seen as alternative ways for dealing with transport problems. Both options are pursued in parallel but not really in combination with each other. The same is true for private and public transport, which are seen as separate, instead of symbiotic. Transport authorities have not yet embraced the new perspective of integrated mobility. No systematic trajectories for experimenting with and learning from new transport technologies and concepts are visible in the Dutch transport technology policy. There have been interesting initiatives but these have never led to concrete programmes and pathway policies pursued as part of a wider transition agenda. One explanation for this is that there is no vision of sustainable transport that is guiding decision makers.

In order to make transport more sustainable, the current mode of governance has to change. For this the policy process should be broadened and sustainable transport should be made a societal goal for which societal support and resources are mobilized. A modulation approach exploits windows of opportunity and seeks to modulate ongoing dynamics in the required direction – here the direction of climate change protection. For this it is important to have an idea of the relevant trends and expectations of key actors within the domain of transport and mobility.

Although authorities cannot plan for sustainable transport, there are many ways through which public authorities can make transport cleaner and safer, and reduce energy use and CO_2 emissions. Besides making the existing transport system more sustainable there should be an effort towards system innovation involving radical change in the way in which we satisfy our mobility needs. Climate protection benefits could be pursued as part of the endeavour for system innovation. System innovation requires a new approach, of process management aimed at modulation of dynamics and creating path dependencies in the right direction – the modulation approach. Among the suggestions for modulation policies for achieving GHG emission reductions in transport are the following.

Given the inertia in transport systems and uncertainty about what solution is best from a sustainability point of view and user point of view, policy suggestions are:

- Engage in the use of social experiments and create niches for promising technologies (strategic niche management).

This raises the question of what technologies we should experiment with. The answer, given in the literature on strategic niche management is: pathway technologies that help to bridge the gap between the current regime and a new (sustainable) one, and thus help to escape lock-in. This leads to the second suggestion:

- Identification and active stimulation of pathway technologies.

This is well accepted in the transport technology policy: electronic vehicle identification, automatic vehicle control, interoperability and global positioning systems are key technologies for system innovation. To these we can add: electric propulsion and transport information, and booking and reservation systems, which have a great potential for achieving environmental sustainability benefits in the long term when they are part of an integrated mobility system. They are supported by public policies and there has been investment in these technologies by industry but there is a gap between research and diffusion.

To increase the chance that a transition will occur and ensure that the path chosen is the best one, we should explore different paths and the possibilities for cross-linkages and cumulative benefits. This leads to a third suggestion:

- Focus on routes of niche accumulation that may lead to regime changes.

There cannot be transition without a transition path. There is a need to identify possible paths and explore these. We should evaluate present transport regimes and the possibilities of shifting them in desirable directions and identifying opportunities to influence niche branching and niche piling. The focus should be on experimenting with a wide range of niche technologies, which in the long term could serve as stepping-stones for a new transportation regime and be set up in such a way that both suppliers and users learn about new possibilities. The next suggestion, therefore, is:

- Modulate 'promise-requirement' cycles of perceptions and expectations.

New technologies hold promise but are still poorly developed in terms of user requirements. This calls for the need to stimulate promise-requirement cycles and the attendant resource-mobilization activities.

- Development programmes for system innovation such as integrated mobility and mobility management.

Opportunities for system innovation producing sustainability benefits should be explored and exploited. An example of system innovation in transport is integrated mobility or chain mobility: the multiple uses of aligned transport services, which reduce congestion and lead to fewer emissions and accidents.

- Transition management as an integrative framework to achieve greater coherence in policy action.

The above actions should not be pursued as isolated actions. They are best undertaken as part of a transition programme with development rounds in which progress is assessed and goals and instruments are evaluated and adjusted.

Institutional Change: Implication for Climate Change Policy

Unlike the US, where climate policy will develop within a remarkably stable institutional environment, the long-term prospects for climate control actions in Europe are certain to be profoundly influenced by the course of institutional change. An analysis of institutional change in Europe shows that time and again it is powerful social, economic and political driving forces that determine the course and timing of institutional change rather than the operations of the institutions themselves. For instance in the period up to 1987, broad social and political consensus on the need to mitigate increasingly conspicuous environmental problems led to the EU adopting some 200 legal measures concerning the environment, despite the fact that the original Treaty of Rome provided no explicit legal basis for the Community to regard the environment as a legitimate object of Community action. The key issue is therefore how the driving forces behind the continuing institutional revolution in Europe will shape the boundary conditions that largely determine the future course and substance of climate control actions. Bennett (Chapter 15, this volume) suggests that four forces may prove to be particularly influential in this respect during the coming decades: globalization, EU enlargement, climate change research and changes in social values and individual perceptions.

Globalization – the process through which the markets and operations of companies become increasingly international – will have far-reaching impacts at all institutional levels. It will drive the process of harmonizing the economic policies and legislation of the main trading blocs – the EU, North American Free Trade Agreement/Free Trade Area of the Americas (NAFTA/FTAA) and parallel initiatives in Asia and Africa – and thereby reduce the scope for autonomous EU policy on many environmental issues. But this development is also likely to feed countervailing needs for, first, more scope for local, national or regional differentiation with respect to trade regulations and instruments where this is necessary in the interests of environmental protection and, second, more effective international enforcement regimes.

For Europe itself, the greatest institutional impacts during the next two decades will in all probability follow from the EU enlargement process. The greater diversity, institutional complexity and implementation challenges, which are the inevitable consequences of possible enlargement to 21 member states and in the longer term possibly 35, will drive EU policy making away from the traditional practice of negotiating highly specific and detailed regulations and directives. Instead a far greater emphasis will be placed on framework measures that lay down basic rules and targets for a particular policy objective but which allow the member states a greater degree of discretion in how the objectives are achieved and which instruments are applied for that purpose. For environmental policy, this implies a shift to the formulation of locally or regionally appropriate environmental and ecological quality and performance targets rather than detailed emission or technological standards. Where feasible, groups of member states may establish particular forms of flexible cooperation, for example with regard to the use of certain economic instruments.

It is probable, if further research were to confirm the more pessimistic viewpoints and a number of conspicuous natural disasters were to be attributed to climate change that the impact on public, corporate and political perceptions will be sufficient to drive a strengthening of international mechanisms for dealing with global-commons problems, or the creation of a dedicated and substantive global climate regime. Consumer pressure will also be such as to force business to demonstrate its environmental responsibility through initiatives that reduce the climate impact of branded products through innovations in product design and manufacture.

Perhaps the most interesting and potentially the most volatile driving force for institutional change in Europe is public perception. Profound longer-term impacts on institutions could result from the increasing need by individuals, groups and organizations to exert pressure on public and private institutions on matters of concern; a tendency supported by their expanding capability to do so through developments in information and communication technology. The question arises of how these developments will interact with the cultural preference in the Central and Eastern European countries for strong political institutions, particularly if and when these countries make up a substantial proportion of the EU member states. Changing public perceptions on democratic accountability and the legitimacy of EU institutions could force radical changes in the architecture of European governance. To be sure, these are foreseeable opportunities that are likely to be created by the institutional impact of driving forces. In that sense they represent a surprise free scenario. But Europe's future will not be surprise free. The importance of developing response strategies as a means of exploiting events that may not be predictable but can at least be anticipated cannot be overemphasized.

4 IN SUM

We sum up by pulling together a number of the observations and ideas that are brought forward in this book.

All authors take for granted that the flexible instruments defined in the Kyoto Protocol will be implemented through international exchange of emissions or emission reductions between private parties, thus exploiting the potential for lowering control costs to the maximum. However, lower costs, while keeping total emissions below the climate gas ceiling set in Kyoto will not be achieved without adequate implementation structure and in particular not without well-designed flexible instruments. The results of past COPs indicate that the parties to the Protocol become increasingly aware of the necessity to bring transparency of parties' emission achievements as well as compliance incentives for parties into the Protocol.

In the book it is argued that the internationally working instruments are inseparably linked to how the national instruments of climate policy at home have been designed and how well they function. International emissions trading between private parties requires for its success a well-monitored and enforced national cap-and-trade programme of participating parties. Credit trading at national level creates flexibility in national control programmes, relying on performance standards. Where such national cap-and-trade and credit-trading programmes exist the stage is set for international emissions trading. Joint implementation can also make use of the already existing infrastructure for baseline setting, monitoring, auditing and enforcement. The clean development mechanism has to cope with the problem that such an instrument infrastructure will usually be lacking in the host country, making it more difficult to achieve efficiency and environmental integrity simultaneously.

Unresolved questions still remain as to where and how much international coordination is required for smooth operation of the flexibility mechanisms. For example, should it be left to the discretion of the national authorities to decide whether to auction allowances in a national programme or to hand them out for free? Economic theory may state that in a perfect market the distribution rule does not affect the market outcome, yet it seems highly probable that political decision makers will wish to create an 'even playing field' and therefore prefer to coordinate their permit distribution policies. Less controversy exists about the necessity of international coordination to avoid a situation where a party decides unilaterally to restrict the import or export of allowances. The Kyoto Protocol's flexible instruments have to fit with the national instrument and vice versa. Introduction of cap-and-trade programmes and even credit-trading programmes may require an overhaul of parts of existing environmental legislation. One of the positive spillovers of global climate change policy might be that it speeds up or even forces innovation of traditional, inflexible approaches in national environmental policies.

The Kyoto Protocol is one of the first steps on the long road towards mitigation of climate change with increasingly stringent emission targets. This will not be an easy and straightforward development. A first signal is the United States' opting out of the Protocol. It is far from clear when and how the US will return to the negotiation table. The European Union has shown itself in the past years a zealous proponent of climate change mitigation; but how will the stage be set after the enlargement of the European Union from 21 to 35 member states? Future global climate change policy will also depend crucially on future findings of scientific research and whether it will influence public perception into a more pessimistic or optimistic direction.

The next decades will be a period of transition. More and more nations have to be brought in as participants in the international agreement on climate change mitigation. Participation will be encouraged by gradually adjusting the distribution of emission targets among parties from being based on historical emissions to a distribution reflecting general accepted criteria of equity. Technical innovations will have an important role to play in preparing the ground for increasingly ambitious control of climate gas emissions. It raises the question of an evolutionary development in instruments to support such innovations and their quick diffusion. The question involves among others the demarcation of responsibilities between the government and the private sector. A leading role for the government is certainly required for the innovative policy approach of transition management which tries to avoid the lock-in position in technologies of the past and create conditions for break-outs towards new sustainable technologies and their social use.

The studies were first presented at the 'Institutions and Instruments to Control Global Climate Change' conference in Maastricht in June 2001. Some of the chapters have been updated to incorporate changes, for example in legislation, which occured after this date.

We wish to conclude by expressing our gratitude to the sponsors of that conference: the Netherlands HDP committee (HDP commissie) of the Royal Netherlands Academy of Arts and Sciences (Koninklijke Nederlandse Academie van Wetenschappen/KNAW) and the Dutch national research programme on global air pollution and climate change (Nationaal OnderzoekProgramma mondiale luchtverontreiniging en klimaatverandering/NOP).

2. The Kyoto mechanisms and the economics of their design

Luke Brander

1 INTRODUCTION

The Kyoto Protocol to the United Nations Framework Convention on Climate Change (UNFCCC) establishes a legally binding obligation on Annex I countries[1] to reduce emissions for six greenhouse gases (GHGs) collectively to 5 per cent below 1990 levels by the years 2008-2012. The differentiated obligations or quantified emissions limitation and reduction commitments (QELRCs) can be expressed as a quantity of permissible emissions or assigned amounts (AAs). Under the agreement there are three instruments by which Annex I countries can obtain part of their GHG reduction commitment from non-domestic sources. These are emissions trading, joint implementation and the clean development mechanism. The motivation for the inclusion of these instruments is to achieve the stated emissions targets at a lower global cost.

Emissions trading (ET) allows the trading of assigned amount units (AAUs) between Annex I countries. Joint implementation (JI) allows Annex I countries to obtain emissions reduction units (ERUs) from investments made in additional GHG-reducing activities in other Annex I countries. The clean development mechanism (CDM) allows Annex I countries to obtain certified emissions reduction (CERs) from investments made in additional GHG-reducing activities in non-Annex I countries.[2]

The relative importance of the three Kyoto mechanisms and domestic action in providing emission reductions to fulfil the Kyoto commitments is largely determined by their respective marginal costs of abatement. The relative marginal abatement costs of the Kyoto mechanisms are in part determined by underlying regional economic and technological characteristics, but also by the design and institutional structure of each mechanism, that is, the cost implications of the design of each instrument. Through the political negotiation process, the design of the mechanisms reflects additional goals than simply the reduction of GHG emissions at minimum cost. For example, proposed limitations on the use of the mechanisms reflect the idea that Annex I countries should shoulder a higher burden due to their historic responsibility for high GHG concentrations. The range of proposals for the design of the CDM is particularly broad and reflects the development concerns of less developed countries. The resulting advantages and disadvantages of the Kyoto mechanisms not only determine their aggregate use in meeting Kyoto commitments and the efficiency of this outcome but also the respective use of each mechanism, which has development and distributional impacts.

The objective of this chapter is to examine the implications of the alternative design options for the Kyoto mechanisms for their relative use and for specific groups of parties (that is, Annex I buyers of credits, Annex I sellers, and non-Annex I sellers). The relative use of the Kyoto mechanisms in an unrestricted international market for emission credits is taken as a starting point for this analysis. Clearly such use will depend on the complete set of designs taken in combination. For the purposes of clarity in the analysis, however, it is only possible to examine a few, but usually one, design issue at a time. To determine the full impact of the design of the Kyoto mechanisms on different groups of parties, both the quantity and price effects need to be considered. For example, supplementarity restrictions will decrease the quantity of credits traded and decrease the price. The net effect on buyers of credits is unclear without examining the elasticities of demand and supply. This is beyond the scope of this chapter but where possible the results of relevant models will be used to clarify these effects.

The structure of the chapter is as follows: Section 2 outlines the international market for emission reductions. Section 3 presents quantitative evidence from a number of modelling studies on the relative importance of the Kyoto mechanisms and domestic action in fulfilling QELRCs. Section 4 provides a discussion of the most important, as yet, undefined characteristics and mechanics of the Kyoto Mechanisms, including supplementarity requirements, inclusion of hot air in ET, the adaptation tax, inclusion of sinks, crediting periods for the project-based mechanisms, compliance rules and liability provisions, and transaction costs. Section 5 provides a brief analysis of the impact of US non-ratification of the Kyoto Protocol, and Section 6 concludes.

2 INTERNATIONAL MARKET FOR EMISSION REDUCTIONS

As stated in the introduction, the starting point for the analysis in this chapter is a fully functioning international market for emission credits. Such a market is represented in Figure 2.1. The demand curve for emission credits from the Kyoto mechanisms represents the marginal abatement cost of domestic action. The total requirement for emission reductions is determined by the extent to

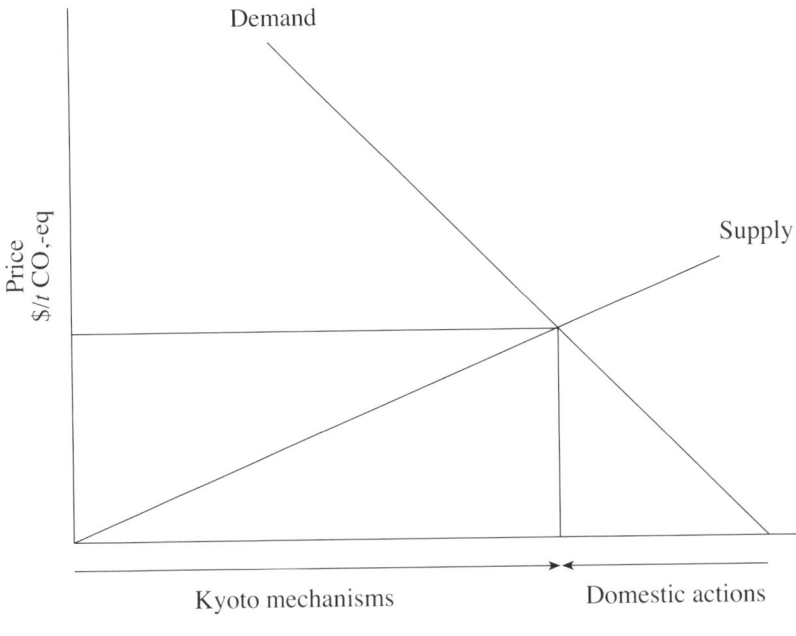

Figure 2.1 Supply and demand for emission reductions from the Kyoto mechanisms

which business-as-usual emissions in Annex I countries exceed their respective emissions limitation and reduction commitments. The supply curve of emission credits from the Kyoto mechanisms is composed of the marginal cost curves of ET, JI and CDM.[3] Emission abatement options can be expected to be taken up in order of cost, so the share of reductions ascribed to each mechanism depends on their relative cost. The general expectation is that, setting aside 'hot air', marginal abatement costs are lower under the CDM than under ET or JI.[4] These relative marginal abatement costs are in part determined by underlying regional economic and technological characteristics, that is, the level of infrastructure and technology currently in place.

In a fully functioning international market for emission credits, the marginal cost of emissions abatement will be equal across the three mechanisms and domestic action. Emission credits from each of the mechanisms and domestic action are perfectly substitutable and tradable – termed 'fungibility'. The respective use of CERs, ERUs and AAUs is determined purely by cost per tonne of CO_2-eq.[5] Buyers and sellers of emissions credits do not exercise monopsony and monopoly power, respectively. No distinction is made between the credits from each mechanism or indeed the source of credits, that is, a tonne of carbon is a tonne of carbon. On the demand side, a distinction between credits could be made if buyers were to act as 'ethical' investors, for example stipulating that they will only hold emission credits that are from renewable energy projects or alternatively that they will not accept emission credits from sinks or nuclear projects. In this case, characteristics such as the source of credits would affect their price. We assume, however, that this is not the case and that the marginal cost of mitigation will be equal across all three mechanisms and equal to a single price for emission credits.

It should be noted that, even with a fully functioning international market for emission reductions, the design of the Kyoto Protocol is not optimal in terms of achieving lower atmospheric GHG concentrations at minimum cost (Nordhaus and Boyer 1999). The use of the Kyoto mechanisms allows 'where' and 'how' efficiency (emission reductions are allocated across countries and across GHGs to achieve given global warming potential reduction at minimum cost) but due to the arbitrary nature of reduction commitments for the first budget period,[6] 'when' (efficient allocation of emissions over time) and 'why' efficiency (emissions are reduced up until the point where the cost of additional reductions equals the benefits) is not achieved. Although the design of the Kyoto Protocol is not optimal, allowing unrestricted use of the Kyoto mechanisms produces an outcome that is not overly far from it (Nordhaus and Boyer 1999). This is particularly the case if emission reduction commitments for future budget periods are set with regard to 'when' and 'why' efficiency considerations.

3 QUANTITATIVE EVIDENCE ON THE ROLE OF THE KYOTO MECHANISMS

A large amount of modelling work has been undertaken to examine the scale of emission reduction requirements and the costs of compliance with Kyoto commitments through a wide range of approaches and under varying assumptions.[7] The focus of most of this work has been on the cost implications of the Kyoto Protocol and the distribution of emission reductions within Annex I. Relatively few of the resulting papers have examined the impacts of alternative design options and reported explicitly on the relative roles of the Kyoto mechanisms.

Table 2.1 presents the results of four studies that give quantitative estimates of the relative use of the Kyoto mechanisms and domestic action in meeting estimated emission reduction requirements for Annex I in 2010. The results are taken from modelling scenarios that correspond as closely as possible with the fully functioning international emission credits market described above.

Table 2.1 Use of the Kyoto mechanisms and domestic action in meeting the estimated Annex I reduction requirement, 2010

Source	Annex I reduction requirement Mt/CO$_2$-eq.	ET Mt/CO$_2$-eq.	JI Mt/CO$_2$-eq.	CDM Mt/CO$_2$-eq.	Domestic action Mt/CO$_2$-eq.
ECN (2000)	2692	90[a]	254	1534	814
Zhang (2000)	2277	576	–	1071	631
MacCracken et al. (1999)	3861	1309	–	1665	887
Haites (2000a)	4166	1195	100	1689	1182

Note: [a]ET within Western Annex I region only.

Adjustments have been made to the results of the various studies to allow them to be directly compared. Where necessary, GHG units have been converted into carbon dioxide equivalents as opposed to carbon equivalents,[8] and quantity estimates for the first budget period as a whole have been converted to a single representative year (2010).

Estimated Annex I reduction requirements vary significantly, from 2277 Mt/CO$_2$-eq. (Zhang 2000) to 4166 Mt/CO$_2$-eq. (Haites 2000a).[9] The estimated volume of emission reductions under each mechanism varies across studies accordingly. In terms of the proportion of Annex I reduction requirements met through each mechanism there is more consistency (see Table 2.2). The CDM is the largest contributor of emission credits in all the models, the range across studies being 41 to 57 per cent. The high estimate from the ECN study is partly

Table 2.2 Kyoto mechanisms and domestic action as a percentage of the Annex I reduction requirement, 2010

Source	ET	JI	CDM	Domestic action
ECN (2000)	3[a]	9	57	30
Zhang (2000)	25	–	47	28
MacCracken et al. (1999)	34	–	43	23
Haites (2000a)	29	2	41	28

Note: [a]ET within Western Annex I region only.

due to the modelled restriction on ET. This restriction will, to varying extents, increase the proportions given to the other mechanisms and domestic action. The second most important source of emission credits is emissions trading (25–34 per cent), followed closely by domestic action (23–30 per cent). Joint implementation accounts for a low share of emission credits in the two models in which it is explicitly considered (9–2 per cent). In the other two models, JI emission reductions are included with ET and so inflate the proportion of reductions that are met through ET.

Variations across the models in the estimated quantity of emissions reductions required to meet QELRCs and the slope of marginal abatement cost curves are due to a number of factors, including the estimation of business-as-usual emissions, methodology used in assessing mitigation costs, and treatment of the complexities of the Kyoto Protocol. Business-as-usual emission estimates are sensitive to projections of economic growth rates, potential energy efficiency improvements and technological change. The large uncertainties regarding the estimation of marginal abatement cost curves arise from difficulties in accounting for the development of emission reduction technologies/rate of technological change, the differences in top-down and bottom-up approaches, inclusion of no-regret possibilities, assumptions on the efficiency of economies, costs included in the analysis and 'double dividends'.[10]

The models treat the complexities of the Kyoto Protocol with different levels of detail, particularly regarding the inclusion of sinks and non-CO_2 GHGs in emissions inventories and reduction potentials,[11] and the treatment of transaction costs specific to the Kyoto mechanisms.

Setting aside the considerable differences in estimated Annex I reduction requirements, which are mostly due to different business-as-usual emission projections, the results presented above show fairly high consistency in terms of the relative importance of the Kyoto mechanisms and domestic action. The results are also consistent with those of other emission credit market models. The Economic Modelling Forum models estimate that 50 per cent of OECD required reductions will take place domestically with Annex I trading only, and 25 per cent with CDM supply included (Weyant and Hill 1999). We take these estimations of the relative use of the Kyoto mechanisms as the background for our analysis of the impact of the outstanding design options on the interaction of the mechanisms.

4 DISCUSSION OF THE KEY DESIGN ASPECTS THAT ARE STILL TO BE DETERMINED

The following unresolved design aspects of the Kyoto mechanisms will influence the extent of their use and also their relative use in meeting the Kyoto commitments of Annex I parties.

Supplementarity

Supplementarity requirements are restrictions on the extent to which the Kyoto mechanisms can be used by Annex I countries in meeting their quantified emission reduction and limitation commitments. The rationale for setting such limits is to force Annex I countries, and particularly the United States, to undertake significant domestic action to reduce emissions, thereby setting the economy on a less carbon-intensive path and encouraging the development of low-emission technologies.

The setting of quantified limits on the use of the Kyoto mechanisms increases the cost of compliance for Annex I parties with high domestic abatement costs, that is, those parties that would otherwise rely most heavily on the Kyoto mechanisms. In other words, such restrictions prevent 'where flexibility' and do not allow GHG mitigation to take place at minimum cost.

There are two broad approaches to setting supplementarity requirements. One is to set quantitative limits on the use of emission permits purchased from abroad, and the second is to create a set of criteria that can be used to judge whether a sufficient degree of domestic action has been taken (Hourcade and Grubb 2000). Most European countries take the view that supplementarity should be defined quantitatively in the form of a ceiling on the use of the mechanisms. Ceilings could be defined separately for each mechanism or for all three mechanisms combined. The text of the Kyoto Protocol implies some level of supplementarity requirement and possibly not the same restriction for each mechanism.

- Article 17 states that ET 'shall be supplemental to domestic actions for the purposes of meeting quantified emission limitation and reduction commitments'.
- Article 6.1(d) states that the use of emission reduction units 'shall be supplemental to domestic actions for the purposes of meeting QELR commitments'.
- Article 12.3(b) states that Annex I parties may use CERs to contribute to compliance with 'part' of their QELRCs.

Differences in wording generally implies differences in intent, and it can be argued that the word 'part' implies a more stringent restriction on CDM than the word 'supplemental' for ET and JI (Michaelowa 1999).

The most recent set of proposals from the chairman of the UN climate negotiations[12] states: 'Annex I Parties to meet their emission commitments chiefly through domestic action since 1990'. This statement appears to support the inclusion of supplementarity restrictions but sets no clear limit on the use of the Kyoto mechanisms and does not mention differentiated restrictions between the mechanisms.

Most proposals for supplementarity restrictions apply a single limit on the combined use of the Kyoto mechanisms.[13] Haites and Yamin (2000) argue that setting separate supplementarity limits would only result in credits of one kind being substituted/traded for credits of another mechanism up to the point that all three limits are reached. They argue that it is therefore only necessary to set an aggregate limit.

The negative impacts of supplementarity restrictions will not be evenly spread across trading parties, although in general both Annex I and non-Annex I parties will be negatively affected. Supplementarity requirements in the form of import restrictions on emission credits result in a price differential between the Kyoto mechanisms and domestic action. Exporters of permits (some economies in transition (EITs) and non-Annex I countries) will export less, which leads to a lower price for traded permits. Therefore these exporters experience a terms of trade loss – lower export price and lower exports. Net importers (Western Annex I) now enjoy a lower import price but lower imports, and also a higher price and level of domestic action.

The imposition of supplementarity restrictions will have asymmetric cost impacts across permit-importing countries. It is even possible for a Western Annex I country, for which imports of permits are below its restriction level, to gain from the lower import price, increase imports of permits and reduce domestic action. In general, however, net importing countries will face increased costs from a supplementarity restriction (Bollen et al. 1999). In addition it is possible that supplementarity restrictions will exacerbate the carbon leakage problem to both EITs with hot air and non-Annex I countries.

Hot Air

The term 'hot air' refers to the quantity of surplus assigned amounts received by a number of Annex I countries, particularly the Russian Federation and the Ukraine, which result from economic collapse in those countries rather than intentional mitigation efforts.

The scale of emissions trading depends to a large extent on the volume of 'hot air' and estimates vary greatly: The International Energy Agency (IEA 2000) estimates that hot air in 2010 will be 623 Mt/CO_2-eq.; the Massachusetts Institute of Technology (MIT) estimates 407 Mt/CO_2-eq.; Haites (2000a) estimates a range from 11 to 807 Mt/CO_2-eq.; Figueres (1998) uses an estimate of 605 Mt/CO_2-eq.; Zhang (2000) estimates a range of 257 to 1371 Mt/CO_2-eq.

The quantity of hot air available depends on the economic growth rates of the EITs, developments in energy efficiency and fuel mix, and methods of accounting for land-use change. It will not be possible to determine the exact quantity of hot air traded through ET as some of the AAUs sold by EITs will be the result of genuine domestic mitigation efforts. Whether or not the hot air problem persists into subsequent budget periods is unclear. This depends on the selection of base years with which to set QELRCs for subsequent budget periods and on emissions growth in the EITs. Given the considerable scope for energy efficiency improvements in the energy sector and industry in the EITs it is not necessarily the case that emission levels will reach 1990 levels in the near future, even with economic recovery.

If hot air is permitted to be traded then ET is likely to crowd out JI to some extent. This will also be the case if an easy system of ET is designed, as it will be easier for low-cost mitigation options in Annex I countries to be developed by the host countries themselves rather than through JI.

The argument against allowing hot air to be traded is that it would allow total Annex I emissions to be higher than would otherwise be the case, although in fact equal to QELRCs. If the inclusion of hot air in ET is not allowed but parties with surplus AAUs are allowed to bank the surplus for use in future budget periods, the quantity of GHG emissions would be the same in the long run, albeit with a different distribution over time, that is, lower emissions in the present and higher emissions in the future (MacCracken et al. 1999; Haites 2000a). Manne and Reichels (1999) show that if hot air can be banked but not traded, the emissions in countries with hot air are higher than in the absence of a prohibition until 2040. Note that the banking of hot air, either prior to the first budget period or for use in subsequent budget periods, is not part of current proposals.

Hourcade and Grubb (2000) review a number of proposed restrictions on the use of hot air. One of these proposes that trading of hot air should be allowed in proportion to the volume of credits from the CDM and sinks. Whereas most restrictions have an implicit effect on the interaction of the Kyoto mechanisms, this approach partially sets their respective use and would help to prevent the crowding out of the other mechanisms by zero-cost hot air. This may be desirable if demand for the Kyoto mechanisms were low, for example due to non-ratification of the Kyoto Protocol by the US. A restriction on the use of hot air would allow the development of the other mechanisms and their institutions.

Setting restrictions on the use of hot air will, however, have the effect of increasing compliance costs for Annex I buyers of credits and removing a source of revenue for EIT countries with hot air.[14] Non-Annex I countries are likely to gain significantly as both the price of emissions credits and use of the CDM are higher.

An issue similar to that of hot air is the potential supply of emission credits from developing countries that are unrelated to emission reductions, this has been termed 'cool air'. The cool air proposal arises from an attempt to include developing countries in the climate regime without requiring them to accept emission limitation commitments.[15] Cool air has been defined as the quantity by which national emissions of GHGs fall below some minimum per capita emissions level (Gupta 1999). This surplus of emission rights can then be credited and traded, possibly through ET, for use by Annex I countries in meeting their commitments. The inclusion of cool air in ET would further increase the quantity of zero-cost emission credits and would add to the potential crowding out of JI and CDM. As with hot air, cool air does not represent any real reduction in GHG emissions and, through its transfer to Annex I countries, allows total emissions to be higher than they would otherwise be (Hourcade and Grubb 2000).

Adaptation Tax

Article 12.8 of the Kyoto Protocol requires the Conference of the Parties serving as the Meeting of the Parties (COP/MOP) to ensure that 'a share of the proceeds from certified project activities is used to cover administrative expenses as well as to assist developing country parties that are particularly vulnerable to the adverse effects of climate change to meet the costs of adaptation'.

There are a variety of proposals for an adaptation tax. It could be in the form of a percentage of the value of CERs generated by the project or a percentage of the CERs themselves. There is no indication given in the Kyoto Protocol as to what this percentage should be. The original Brazilian proposal for a Clean Development Fund was for 10 per cent of the revenue from penalties for non-compliance to go to adaptation. UNEP (1998) proposes that 50 per cent of the CERs go to the Annex I partner, 20 per cent to the non-Annex I partner, 20 per cent to the host government, 4 per cent to the adaptation assistance fund, and 1 per cent to administrative expenses.

It is unclear why the non-Annex I partner and host government should receive CERs other than as an incentive to support the aspects of the project that relate to reducing GHG emissions. The benefits otherwise received by the host entities/governments are in terms of technology transfer, contribution to investment costs (for whatever the project may be) – the size of this contribution is determined by negotiations between investor and host, and does not need to be controlled by a CDM institution – and external benefits (for example, reduced air pollution, lower energy costs). If CERs were to be shared between participants, this may have implications for how other benefits of the project are shared (for example, the financial profits of a power plant) and also the risks of a project (for example, host entities would then also face the risks of the project not producing creditable emission reductions).

The most recent proposal by the chairman of the UN climate negotiations (UNFCCC 2001) proposes that the Adaptation Fund should be established under the Global Environment Facility (GEF) as a trust fund. The fund's finances are to be generated through a levy (share of proceeds) of 2 per cent on the CERs generated through CDM projects. When compared with the administrative costs of comparable institutions 2 per cent is on the low side. For example, the administrative expenses of the World Bank, the Montreal Protocol Fund and the GEF are in the range of 3-15 per cent of total disbursements. It should be noted, however, that total disbursements are not the same as value of CERs generated. It is also the case that administrative costs are to a large extent fixed costs and so large-scale projects will be paying more than their actual administrative costs. This may help reduce the observed advantage of large-scale projects and have an impact on the geographical distribution of CDM investments.

A tax on CDM alone will reduce the use of the CDM and increase the price of emission credits. Annex I sellers of credits will gain from both the higher price and higher volume of trade. Annex I buyers of credits face higher compliance costs, and non-Annex I parties attract fewer mitigation investments. To raise a given amount of revenue and avoid a possible shift of investments away from the CDM it may be preferable to apply a smaller levy to all three mechanisms rather than a relatively large levy only on the CDM. In order to encourage investment and involvement in the least-developed countries (LDCs) it has been proposed that CDM projects in LDCs should be exempt from the adaptation tax. Levying a charge on all three Kyoto mechanisms would also contribute to the goal of supplementarity, by raising the costs of the mechanisms relative to domestic action.

In addition to the Adaptation Fund, a Special Climate Change Fund has been proposed to finance additional 'activities, programmes and measures related to climate change, in the fields of technology transfer, capacity building, economic diversification, energy, transport, industry, agriculture, forestry and waste'. This fund is to be financed through contributions by Annex I parties in the form of financial contributions and/or units of assigned amounts. A minimum amount of US$1 billion per year is proposed for this fund. Target contributions are proportional to Annex I parties' CO_2 emissions in 1990. In this case the US, which was responsible for 25 per cent of global emissions in 1990, might pay about $250 million (Environment News Service 2000).

Inclusion of Sinks

The inclusion of sinks as a source and store of GHG emissions has impacts on both the demand for and supply of the Kyoto mechanisms. The scale and distribution of GHG emissions or sequestration due to land-use, land-use change and forestry (LULUCF) depends on the accounting system used.[16] The most recent proposals from the chairman of the UN climate negotiations set a number of boundary conditions for LULUCF accounting for the first commitment period. These conditions are intended to limit the use of sinks in both Annex I accounting and in the project-based mechanisms.

On the demand side it is estimated that sinks will provide significant domestic reductions, particularly if the management of cropland and grazing land is included. Most Annex I countries are currently experiencing reforestation and, as net land-use accumulation of carbon is not included in the calculation of baseline emissions, this counts as a reduction in GHG emissions. For the US, for example, based on the Intergovernmental Panel on Climate Change (IPCC) accounting method and the inclusion of forestry management, LULUCF will provide a net uptake of 1,056 Mt/CO_2-eq. per year during the first commitment period – equivalent to about half of the US commitment (Vrolijk and Grubb 2000). On the supply side, the inclusion of sinks in JI and CDM is expected to have even larger impacts. Forestry projects and particularly the prevention of deforestation may provide very low-cost emission credits. By increasing the supply of and reducing demand for emission credits, the inclusion of sinks will decrease the market price for credits and reduce compliance costs. The impact on the interaction of the Kyoto mechanisms, however, is unclear and depends on the magnitude of the impact of sink inclusion on the relative costs of each mechanism and domestic action.

In terms of competition between JI and CDM, sinks are not treated equally. Whereas Article 6.1(b) of the Kyoto Protocol clearly establishes the eligibility of sink enhancement projects for JI, there is no explicit reference to such projects for the CDM. The most recent proposals from the chairman of the UN climate negotiations (UNFCCC 2001) do include LULUCF project activities under the CDM but these are limited to afforestation and reforestation only. This restriction is due to concerns over the permanence, additionality and leakage of land-use activities. Different rules for sink enhancement activities in Annex I and non-Annex I countries clearly put the CDM at a relative disadvantage.

Crediting Periods

For CDM projects, emission reductions can be credited from 2000 whereas for

JI projects the crediting period is only the duration of the first budget period, 2008-12. The scope for CDM projects to generate emission credits over a longer time horizon results in higher cost efficiency (lower cost per emission credit) for a CDM project than for a comparable JI project. In other words, total project costs can be divided over a larger number of emission credits. Setting aside any transaction cost and adaptation tax differences between the two mechanisms, CDM will have a cost-efficiency advantage over JI for the first budget period. The extent of this advantage depends in part on how soon CDM projects are implemented and start to produce CERs. It is intended that the start of the CDM should be as prompt as possible, with the proposed election of the CDM Executive Board at COP7. The crediting advantage of the CDM also depends on how long CDM projects are considered to be additional. If baseline estimations are relatively generous to CDM investors, that is, CDM projects are judged to be additional for a large proportion of project lifetimes, then more emission reductions can be certified per project.

The cost-efficiency advantage of the CDM due to its crediting period will also crowd out ET and domestic action. The volume of hot air traded through ET, however, is likely to be less sensitive to alternative designs and changes in the cost efficiency of other mechanisms because the surplus AAUs are achieved at zero cost and can therefore compete with other very low-cost options.

An alternative view of emission reduction strategies is that Annex I parties will observe the inventory of CDM credits available to them (through secondary trade, bilateral, unilateral or multilateral investments) at the beginning of the budget period, and can then revise their use of the other mechanisms or domestic action accordingly. If there is a large supply of CERs available, the other mechanisms will be used relatively little (Haites 2000a).

It is possible that crediting of JI emission reductions prior to 2008 may be allowed in order to avoid investment decisions being biased towards CDM, although this idea does not feature in current UNFCCC proposals on the design of the Kyoto mechanisms. Even if this option is not taken up, the Kyoto Protocol does not prohibit an Annex I party from promising to issue ERUs for emission reductions prior to 2008. This activity does not reduce environmental integrity because the issued ERUs would still come under the cap for 2008-12. This procedure would encourage earlier investments in JI projects but would require more stringent control of emissions during the 2008-12 period (Haites 2000a).

An additional advantage for the CDM under UNFCCC proposed rules is the possibility of CERs being banked towards meeting commitments in the second commitment period. In fact this increases flexibility in the use of credits from all three mechanisms as surplus ERUs or AAUs could be substituted for CERs to be banked for future use.

Compliance Rules and Liability Provisions

Non-compliance represents a decrease in demand for emission credits and so results in a decrease in the price of credits. Significant penalties for non-compliance with emission reduction commitments are necessary to ensure environmental integrity. The penalty per tonne below the commitment needs to be higher than the emission credit price per tonne in order to prevent non-compliance from being financially attractive (Haites 2000b).

Current UNFCCC proposals for compliance penalties are to reduce assigned amounts in the subsequent budget period for non-compliance in the current budget period. The quantity of assigned amount deducted is proportional to the quantity by which current commitments are exceeded, but increases as the scale of non-compliance increases.[17] Effectively this means that parties can borrow emission reductions from future budget periods at an increasing rate of interest. If this is permissible continuously there is no real penalty imposed – parties can always borrow from the next commitment period to cover a penalty received for non-compliance in the previous commitment period. It may also be the case that emission reduction commitments for subsequent budget periods can be negotiated taking into account any penalty received.[18]

In addition to non-compliance penalties, it is proposed that liability provisions should be introduced to limit the sale of AAUs by an Annex I party in order to avoid overselling. Some proposed liability provisions, such as a requirement that an Annex I party proves compliance before being allowed to sell allowances, would heavily restrict the use of ET. The most recent UNFCCC proposals suggest that a minimum limit for credits retained in the national registry is set at 90 per cent of a party's assigned amount, or five times its most recently reviewed inventory – whichever is lowest. This is a restriction on ET only, as JI and CDM emission reductions are by definition additional and are not included in the portion that is required to be retained. This liability provision reduces the environmental and compliance risk of overselling of AAUs and is also not overly restrictive of ET.

Transaction/Institutional Costs

The institutional design of the Kyoto mechanisms has implications for the level of transaction costs associated with each mechanism. Although some of the discussion above includes transaction cost issues, such as the adaptation tax, it is worth examining some additional transaction cost considerations.

Most models of emission reduction potentials do not explicitly include transactions costs and 'real-world' constraints. In addition to the transaction costs directly related to the accounting and reporting requirements of the UNFCCC, CDM and JI projects are likely to incur high transaction costs related to the countries in which they are situated. Such transaction costs are often difficult to estimate and may not be included in the cost information used to construct marginal cost curves for each mechanism. For this reason the costs of CDM and JI may be underestimated.

ET may have lower transaction costs than CDM and JI as the project-based mechanisms incur costs related to baseline estimation, and verification and crediting of emission reductions.[19] Potentially JI may have lower verification and crediting costs than CDM. Certification of emission reductions under the CDM is to be conducted by an independent body, whereas Article 6 does not specify how ERUs will be certified. Two possible certification procedures for JI are (i) to use the same international review process as CDM or (ii) to allow host governments to monitor and certify the quantity of ERUs produced from JI projects. It is potentially feasible to allow host governments to certify ERUs because unlike the CDM there is no incentive to certify more emission reductions than are actually realized. The crucial difference between the two mechanisms is that the sale of ERUs involves the transfer of AAUs whereas the sale of CERs does not. If an EIT party overcertifies ERUs it may face problems in complying with its commitments. The second option for JI certification may result in lower transaction costs and therefore a relative advantage for JI. Such a crediting process would require significant non-compliance penalties to ensure that overcrediting of ERUs is not profitable to JI host countries.

The impact of higher transaction costs associated with the project-based mechanisms is to reduce their use relative to ET and domestic action. Higher transaction costs raise the market price of emission credits and the compliance cost for Annex I buyers. Haites (2000a) models the impact of different transaction cost levels on the carbon market and shows that the Annex I and non-Annex I sellers of emission credits are better off with higher transaction costs as the increase in market price for credits more than offsets their burden of the increased transaction costs.

5 IMPACT OF THE POTENTIAL US NON-RATIFICATION OF THE KYOTO PROTOCOL

The US announcement that it will not ratify the Kyoto Protocol introduces huge uncertainty to all aspects of the climate change negotiations. Although a number of compromise proposals have already been made – essentially to reduce the US compliance costs – it appears unlikely that the issue of non-Annex I commitments will be resolved at this stage and so the US will not participate in GHG reductions as set out in the Kyoto Protocol. The possibilities for the redesign of the climate change regime are large and it is not the intention of this section to speculate on them. We only consider the possibility that all other parties ratify the Kyoto Protocol as it currently stands.

The withdrawal of the US from the Kyoto Protocol represents a huge fall in demand for the Kyoto mechanisms. Table 2.3 presents three sets of estimates of the respective US and non-US proportions of Annex I reduction requirements. The US accounts for 60-74 per cent of Annex I reduction requirements. This fall in demand will drastically reduce the price of emission credits to the remaining Annex I buyers of credits. Low- or zero-cost emission reduction options can be expected to satisfy non-US Annex I QELRCs. Indeed the estimated quantity of hot air is roughly equal to the estimated rest of Annex I reduction requirement.

This suggests that the relative use of the Kyoto mechanisms and domestic action may differ greatly from the shares presented in Table 2.2, with emissions trading of hot air becoming the most significant source of emission credits. Annex I sellers of credits are, however, likely to be worse off due to the fall in international permit price. Non-Annex I parties are also negatively affected due to the decrease in CDM investments. Implementation of domestic action and the use of the project-based mechanisms may be very low unless supplementarity requirements and restrictions on the use of hot air are introduced. Other design issues affecting the relative marginal abatement costs of the mechanisms, such as the adaptation tax and crediting periods, are likely to be of less importance given the zero cost of hot air credits.

Table 2.3 US and rest of Annex I reduction requirements, 2010

Source	Annex I reduction requirement	US reduction requirement		Rest of Annex I reduction requiremen	
	Mt/CO_2-eq.	Mt/CO_2-eq.	% of Annex I reduction	Mt/CO_2-eq.	% of Annex I reduction
ECN (2000)	2692	1997	74	695	26
Zhang (2000)	2277	1554	68	722	32
MacCracken et al. (1999)	3861	2325	60	1536	40

CONCLUSIONS

The relative advantages and disadvantages of the Kyoto mechanisms - as determined by the various design options for the mechanisms - will impact on the relative use of the mechanisms and the costs of compliance. Without restrictions, the relative role of the Kyoto mechanisms and domestic action will be determined by their respective marginal costs. There is a high degree of consistency in the results of a number of carbon market modelling studies in terms of the proportion of required emission reductions met through each of the mechanisms. The CDM is anticipated to be the largest source of emission reductions in the first commitment period, followed by ET, domestic action and JI. The design of each mechanism and the mechanics of the trading system are not yet fully defined and are likely to reflect additional goals than simply the achievement of minimum cost emission reductions. Qualitative arguments for the effects of current proposals for the design of the Kyoto mechanisms and the distributional impacts for both buyers and sellers of credits have been given. Any restrictions on the use of the mechanisms will increase compliance costs for buyers of credits as a group (Western Annex I parties), although increased compliance costs will not be distributed evenly and some buying parties may benefit under certain restrictions. Sellers of credits may benefit from some restrictions due to resulting increases in the price of emission credits or through any relative advantage gained by the mechanism through which they sell credits. Modelling work is required to quantify and aggregate the various effects in order to determine the net effect on the relative use of the mechanisms and the distributional consequences.

Non-ratification of the Kyoto Protocol by the US represents a fall in demand for GHG emission reductions of between 60 and 74 per cent. The price of emission credits can be expected to fall dramatically as a result. The relative roles of the Kyoto mechanisms will also be affected and in the absence of supplementarity requirements and/or restrictions on the use of hot air, ET may crowd out CDM and JI in the first commitment period.

NOTES

1. UNFCCC Annex I lists the countries that have committed themselves to a quantitative GHG emissions reduction target (OECD members plus most of the Central and Eastern European countries). Annex B of the Kyoto Protocol lists 37 countries with their specific emissions reduction commitments. Annex B includes some countries (Slovakia, Slovenia, Liechtenstein and Monaco) that are not listed in Annex I and excludes some countries (Belarus and Turkey) that are listed in Annex I.
2. CDM investments have the additional requirement of contributing to sustainable development in the host country.
3. When an Annex I country has lower emissions than its total assigned amount, the excess

AAUs are included in the supply of emission credits through ET.
4. The issue of implementation costs differences between the mechanisms is dealt with later.
5. The five non-CO_2 GHGs covered by the Kyoto Protocol are converted into CO_2 equivalents based on a global warming potential for 100 years.
6. Arbitrary in that they are set unrelated to expected damage costs of the GHG concentration trajectory.
7. See particularly the Special Issue of *The Energy Journal* on 'The Costs of the Kyoto Protocol: A Multi-Model Evaluation' (1999).
8. The conversion rate of 1 tonne carbon equivalent to 3.667 tonnes CO_2 equivalent was used.
9. Note that these estimates include the US demand for emission reductions in meeting its first budget period commitments.
10. See Sijm et al. (2001) for a discussion of the differences in mitigation cost assessments.
11. See Reilly (2000) for an overview of the issues and impacts of including non-CO_2 GHGs and carbon sinks in the Kyoto Protocol.
12. The chairman of the UN climate change negotiations put forward a new set of compromise proposals on rules for the Kyoto Protocol on 12 April 2001 (UNFCCC 2001).
13. See Michaelowa (1999) and Hourcade and Grubb (2000) for discussions on possible definitions of supplementarity requirement.
14. It has been argued by a number of authors that the prospect of emissions trading in hot air was vital in persuading the US, Russia and the Ukraine to accept emission reduction commitments (Nentjes and Woerdman, 2000). To restrict the use of hot air therefore threatens the acceptability of the Kyoto Protocol to these parties.
15. The cool air concept has also been proposed as a criterion by which to distribute adaptation fund assistance.
16. Different accounting scenarios result in different estimates of uptake or emissions resulting from LULUCF. Many of the definitions and complexities relating to the accounting of sinks are yet to be resolved. For example, the choice of definition of a forest has a large impact on the amount of land falling within the accounting system. See Vrolijk and Grubb (2000) for an overview of the complexities of LULUCF and the impacts of alternative accounting systems.
17. The proposed penalties in the 'New Proposals by the President of COP6' (UNFCCC 2001) are to deduct from assigned amounts in the subsequent commitment period, a number of tonnes of allowable emissions equal to: (i) 1.1 times the tonnes of excess emissions if the party has exceeded its assigned amount by less than 1 per cent; (ii) 1.5 times the tonnes of excess emissions if the party has exceeded its assigned amount by more than 1 per cent, but less than 8 per cent; (iii) 2.0 times the tonnes of excess emissions if the party has exceeded its assigned amount by more than 8 per cent.
18. Although commitments for Annex I parties for a second commitment period are to be adopted before 2008, it is likely that parties will have a reasonable view of whether they will meet their commitments and the scale of any non-compliance penalty that they will receive.
19. It is argued by Woerdman (forthcoming), however, that ET does not inherently incur lower transaction costs, particularly if baselines for CDM and JI projects are standardized and if the ET market does not conform to the perfect competition paradigm.

REFERENCES

Bollen, J., A. Gielen and H. Timmer (1999), 'Clubs, ceilings and CDM: macroeconomics of compliance with the Kyoto Protocol', *The Energy Journal*, Special Issue on 'The Costs of the Kyoto Protocol: A Multi-Model Evaluation', 177-206.
ECN (2000), 'Kyoto mechanisms: the role of joint implementation, the clean development mechanism and emissions trading in reducing greenhouse gas emissions', Energy Research Centre, Petten, The Netherlands.

Environment News Service (2000), http://www.ens.lycos.com/ens/apr2001/2001_-04-12-03.html.
Figueres, C. (1998), 'How many tons? Potential flows through the clean development mechanism', Working Paper, available at http://www.csdanet.org/wp.figueres.html.
Grubb, M. (2000), 'Quantifying Kyoto – A review of model results and sensitivities', Paper presented to RIIA workshop 'Quantifying Kyoto', Royal Institute of International Affairs, Chatham House, London, 30-31 August.
Gupta, J. (1999), 'North–South aspects of the climate change issue: towards a constructive negotiating package for developing countries', *UNFCCC Strategies for Developing Countries*, **8** (2).
Haites, E. (2000a), 'Institutional features of the Kyoto mechanisms and the COP-6 decisions', Marganee Consultants Inc., Toronto.
Haites, E. (2000b), 'The Kyoto mechanisms and global climate change: coordination issues and domestic policies', prepared for the Pew Centre on Global Climate Change, Arlington, VA.
Haites, E. and F. Yamin (2000), 'The CDM: proposals for its operation and governance', *Global Environmental Change*, **10** 27-45.
Hourcade, J.-C., and M. Grubb (2000), 'Economic dimensions of the Kyoto Protocol', in J. Gupta and M. Grubb (eds), *Climate Change and European Leadership: A Sustainable Role for Europe?*, Dordrecht: Kluwer Academic Publishers.
Janssen, J. (2000), 'Will joint implementation survive international emissions trading? Distinguishing the Kyoto mechanisms', FEEM Working Papers.
Jepma, C. (1997), 'Banking', *Joint Implementation Quarterly*, **3** (3).
Lefevere, J. (2000), *Defining and Distributing the 'Share of Proceeds' under the Clean Development Mechanism*, Washington, DC: Centre for Clean Air Policy.
MacCracken, C.N., J.A. Edmonds, S.H. Kim and R.D. Sands (1999), 'The economics of the Kyoto Protocol', *The Energy Journal*, Special Issue on 'The Costs of the Kyoto Protocol: A Multi-Model Evaluation', 25-71.
Manne, A.S. and R.G. Reichels (1999), 'The Kyoto Protocol: a cost-effective strategy for meeting environmental objectives?', *The Energy Journal*, Special Issue on 'The Costs of the Kyoto Protocol: A Multi-Model Evaluation', 1-24.
Michaelowa, A. (1999), 'Clean development mechanism and joint implementation – which instrument is likely to have a higher impact?', Working paper.
Nentjes, A. and E. Woerdman (2000), 'The EU proposal on supplementarity in international climate change negotiations: assessment and alternative', ECOF (Department of Economics and Public Finance) Research Memorandum No. 28, Groningen: University of Groningen.
Nordhaus, W.D. and J.G. Boyer (1999), 'Requiem for Kyoto: an economic analysis of the Kyoto Protocol', *The Energy Journal*, Special Issue on 'The Costs of the Kyoto Protocol: A Multi-Model Evaluation', 93-130.
Parkinson, S., K. Begg, P. Bailey and T. Jackson (1999), 'JI/CDM crediting under the Kyoto Protocol: does "interim period banking" help or hinder GHG emissions reduction?', *Energy Policy*, **27**, 129-36.
Reilly, J. (2000), 'The Kyoto Protocol and non-CO_2 greenhouse gases and carbon sinks', Paper presented at RIIA workshop, 'Quantifying Kyoto', Chatham House, London, 30-31 August.
Sijm, J.P.M., L.M. Brander, O.J. Kuik and S.N.M. Rooijen (2001), 'Cost assessments of mitigation options in the energy sector', Forthcoming.
UNFCCC (2001), 'New proposals by the President of COP6', http://www.unfccc.int/sessions/cop6_2/unfccc_np.pdf.

Vrolijk, C. and M. Grubb (2000), 'Quantifying Kyoto: how will COP6 decisions affect the market?', report of RIIA workshop 'Quantifying Kyoto', Chatham House, London, 30-31 August.

Weyant, J.P. (1999), 'Emissions trading: reducing the cost of implementing the Kyoto Protocol', *Energy and Environment*, **10** (5), 539-48.

Weyant, J.P. (2000), 'An introduction to the economics of climate change policy', prepared for the Pew Center on Global Climate Change, Arlington, VA.

Weyant, J.P. and J.N. Hill (1999), 'Introduction and overview, *The Energy Journal*, Special Issue on 'The Costs of the Kyoto Protocol: A Multi-Model Evaluation', pp. 1-24.

Woerdman, E. (2000), 'Implementing the Kyoto Protocol: why JI and CDM show more promise than international emissions trading', *Energy Policy*, **28**, 29-38.

Woerdman, E. (forthcoming), 'Emissions trading and transaction costs: analysing the flaws in the discussion', *Ecological Economics* (accepted for publication).

Zhang, X.Z. (2000), 'Estimating the size of the potential market for the Kyoto mechanisms', http://ideas.uqam.ca/ideas/data/Articles/dgrrugccs1999q4-4.html.

3. Alternative design options for emissions trading: a survey and assessment of the literature

Jan-Tjeerd Boom and Andries Nentjes

1 INTRODUCTION

The Kyoto Protocol of 1997 sets ceilings for the emissions of greenhouse gases of Annex B parties to be achieved in the commitment period 2008-12. For some parties, such as the US and the European Union (EU), keeping emissions below their assigned amounts will imply high marginal costs whereas others, for example Russia and Ukraine, can realize their emission targets with little economic effort. Article 17 of the Kyoto Protocol introduces flexibility by allowing international emissions trading in greenhouse gases. Parties can avoid high marginal cost of emission reduction by buying their assigned amounts from countries which accept an equivalent decrease of their assigned amounts since they are able to expand their emission control at relatively low marginal cost.

In the political discussion, international emissions trading is seen as transactions between national governments, increasing or decreasing their assigned amounts. After the revision of the national emission targets, adequate national policies and instruments should be designed and implemented to realize the revised targets. On the other hand, economists who have reflected and written on international emissions trading have pointed out that the flexibility of the Kyoto Protocol and cost savings would be much higher if international emissions trading between private parties is made feasible: private parties have better information than governments on their emission control costs, as well as the incentive to keep costs as low as possible.

Private party trading means that a firm in country X can buy or sell emission permits from or to, respectively, a firm in country Y. It sets the stage for a truly international market for tradable emission permits, the tasks of national governments being restricted to registering the concurrent changes in assigned amounts and to monitoring and enforcing compliance of national firms.

This type of trade in emission allowances between private parties under

Article 17 of the Kyoto Protocol should be clearly distinguished from joint implementation (JI) as defined in Article 4. Although JI also involves a transaction between private parties residing in Annex B countries, allowing one party (in the donor country) to increase its emissions thanks to the extra emission reduction by the other party (in the guest country), its design is basically different and its economic impact as well.

In this chapter we shall discuss the design of two flexibility mechanisms; permit trading and credit trading. With permit trading, a cap is placed on firm and total emissions, after which firms are allowed to trade emission allowances. Credit trading on the other hand is based on relative standards with no absolute cap on emissions. In our analysis we concentrate on emissions of carbon dioxide and on the type of design that offers the legal setting for transactions between private parties. Quite understandably the literature on the subject focuses on the international dimension. Yet adequate functioning of the international flexibility mechanisms depends on how well the national instruments are designed and implemented. This simple truth is often overlooked. In this contribution we shall therefore concentrate on the national basis of international permit and credit trading.

Among economists (see Ellerman 1998; Bohm 1999; Hahn and Stavins 1999; and Zhang and Nentjes 1999) there is a consensus that international permit trading between private parties makes sense only if it is embedded in well-enforced national schemes of tradable permits. Consequently international emissions trading is basically private party trading within internationally linked national schemes of tradable emission permits. Various designs of national permit-trading schemes have been proposed in the past few years. We shall give a survey in Section 2 and discuss their strengths and weaknesses, selecting what we consider to be the major issues. The link between national cap-and-trade schemes and international emissions trading is addressed in Section 4.

Although interest in tradable permit schemes is growing, direct regulation is still the dominant national instrument of environmental policy - for air pollution usually in the form of performance standards. To meet the bottlenecks caused by the rigidity of direct regulation, flexibility has been introduced in the US by allowing trade in emission reduction credits. Whereas it is normally argued that credit trading has to be based on explicit abatement arguments, we show in this chapter that this need not be. Actually, credit trading can be organized in much the same way as permit trading. This also extends to the international level, where international credit trading can be set up by linking national trading schemes. The result is that international credit trading does not fall under Article 4 of the Kyoto Protocol, but under Article 17, which defines emissions trading. The particularities of national credit schemes and the major differences with cap-and-trade programmes will be discussed in Section 3 and joint implementation as international credit trading is the subject of Section 4. Section 5 concludes.

2 NATIONAL PERMIT TRADING WITH A CAP

The basic elements of a scheme of tradable emission permits at the national level are the following: set a ceiling on total emissions of the group of participants in the scheme; distribute the emission permits among participants; allow trade; monitor (transfer of) emission permits and of actual emissions and enforce compliance with the scheme. Differences between proposed schemes arise from differences in how these elements have been worked out.

Design of a National Permit-trading Scheme

Since carbon dioxide is released by burning the carbon contained in fossil fuels, emissions can be controlled by restricting the use of (carbon in) fossil fuels. In the discussion on appropriate design of tradable carbon permit schemes a major issue is at what level to organize it.

Four basic designs of permit trading can be distinguished (Jepma et al. 1998 and Hargrave 1998): upstream, downstream, hybrid and mixed approaches. In an upstream scheme, the producers, processors and transporters of fossil fuels are regulated. In a downstream scheme, the consumers of fuels can trade emissions and in a hybrid scheme large consumers of fossil fuels are directly regulated, while the remainder of fuel consumption is regulated through an upstream scheme. In a mixed scheme, large emitters are regulated through a tradable permit system, while small emitters are regulated through some other instrument. These four schemes will be discussed below.

Several criteria have to be taken into consideration in the evaluation of the four design options (Hargrave 1998; Hargrave et al. 1999):

- *Environmental effectiveness* The larger the coverage of total carbon dioxide emissions of the scheme, the greater the certainty that the emission level set by the government is realized.
- *Economic efficiency* Economic efficiency depends on the coverage of emissions in the scheme and on the number of sources captured.
- *Effects on competition* Competition can be distorted when competitors do not face the same marginal costs of abatement.
- *Administrative burden* Intricate systems increase the costs of setting up and maintaining the trading scheme. The administrative burden depends on (Hargrave 1998):
  a. The number of regulated sources. The larger the number of sources,

the more information is needed in the setting up of the system.
b. The availability of needed data. If the data is readily available, the previous point becomes less important.
c. The level of reporting requirements and the level of monitoring needed. If reporting requirements are very intricate, the costs for the regulated sources are high. High levels of monitoring mean high costs for the monitoring authority.
d. Proper accounting. Ideally, firms are only required to hold emission quotas for emissions of greenhouse gases, and only for domestic emissions.
- *Relationship to existing policies and measures* When not all emissions are captured by the trading scheme, but some are regulated through other instruments, the interaction between the systems can have both environmental and economic consequences.

Upstream

In an upstream approach, the emitters of carbon do not receive permits, but instead the suppliers of fossil fuels have the obligation to cover their fuel sales (in terms of carbon) with permits. Producers, processors and distributors of oil, coal and natural gas receive permits for free or buy them at an auction. After the initial distribution, they can trade the permits among one another. The price of permits, which will result in the market, will be passed on to the consumers through the price of fossil fuels. Hence, the consumers pay a kind of 'carbon tax'. The result is that fuels with higher carbon content, such as oil and coal, will rise more in price than those with a low carbon content, such as natural gas.

Two main advantages of an upstream approach can be identified (Hargrave 1998; Jepma et al. 1998; and Bohm 1999). First, an upstream approach would comprise virtually all fossil fuel use, and thereby carbon emissions. A second advantage is that there are relatively few regulated entities in such a system and only their carbon sales have to be monitored and compared with the permits the suppliers have acquired. Therefore, the administrative costs of an upstream system will be low.

Several disadvantages of an upstream system are mentioned in the literature. The main problem is the low political acceptability of such a scheme. Since there are (almost) no options for the suppliers of fossil fuels to reduce the carbon content of the fuel, the carbon cap is actually a fuel cap. According to Hargrave (1998), this may induce strong resistance from producers since it will affect their profits. This is, however, not a very strong argument since any measure to curb greenhouse gas emissions will affect the producers and suppliers of fossil fuels. The resistance against an upstream system is more likely to be connected with the way of distributing the permits.

If in an upstream system the permits were to be grandfathered, the receivers of the permits would be rewarded with a large rent. Since they will transfer the main part of the costs of the permits to the end users, the permit holders do not pay for the reduction of emissions in any direct way (Cramton and Kerr 1998). In this way, a small group of producers and transmitters of fossil fuels receives large rents without incurring costs. Although grandfathering might change producers into enthusiastic supporters of the scheme (see Dijkstra 1999 and Svendsen 1998), it will have the same distributional impact on end users as a carbon tax. However, it is not politically acceptable that such large rents are distributed to a relatively small group of large companies (Cramton and Kerr 1998; Hargrave 1998; and Woerdman 2000). This leaves open the option of auctioning the permits. However, this will meet resistance from the affected sectors: the suppliers as well as the end users. Since there are relatively few suppliers that will be involved, they will be very effective at organizing themselves (see Olson 1965). It is therefore likely that they will have an influence on the policy outcome. The same is true of the few very large energy-intensive end users.

Besides this principal political bottleneck, several other disadvantages of upstream trading are mentioned in the literature. One of the arguments is that consumers have no incentive to reduce their emissions of carbon other than through the reduction of fuel use. Hence, techniques aiming at removing carbon after the fuel has been used (end-of-pipe technologies) will not be developed (Hargrave 1998). In our view, this could easily be solved by introducing refunding for carbon removal after fuel use. Another somewhat peculiar argument runs that an upstream approach may not provide as great an incentive for energy efficiency and fuel switching as a downstream approach. The argument is based on the assumption that energy consumers do not respond to price signals in the same way as to quantity signals. Hargrave (1998), however, does not give any references to confirm this view. Hence, it is not certain whether this is a certified fact, or just an opinion. Finally, it is mentioned that the low number of participants in the scheme increases the possibilities for market power by one or a few participants (see also Jepma et al. 1998). It has been estimated, though, that market power would not be a problem in the US, since the number of regulated firms in an upstream system would be about 2000, with the largest firm having a market share of about 6 per cent (Cramton and Kerr 1998). According to Koutstaal (1997), even in a small country like the Netherlands there would be 40 to 50 traders in an upstream scheme, which should be sufficient for a viable permit market.

We conclude that upstream schemes have the attractive features of wide coverage and low administrative costs. However, they meet the political obstacle that grandfathering as well as auctioning of the permits will meet strong resistance from interest groups.

Downstream

In a downstream system, all emission sources are required to hold emission permits and the emissions of all sources will be monitored. The government can either distribute the permits for free (grandfathering), or auction the permits. In all permit-trading schemes so far, the permits have been grandfathered (see UNCTAD 1998 and Stavins 2000). The reason for this is that it enhances the political acceptability of the scheme.

It is more acceptable to grandfather permits in a downstream system than in an upstream system. In a downstream system the consumers receive the permits and thereby the rents following from grandfathering, but they also have to bear the burden of emission reductions. Grandfathering the permits will make the scheme much more acceptable for the regulated parties than distribution through an auction (Hargrave 2000). Apart from grandfathering, a second factor, which pertains in particular to the US, is familiarity with the system (Festa 1998). In the US, several kinds of downstream trading (credit and permit trading) have been used (see Tietenberg 1989; Klaassen and Nentjes 1997; Svendsen 1998; and Schmalensee et al. 1998 for some assessments). However, downstream trading is also known in Europe, for example in the milk production and fishing quotas in the EU and manure production quota in the Netherlands (Boom et al. 1998 and Stavins 2000). Familiarity with the system will make it easier to explain and might diminish resistance from politicians and bureaucrats. In a downstream system, all individual emission sources are regulated. This means that there are many traders facing the same price, trading will be regular, ensuring that new information is dispersed quickly through the market and there is little risk of market power (Jepma et al. 1998). The size of the market will also make the development of derivatives such as options possible, which will make the market even more efficient.

Besides these, some other advantages of downstream trading are mentioned in the literature. According to Festa (1998) a downstream system is a greater stimulus for innovation than an upstream system. Festa asserts that price signals are not always sufficient motivation for consumers to implement profitable energy savings but that quantitative signals are sufficient motivation. Again, there is no evidence provided to support this claim. Another argument related to innovation, is that a downstream system where emissions are monitored at the source also provides incentives to remove carbon, after emission use (Festa 1998).

Although the literature endows a downstream system with many advantages, some disadvantages are mentioned too. There is general consensus that there are high administrative costs connected with a downstream approach (Festa 1998; Hargrave 1998; Jepma et al. 1998). One of the reasons for this is that the permits have to be distributed to all the sources and their emissions have to be monitored. If the distribution is based on historical emissions per source, the distribution will be very costly, since then the government needs to know the emission data from all sources for some period in the past. In addition, the monitoring of the emissions and checking of compliance by comparing permits and emissions at every source will be extremely expensive.

One remedy to this problem would be to lower the coverage of the system by including only large sources or specific sectors. Other sources and sectors would of course have to be regulated with other instruments. That could contain the problem that the burden of emission reductions would fall solely on the firms placed under a cap and reduce carbon leakage as well (Hargrave 1998).

In summary we conclude that a major argument for downstream systems is the possibility of grandfathering; thus reducing the opposition from interest groups against restriction of carbon use. But on the other side the high administrative cost of a pure downstream system thwarts its feasibility.

Hybrid

In a hybrid system, large polluters are regulated directly as in a downstream system, while other polluters are targeted through an upstream system (Koutstaal 1997). Hybrid trading can be seen as a compromise system between upstream and downstream trading, giving low administrative costs and a reasonable level of political acceptability.

An advantage of a hybrid system is that the number of parties to be monitored and checked for compliance is relatively low so that the administrative costs of the system are low, although higher than in an upstream system. It will also secure a reasonable number of traders, making the system more efficient and flexible (Jepma et al. 1998). At the same time, the environmental effectiveness of the scheme is high because all sources are covered: the large ones directly through the downstream part and the small ones indirectly through the upstream part of the system.

One of the major issues in a hybrid system is the distribution of the permits. The permits of the emission sources regulated directly through tradable permits can be grandfathered. These firms will have to bear the costs of emission reductions and therefore political resistance against grandfathering will be low. The permits for the fossil fuel suppliers have to be auctioned since handing them out for free to fuel suppliers without any real effort in return will not be accepted politically.

A challenge connected with a hybrid system is to avoid double counting (Hargrave 2000). Fuels consumed by sources included in the trading programme must be exempt from the indirect fuel tax that is put on the fuel price by the producers through the upstream system.

Basically the hybrid scheme is a compromise of upstream and downstream elements. In terms of advantages and disadvantages it chooses the political middle of the road of average political acceptability (lower than downstream and higher than upstream) and average administrative cost (lower than downstream, higher than upstream).

Mixed

In a mixed system, large polluters are regulated through a downstream tradable permit system, while other sources are regulated through some other instrument, such as performance standards or taxes. The advantages of such a system are that it gives an effective and efficient policy for large emitters. At the same time, it also includes small emitters (Jepma et al. 1998). Furthermore, a high coverage of total carbon dioxide emissions is possible in such a system. The instrument that is most likely to be used to regulate the small emitters is a tax. Taxes are relatively easy to monitor, whereas the monitoring system needed for standards would be as high as in a downstream system. Hence, they would not give lower administrative costs.

The disadvantages of a mixed system stem mostly from the fact that two or more instruments are used simultaneously (see Hargrave 1998). First of all, it would bring high administrative costs because additional programmes have to be set up besides the permit-trading system. In our view this is a flawed argument. Administrative cost will be lower than in the case of pure downstream permit trading. Second, if these other measures lead to other marginal costs of abatement in the other industries, inefficient allocation of emission control between sectors and carbon leakage may still take place. A third problem with other measures is that they do not guarantee that the countrywide emission ceiling is met.

Even though many problems are associated with a mixed system, it is the one preferred by the European Commission (2000). Remarkably, the Commission does not even mention other design options for emissions trading. The main reason why it prefers a mixed system is that in this way companies will be able to trade on an EU permit market, which in turn will give them the opportunity to prepare for international emissions trading. The problem of carbon leakage is dealt with in one sentence: 'the potential competitive distortions caused by leaving out some sectors, or smaller sources within the covered sectors can be limited by ensuring that equivalent policies and measures are imposed on sectors and sources not covered by the trading system' (European Commission (2000, p. 13). Here the Commission seems to think that it is easy to set 'equivalent policies and measures'. However, it is not likely that an equal price can be created through other measures. Moreover, it will be impossible to adapt these measures as the price of permits changes over time.

Most of the authors mentioned above agree that the upstream design will outperform the other alternatives (Bohm 1999; Cramton and Kerr 1998; Festa 1998; Fischer et al. 1998; Hargrave 1998, 2000; and Jepma et al. 1998). The reason for this is that an upstream design gives a high coverage of emissions and low administrative costs. It is precisely on these points that a downstream system performs badly. Even though a downstream system will have more parties participating in the permit market and might therefore be more efficient than an upstream system, it is assessed that this cannot outweigh the above-mentioned problems. There is also some support for a hybrid system, although the administrative costs for preventing double counting can be high. A mixed system is almost unanimously rejected (except by the European Commission) because of the problems that arise when combining different instruments. There is, however, an alternative to the designs above that the authors do not discuss.

Alternative design
In another strand of the literature on CO_2 emissions trading, an alternative design is mentioned, which combines elements from upstream and downstream systems (see Koutstaal 1993, 1997; Zhang and Nentjes 1999; Duijse et al. 1998; and Nentjes and Rietveld 2000). In this system, allowances are grandfathered to sources (big and small), as in a downstream system, but compliance is monitored upstream at the level of producers and importers. This approach seeks to improve political acceptability by extending grandfathering to small sources and to avoid high administrative cost by concentrating monitoring of compliance on the few firms operating upstream.

Permits for large-scale fuel users could be grandfathered proportional to their carbon use in a reference year. Note that this is the usual approach when quantity is rationed and quotas are distributed among firms. Grandfathering to small users can be done on a general basis: it is proportional to CO_2 emissions resulting from average fuel use per adult person in a reference year. In this way the administrative cost of establishing the fuel use of every person or household in a reference year is avoided. It can be expected that it will be politically more acceptable to grant permits for a basic good, such as fuel for consumer households, on an egalitarian base rather than proportional to past use.

The scheme requires that an end user who purchases fossil fuels 'pays' for the emissions by handing over emission permits to the fuel distributor. The distributor in turn can only buy fuels if he/she transfers the adequate number of permits to his/her supplier. In this way, the permits end up in the hands of producers and importers of fuel. They have to demonstrate, at the end of the year, that their sale of fuels (potential CO_2 emissions) is covered by an equivalent number of permits in their possession. The number of permits they actually can receive is restricted by the quantity of permits grandfathered to sources. The scheme is to a large extent self-enforcing: the seller of fuels has an interest in receiving the right number of permits from the purchaser. Since checking compliance is restricted to the relatively few upstream firms and the scheme is self-enforcing at the downstream levels, the administrative costs of monitoring and enforcement are kept low.

The administrative cost of initial distribution and the transaction costs of permit transfer and trade between downstream sources can be kept low by delegating implementation to a national agency and organizing the scheme in the form of carbon accounts and making permit transfers using pin-card technology. All participants are registered at the national agency. At the beginning of the year end users receive their permits for that year on their account. The agency also sends them a carbon pin card. When purchasing fuels, for example after filling up at a petrol station, the end user uses his/her pin card to transfer the amount corresponding with the carbon content of the acquired fuel to the permit account which the distributor holds at the national permit agency. Transfer of carbon permits to the account of the gas and electricity distributor can be synchronized with paying gas and electricity bills. Permit trade between end users can be facilitated by placing machines at strategic points where they can electronically increase or decrease their carbon account by buying or selling at the current carbon permit price. The machines are exploited by companies who trade professionally in carbon permits. The current market price arises from the transactions of and between the permit-trading companies. At the end of the year the national agency establishes the balance of the permit account for every user unit. This is equal to: grandfathered permits (via pin card or account) plus the purchased permits minus the permits sold minus the permits used and transferred. This balance can be positive, but not negative. The positive balance is added to the permit account for the next year.

The implementation costs consist of the registration of the participants as well as the yearly allocation of permits and mailing of pin cards. We estimate this roughly as a few euros per participant. The cost of permit transfers between accounts will be comparable to the cost of money transfers between bank accounts. Monitoring focuses on the limited number of fuel importers and producers. Valued at a carbon price of €40 per ton of CO_2, a reasonable estimate of administrative cost is 1 per cent of total carbon value (Nentjes and Rietveld 2000). The transaction costs of purchasing additional permits or selling excess permits will not be higher than 2 per cent per transaction as we know from experience in the US.

The flexibility of the scheme arises from the possibility of trading permits freely, thus allowing firms and households to adjust the number of permits to their actual CO_2 emissions. Families living in small, well-insulated apartments and without a car will end up with a permit surplus at the end of the year, which can either be sold or banked to cover emissions next year or later. This feature of grandfathering permits to all fuel users would increase the political support for the scheme compared to other designs. Simultaneously, the administrative costs, although higher than with an upstream system, are kept low, thus avoiding the major bottleneck of the pure downstream scheme.

CREDIT TRADING

Credit trading was originally developed in the 1980s in the US in the Environmental Protection Agency (EPA) emissions trading programme to introduce flexibility in a stringent scheme of direct regulation of emission standards for sources. The original spatial level for credit trading was the region, since it was developed for non-uniformly dispersing pollutants. In the case of a national climate change policy based on performance standards, the credit-trading scheme could be implemented on a national scale. The legal framework for such a type of national credit trading is already in place in the US and other countries could follow the example.

Credit trading is an instrument that cannot be used on its own but must always be combined with some other instrument that sets a baseline for emissions. The most commonly used instrument is some form of relative standard that does not place a cap on total emissions.[1] Such relative standards can specify the allowed emissions per unit of output or per unit of some input. The allowed emissions per year are calculated by multiplying allowed CO_2 emissions per unit of output (input) by the level of output (input) that year. In this way, the emission baseline is determined *ex post* and need not be estimated beforehand. Although *ex post* estimation of the baseline may seem a large departure from the usual procedure, it does not differ so much from it in the end. Even when the baseline is estimated beforehand, it is continuously adjusted as new developments arise. Hence, in the end, the baseline is the same, whether one estimates it beforehand and adjusts it, or whether one sets it *ex post*.

After the implementation of the relative standard, firms can start to trade credits. A firm can simply start selling credits whenever it expects that its total emissions will be below its baseline. At the end of the trading period, usually a year, all emission sources have to show that they are in compliance, that is, that their actual emissions are below allowed emissions. Allowed emissions are here defined as the relative standard multiplied by the level of output plus the net purchased emission credits. If firms are not in compliance, they could be given a grace period in which they can purchase credits up to the level of their actual emissions as is the case in the US SO_x trading programme.

It should be noted that with credit trading defined in this way, there is no need for explicit abatement projects to create credits. When a change in production leads to lower emissions per unit of output for example, the firm will stay below its baseline. This can hardly be defined as a project. However, the firm would receive credits for this. A consequence of this is that credit trading does not necessarily lead to reductions in emissions by the seller. If the government sets the relative standard higher than the emissions per unit of output would be, the firm can sell credits without reducing emissions. It is clear that this definition of credit trading diverges from the common descriptions of it. However, project-based credit trading can be seen as a restricted version of the general credit-trading design that we give here. Most experience with credit trading is with the project-based kind. However, in the Netherlands, two general credit trading schemes are proposed. In 2001 the Minister of the Environment submitted to Parliament a proposal for NO_x emission trading (for all stationary sources larger than 20 MWth, about 200 firms) which seeks to achieve the NO_x emission target of 55 kilo tonnes in 2010 by setting a performance standard of 50gr NO_x emissions per GJ energy input in 2010 (Kamerstuk 26578 no. 3, Vergaderjaar 2000–2001). Also for CO_2, an emissions trading scheme is discussed. This will be based on an energy efficiency standard set through negotiations with industry. Hence, in the NO_x trading scheme the basis is a relative input standard, while the basis for the CO_2 trading scheme is an output standard.

As will be clear, the above description of credit trading has much in common with permit trading with a cap. The only difference is the instrument that forms the basis of the trading system. With permit trading this is a ceiling on the total emissions of a firm. With credit trading, this is some relative standard that does not set an absolute ceiling on emissions, but a relative one; emissions are allowed to vary with output (or some input).

The disadvantages of credit trading can almost be characterized as the cardinal sins in economics: low effectiveness and low efficiency. The low effectiveness of the system is caused by the low effectiveness of the relative standards that are used as a basis for credit trading. Firms that enter the industry and expanding firms get a licence to emit for free up to the level set by the relative standard. In the case of unexpected rapid economic growth the emission target for the group of sources will be exceeded even though compliance at the firm level is perfect. In the proposed Dutch NO_x emissions trading scheme such a development should be signalled by an evaluation in 2006. If deemed necessary the performance standard will be made more stringent, but not lower than 40 gr NO_x per GJ. This makes it clear that the credit-trading scheme requires more central planning and intervention than a cap-and-trade scheme without the certainty that it will prevent too lax or too late adjustment of performance standards. In cap-and-trade schemes the whole problem is avoided by simply not allowing additional emissions for free once the allowed emissions have been distributed.

Apart from effectiveness, the efficiency of credit trading is also to be criticized. Just as in permit trading, a credit-trading system will improve cost efficiency for the trading firms as compared to no trading. Firms with low emission reduction costs will abate more than is necessary, and can sell credits to firms with high abatement costs. The result is that the marginal abatement costs of the trading firms will be more equal. However, a credit-trading system will not be as efficient as permit trading. With credit trading there is an imbalance between emission reduction through a reduction in production and through other measures. More precisely, production is too high under credit trading based on performance standards, leading to higher marginal abatement costs (Ebert 1998, 1999 and Dijkstra 1999). The reason for this is that by regulating emissions through relative standards, firms cannot comply by reducing total emissions, but only by reducing emissions per unit of output or input. Hence, reducing emissions by reducing output will not result in better compliance. By regulating in such a manner, one efficient possibility of reducing emissions, reducing output, is excluded. It will be clear that this can never lead to the most efficient outcome.

Under relative standards, and thereby credit trading, firms that expand production receive additional emission allowances and firms that reduce production lose emission allowances. The same happens for firms that enter or exit the market. They respectively receive emission allowances for free or lose them on exit. Hence, a firm that wants to leave the market cannot sell its credits. Although this may lead to lower emissions if the other firms do not react by increasing production, it may also give an incentive for inefficient firms to stay in the market. With permit trading, firms with very low profitability would stop production and sell their permits since this will maximize profits. However, with credit trading, reducing output does not generate credits and therefore, firms with low profitability have no incentive to terminate production. In this way, a low-cost option to reduce CO_2 emissions is foreclosed.

The major advantage of credit trading is that it has a high political acceptability (Boom and Svendsen 2000a,b and Boom 2001a). Resistance from industry will be low since credit trading based on relative standards gives firms maximum flexibility. This system allows them to increase emissions with output and provides additional flexibility through the possibility of emissions trading. Furthermore, the emission allowances are distributed for free to the emission sources. Another important factor, especially to export-oriented industry, is that the price of the goods produced in the regulated industries will be lower under credit trading than under permit trading. In this way, industry will have an advantage over foreign competitors regulated through taxes or tradable permits. Precisely this aspect is clear in the proposals for CO_2 emissions trading in the Netherlands. Here the so-called 'sheltered' sectors (those not facing foreign competition) will be regulated through an upstream system based on tradable permits, while the 'exposed' sectors (those facing foreign competition) will be regulated through a credit-trading system based on relative standards.

For all parties involved in designing and participating in the scheme, mainly politicians, civil servants and industry, credit trading has the advantage of familiarity. The basis of credit trading, relative standards, is well known to all groups. It is likely that the most preferred instrument is the one that deviates least from the existing policy (Lindblom 1959).

In many analyses, credit trading is associated with high transaction costs (UNCTAD 1998). Looking at the history of credit trading, this also seems to be vindicated by the practical experience with the instrument. However, in the design of credit trading outlined above, transaction costs will be as low as with permit trading. Transaction costs will first be high when credit trading is based on abatement projects where the baseline is determined beforehand. In that case, the baseline has to be adjusted continuously and the project needs to be monitored constantly to ensure that the projected abatement level is realized. In the credit-trading system described above, such projects are not necessary and nor is *ex ante* estimation of the baseline.

Another commonly held belief is that the preparation time for credit trading is long. Two points have to be taken into account here: the preparation time for setting up the programme and the preparation time for individual trades. At least within Europe, permit trading is a radical break with the past in two ways. First of all, permit trading is based on ceilings on total emissions. This is not a very common way of designing environmental policy. More often, relative standards or taxes are used. With both these instruments, total emissions are allowed to increase. Hence, emission ceilings close the open access to emission space. On top of that, emissions trading is allowed, which also constitutes a break with the past. The process of putting both subjects on the political agenda for discussion and deciding on implementation will take time. With credit trading, the underlying instrument does not constitute a break with the past. Therefore, we expect that less time is needed to implement it. However, credit trading also constitutes a break with the past in that it allows for emissions trading. For both types of private emissions trading, making the necessary adjustments in legislation will be substantive and very time consuming. For example, it has been pointed out that private trading does not comply with the EU rules on environmental regulation. Specifically, it would not lead to emissions that are as low as reasonably achievable (the ALARA principle) and that it would not lead firms to implement the best available techniques (the BAT principle). One can, however, argue that emissions trading does just that by equalizing the marginal costs of abatement. Since credit trading will be based on well-known instruments, the preparation time for setting up the programme will be shorter than with permit trading. If permit and credit trading are both based on sound national policies, and credit trading is designed in the way outlined above, there will not be a large difference in the transaction costs per individual trade. Therfore, we do not expect important differences in transaction costs in the case of national schemes. In Section 4 we shall discuss how far this also holds for international application of the flexibility mechanisms.

The upshot of the above discussion is that although performance standards completed with credit trading are less effective and less efficient than cap-and-trade schemes they may meet less resistance from dominant interest groups, which makes the schemes politically more feasible.

4 INTERNATIONAL PERMIT AND CREDIT TRADING

International Permit Trading

Article 17 of the Kyoto Protocol allows international emissions trading with a cap. Although the Protocol only talks of emissions trading between parties, that is, transactions between the governments of the Annex I countries, there is wide consensus that it can include emissions trading with a cap between private parties (Bohm 1999; Hahn and Stavins 1999; and Zhang and Nentjes 1999). International emissions trading between private parties requires that caps have been established for participating firms. The caps on total emissions have to be set by national authorities and apply to the firms in their respective jurisdictions. Monitoring of firms' emissions, ownership of permits and their transfer, as well as enforcement are also tasks of national authorities in the first instance. The requirements imply that international cap-and-trade schemes cannot work unless national cap-and-trade schemes have already been established. They may comprise the whole economy or selected sectors or a group of sources only. International emissions trading is basically the international linkage of national emissions trading schemes. Once the unit of trade has been defined, for example, one tonne of carbon, firms can trade internationally.

The national agencies to which the implementation of the national private trading scheme has been entrusted, should inform one another about transfrontier permit transactions. Suppose a Dutch firm buys permits from a firm in Denmark. The agency in Denmark registers the reduction of permits of the Danish firm and must ensure that the Danish firm complies by not emitting more than its reduced number of permits allows. The Danish agency also registers that the international transaction has reduced Denmmark's assigned amounts. The Dutch agency registers the increase in the permit account of the firm in the Netherlands' assigned amounts. Coordination here means that the two agencies inform each other and check whether the number of permits sold in Denmark equals purchases in the Netherlands, to ensure consistency in the transfer of assigned amounts. Each agency is responsible for compliance with the after-trade emission ceiling in its own country. Of course the United Nations Framework Convention on Climate Change (UNFCCC) institutions set up for compliance monitoring also have to be informed in due course on the change in assigned amounts of countries, caused by international emissions trading of private parties.

The alternative for international private party emissions trading, conceived as internationally interlinked national cap-and-trade schemes, would be that national authorities define a cap for sectors or groups of firms and place them directly under the control of an international emissions authority which would have to work out and supervise the international emissions trading scheme. It is highly unlikely that national states would be willing to give up so much of their national sovereignty in the face of the possibility of linking national cap-and-trade schemes.

If it is accepted that national authorities enforce compliance of private parties engaged in national and international permit trade, this implies that the private seller of permits is liable in the case of non-compliance. If the seller's emissions exceed the reduced quantity of permits he/she possesses, the national authority has to apply sanctions. Introducing buyer liability for private parties means that the legislator doubts the ability of the emission authority to enforce the scheme and therefore needs buyer liability. Buyer liability would ensure that buyers assess the 'quality' of the permits they buy (will they be covered by genuine emission reductions or not?).

The discussion on seller versus buyer liability (see Yamin et al. 2001; Zhang 2000) is only relevant in so far as liability between parties, that is, national governments that have signed the Kyoto Protocol as an Annex I party, is at stake. Seller liability would conform with the system we have expounded in this chapter. Buyer liability seems to make sense if there are sound reasons to distrust the capability or willingness of some national authorities to enforce their national schemes properly. However, buyer liability would complicate the system of internationally linked national cap-and-trade schemes. A country that has been a buyer and sees its assigned amounts reduced, because it has purchased from a selling country that has not reduced its emissions sufficiently, is under an obligation to tighten the national cap. In the case of grandfathering this would mean an unexpected reduction of permits for firms, thus creating an additional source of uncertainty for firms in the cap-and-trade scheme.

In our view a more appropriate approach to cope with the problem of inadequate monitoring and enforcement by some countries is to allow participation in international private party emissions trading only for countries with certified international cap-and-trade schemes. The criterion used in certification should in particular specify the requirements of registering permit ownership, monitoring emissions, establishing compliance and application of sanctions (see Zhang and Nentjes 1999 and Boom and Nentjes 2000). If a country fulfils these criteria, it is likely that it will comply with its commitment. One can imagine that under the supervision of the UNFCCC, the criteria are drafted and implemented much like the European Union drafted and applied its criteria for participation in the euro-scheme. This solution presupposes that only the seller country is liable. Otherwise, setting up the criteria would have no meaning.

Since not all Annex I countries would meet the requirements for carbon permit trade an international system of linked national tradable permits schemes might initially start with only a handful of countries. Consequently, the 'full efficiency' of the scheme will only be achieved between the subset of participating countries. The Annex I countries not qualifying for participation in the emissions trading scheme, can still trade emissions through joint implementation. A start with a small number of states does not preclude subsequent expansion to include other qualified countries according to the rules of procedure agreed before trading begins. Such an expansion will bring more emission sources into an international permit trading scheme and increase the scope for efficiency gains.

Although the requirements seem rather harsh, it should be noted that they are nothing more than the requirements for prudent national environmental policy. When a country does not satisfy these requirements it will not even be able to either monitor its domestic sources or enforce environmental policy on them, or both. Hence, in that case any domestic environmental policy will fail. In the case of tradable permits the outcome would be disastrous.

International Credit Trading

Similar to cap-and-trade schemes, international credit trading can be crafted upon national credit-trading programmes. National schemes are interlinked internationally by allowing credit trade with private parties in other Annex I countries with well-established national schemes of credit trading. Note that national credit trading normally will be based on national performance standards, allowing limited, free emissions to entrants and expanding firms. Emission reduction credits are earned by emitting less than allowed emissions calculated by multiplying emission standards with the appropriate measure of capacity.

International credit trade is feasible and efficient if conditions are met similar to those relevant for international cap and trade: credits should be expressed in the same unit or different with fixed conversion factors, national agencies should register international credit transfer and inform one another and monitor and enforce compliance of the sources under their jurisdiction. Participation in the scheme would be reserved for countries with certified national credit schemes. The answer to the question on which issues international coordination or even harmonization is necessary are similar to those for the cap-and-trade programme, although the controversy on auctioning versus grandfathering is avoided, since performance standards have always implied grandfathering.

Does the Kyoto Protocol allow international credit trading and if so which article(s) does or do apply here? In our view Article 17 of the Protocol does not apply exclusively to cap-and-trade schemes, but to credit trading as well. The article defines the transfer of parts of assigned amounts between parties' governments; it says nothing about how the underlying international exchange between private parties should be organized. It is clear that the article refers to emissions trading between parties that have committed to an emission ceiling. However, nothing is said to suggest that trade between private parties should be placed under a cap too. The interpretation is that the parties are responsible for meeting the emission ceilings and not the individual firms. It could be argued, and some authors have done so (Yamin et al. 2001; Hahn and Stavins 1999; Janssen 2001), that joint implementation is a type of baseline-and-credit trading. One might be tempted to go one step further and see international credit trade as an activity covered by Article 6 – the joint implementation article. At first sight it seems that despite the similarities between joint implementation and credit trading as specified in Section 3, there are also major differences, making the application of Article 6 to international credit trading less appropriate.

Article 6 specifically mentions that joint implementation should be based on investment projects that lead to emission reductions. Credit trading as defined above, however, does not fit this description. With joint implementation, a party (a national government) can earn emission reduction units (allowing it to raise its emissions above its Kyoto emissions commitment) by reducing emissions below a baseline in a project carried out in the jurisdiction of another party; the emission reduction units will lower the transferred emission ceiling of that other party. The consensus is that joint implementation projects will usually be carried out by private parties in the donor and the host countries. Article 6 states that approval of the project by the parties is required. A plausible interpretation is that the emission reduction units or credits of a joint implementation project are calculated *ex ante*, even before the investment is carried out. They have the property of future (expected) emission reductions over the commitment years 2000 to 2012. Approval and certification of the emission reduction units is required *ex ante*. The parties, that is the governments of the host and donor countries should then *ex ante* agree on the emission reduction units to be transferred from the host country to the donor country.

In our design of international credit trade, the credits need not come from specific emission reduction projects. If a reduction in production leads to lower emissions per unit of product, a lower demand of the product may lead to overcompliance by the firm. In that case, the firm could receive credits since it stayed below the relative standard. However, this would hardly qualify as an emission reduction project. Furthermore, in our design trade in future emissions reduction between private parties is possible, but only registration of the transfer is required. Future emission reduction units are transferred between the private parties and simultaneously between the parties. *Ex ante* approval is not needed. *Ex post*, for example on an annual basis starting in 2008, compliance will be checked at the firm level: whether actual emissions exceed the allowed emissions that have been calculated on the base of the performance standard multiplied by output and transferred credits. When the enforcement regimes in both countries are adequate the legal entities in both countries will comply.

Another difference with joint implementation is that credit trading does not have to lead to genuine emission reductions. For example, the government could set the relative standard higher than the actual emissions per unit of output will be. In that way, the firm can receive credits for staying below the standard without having to reduce emissions. This also implies that credit trading does not necessarily exclude trading in hot air. Only if trading is explicitly based on projects where the baseline is defined on expected emissions, will the permits or credits always be backed by genuine emission reductions. In all other cases, this does not have to be so.

A further question is whether and how schemes of private party emissions trading with a cap (nationally and internationally) can be combined with private party credit trading on an international scale. How do the two schemes interact and what are the consequences for effectiveness and efficiency (see also Nentjes and Rietveld 2000, pp. 183–5)? Some countries may want to implement the two schemes for different sectors, as is the case in the Netherlands. It is even more likely to meet such a combination at the international level. As Boom (2001b) shows, countries may have a preference for one of the trading schemes depending on the trade balance in the goods market and on the difference between the domestic and international price of credits and permits (see also Ulph 1996 and Dijkstra 1998). Some countries will therefore implement a credit-trading system, while others will implement a permit-trading system. A discussion of all the economic implications of continuing credit trade and permit trade requires a separate study and we shall therefore not pursue that line of research here.

5 CONCLUSIONS

In this chapter, we have presented blueprints for two international emissions trading schemes between private parties. Both schemes have in common that a well-functioning international flexibility mechanism requires implementation of well-defined and enforced national schemes: either national cap and trade, or performance standards complemented with credit trading. On this basis, international emissions trading can be organized as internationally linked national schemes with seller liability between private parties and Between parties for assigned amounts. A well-functioning, transparent permit market, or credit market with low transaction costs and low administrative costs, can develop. Internationally linked, well-designed national schemes of permit trading with a cap, or alternatively credit trading, both create flexibility nationally and internationally for carbon users and tend to equalize the marginal cost of CO_2 emission control: a basic tenet of technical efficiency.

The efficiency of permit trading is highly dependent on the design of the domestic schemes of tradable permits. Although a large part of the literature points to an upstream scheme as the best design, such a design implies auctioning of permits, which decreases its political feasibility. An alternative is possible that will potentially perform better. In this alternative design all sources receive tradable permits for free, but the monitoring is done at the level of the suppliers of fossil fuels. In this way, high efficiency is combined with low administrative costs and relatively high political acceptability.

The crucial difference between cap-and-trade and performance standards with credit-trading schemes is how the entry to 'emission space' is organized. Where credit trading is an addition to performance-based direct regulation or covenants, entering firms and expanding incumbent firms can obtain additional emissions for free and firms that terminate their business lose their licence to emit. In permit trading with a cap this is not the case. This may encourage more support from industry for credit trading than for cap-and-trade permit schemes. However, by giving up tradable permits with an emission ceiling as a national instrument, the government sacrifices its control of total emissions and the certainty of realizing the emission goals set in the Kyoto Protocol. In addition, performance standard-based credit trading does not put a price on residual emissions not exceeding the standard. This leads to inefficiently high energy-intensive output, requiring inefficiently high emission abatement.

However, for various reasons, some governments may prefer credit trading to permit trading. It is therefore likely that both schemes will be used and that they will be combined at the international level. Such a combination, however, gives rise to several problems. One is caused by the fact that credit trading does not put a ceiling on emissions. This is already a problem at the national level, but may be aggravated at the international level. Furthermore, a combination of credit and permit trading will lead to inefficiencies.

NOTE

1. Using an emission ceiling as the underlying instrument would transform the system in a permit-trading scheme.

REFERENCES

Bohm, Peter (1999), 'International greenhouse gas emissions trading – with special reference to the Kyoto Protocol', TemaNord 1999:506, Nordic Council of Ministers, Copenhagen.
Boom, Jan-Tjeerd (2001a), 'International emissions trading under the Kyoto Protocol: credit trading', *Energy Policy*, **29** (8), 605–13.
Boom, Jan-Tjeerd (2001b), 'Strategic choice of international emissions trading scheme in an open economy', mimeo, Institute of Economics, University of Copenhagen.
Boom, Jan-Tjeerd and Andries Nentjes (2000), 'Level of international emissions trading: should governments trade, or should firms?', Economic Discussion Paper 4/2000, Department of Economics, University of Southern Denmark, Odense.
Boom, Jan-Tjeerd and Gert Tinggaard Svendsen (2000a), 'The political economy of international emissions trading scheme choice: a theoretical analysis', *Journal of Institutional and Theoretical Economics*, **156** (6), 548–66.
Boom, Jan-Tjeerd and Gert Tinggaard Svendsen (2000b), 'The political economy of international emissions trading scheme choice: empirical evidence', Discussion Paper 00-19, Institute of Economics, University of Copenhagen.
Boom, J.-T., K. van Buiren, P. van Duyse, A. Duizendstraal, E. Dijkgraaf, G.A. van Es, A.W.M. de Groot, H. Heijnes, R. de Jong, J.M.M. Koster, A. Nentjes, M. Varkevisser, J.W. Velthuijsen, D. Wiersma and Z.X. Zhang (1998), 'Market performance and environmental policy: a scenario study for a market oriented environmental policy', SEO-report 460, Foundation for Economic Research, University of Amsterdam, Amsterdam.
Cramton, Peter and Suzi Kerr (1998), 'Tradable carbon permit auctions: how and why to auction not grandfather', Discussion Paper 98-34, Resources for the Future, Washington, DC.
Dijkstra, Bouwe R. (1998), 'The international dimension of environmental policy instruments', ECOF Research Memorandum 18, Department of Economics and Public Finance, Groningen University.
Dijkstra, Bouwe R. (1999), *The Political Economy of Environmental Policy: A Public Choice Approach to Market Instruments*, Cheltenham, UK and Northampton, MA, USA: Edward Elgar.
Duijse, P. van, A. Nentjes, J. Krozer, K. Blok and M. van Brummelen (1998), *Verhandelbare CO_2-emissierechten* (Tradable CO_2 emision permits), Achtergrondstudies 002, VROM-raad, The Hague.
Ebert, Udo (1998), 'Relative standards: a positive and normative analysis', *Journal of Economics*, **67**, 17–38.

Ebert, Udo (1999), 'Relative standards as strategic instruments in open economies', in E. Petrakis, E. Sartzetakis and A. Xepapadeas (eds), *Environmental Regulation and Market Power*, Cheltenham, UK and Northampton, MA, USA: Edward Elgar, pp. 210-32.

Ellerman, A. Denny (1998), 'Obstacles to global CO_2 trading: a familiar problem', Report 42, Massachusetts Institute of Technology Joint Program on the Science and Policy of Global Change, MIT, Cambridge, MA.

European Commission (2000), 'Green Paper on greenhouse gas emissions trading within the European Union', Green Paper, European Commission, Brussels.

Festa, David H. (1998), 'US carbon emissions trading: some options that include downstream sources', Airlie Carbon Trading Papers, Center for Clean Air Policy, Washington, DC.

Fischer, Carolyn and Suzi Kerr and Michael Toman (1998), 'Using emissions trading to regulate U.S. Greenhouse gas emissions. Part 1 of 2: basic policy design and implementation issues', RFF Climate Issue Brief 10, Resources for the Future, Washington, DC.

Hahn, Robert W. and Robert N. Stavins (1999), 'What has Kyoto wrought? The real architecture of international tradable permit markets', Discussion Paper 99-30, Resources for the Future, Washington, DC.

Hargrave, Tim (1998), 'US Carbon Emissions Trading: Description of an Upstream Approach', Airlie Carbon Trading Papers, Center for Clean Air Policy, Washington, DC.

Hargrave, Tim (2000), 'An Upstream/Downstream Hybrid Approach to Greenhouse Gas Emissions Trading', Airlie Carbon Trading Papers, Center for Clean Air Policy, Washington, DC.

Hargrave, Tim, Ned Helme, Tim Denne, Suzi Kerr and Jürgen Lefevere (1999), 'Design of a practical approach to greenhouse gas emissions trading combined with policies and measures in the EC', Working Paper, Center for Clean Air Policy, Washington, DC.

Janssen, J. (2001), 'Does international emissions trading jeopardize joint implementation? Distinguishing the Kyoto mechanisms from economic perspectives', in H. Abele, T.C. Heller and S.P. Schleicher (eds), *Designing Climate Policy: Challenge of the Kyoto Protocol*, Vienna: Austrian Council on Climate Change.

Jepma, Catrinus J., Wytze P. van der Gaast and Edwin Woerdman (1998), 'The compatibility of flexible instruments under the Kyoto Protocol', Report 410 200 026, Dutch National Research Programme on Global Air Pollution and Climate Change, Bilthoven.

Klaassen, Ger and Andries Nentjes (1997), 'Sulfur trading under the 1990 CAAA in the US', *Journal of Institutional and Theoretical Economics*, **153** (2), 384-410.

Koutstaal, Paul R. (1993), *Verhandelbare CO_2-emissierechten in Nederland en de EG* (Tradable CO_2 emission permits in the Netherlands and the EU), Ministry of Economic Affairs, The Hague.

Koutstaal, Paul R. (1997), *Economic Policy and Climate Change: Tradable Permits for Reducing Carbon Emissions*, Cheltenham, UK and Northampton, MA, USA: Edward Elgar.

Lindblom, Charles E. (1959), *The Science of 'Muddling Through'*, Indianapolis: Bobbs-Merrill.

Nentjes, Andries and Piet Rietveld (2000), *Verhandelbare Rechten voor Verkeer en Vervoer als Instrument van Klimaatbeleid* (Tradable permits in transport as

instrument of climate change), Achtergrondstudie, VROM-raad, The Hague.
Olson, Mancur (1965), *The Logic of Collective Action*, Cambridge: Cambridge University Press.
Schmalensee, R., P.L. Joskow, A.D. Ellerman, J.P. Montero and E.M. Bailey (1998), 'An interim evaluation of sulfur dioxide emissions trading', *Journal of Economic Perspectives*, **12** (3), 53-68.
Stavins, Robert N. (2000), 'Experience with market-based environmental policy instruments', KSG Working Paper 00-004, Harvard University, Cambridge, MA.
Svendsen, Gert Tinggaard (1998), *Public Choice and Environmental Regulation: Tradable Permit Systems in the United States and CO_2 Taxation in Europe*, Cheltenham, UK and Northampton, MA, USA: Edward Elgar.
Tietenberg, T.H. (1989), 'Marketable emissions permits in the U.S.: a decade of experience', in Karl W. Roskamp (ed.), *Public Finance and the Performance of Enterprises*, Detroit, MI: Wayne State University Press.
Ulph, Alistair (1996), 'Environmental policy instruments and imperfectly competitive international trade', *Environmental and Resource Economics*, **7**, 333-55.
United Nation Conference on Trade and Development (UNCTAD) (1998), *Greenhouse Gas Emissions Trading: Defining the Principles, Modalities, Rules and Guidelines for Verification, Reporting and Accountability*, Geneva: UNCTAD.
Woerdman, Edwin (2000), 'Organizing emissions trading: the barrier of domestic permit allocation', *Energy Policy*, **28**, 613-23.
Yamin, Farhana, Jean-Marc Burniaux and Andries Nentjes (2001), 'Kyoto mechanisms: key issues for policy-makers for COP-6', *International Environmental Agreements: Politics, Law and Economics*, **1** (2), 187-218.
Zhang, Zhong Xiang and Andries Nentjes (1999), 'International tradable carbon permits as a strong form of joint implementation', in Steve Sorell and Jim Skea (eds), *Pollution for Sale: Emissions Trading and Joint Implementation*, Cheltenham, UK and Northampton, MA, USA: Edward Elgar, pp. 322-42.

4. To design and implement climate change measures and the need to strike a balance between environmental protection and international trade law

David Grimeaud

1 INTRODUCTION

The design and the enacting of an ambitious international climate change regime are difficult and politically sensitive tasks. In particular, the Conferences of the Parties (COP) of the 1992 United Nations Framework Convention on Climate Change (UNFCCC) have clearly shown how complex it can be to reach an agreement on the operationalization of the 1997 Kyoto Protocol (KP).[1] Parties have indeed long been divided over numerous key issues, including the use and accounting of sinks as a means to achieve parts of Annex I parties' quantified greenhouse gas (GHG) emission limitation and reduction commitments (QELRCs), on the amount of QELRCs to be realized at the domestic level (the supplementary issue) and on the features of the future compliance mechanism. *De facto*, a breaking point was eventually reached at COP6 (13-24 November 2000, The Hague) where the inability of the United States (US) and of the European Community (EC) to agree upon such questions led to its suspension.[2] However, several of those conflicts have since then been resolved at COP6 (II) held in Bonn on 16-27 July 2001 and at COP7 (29 October-9 November 2001, Marrakesh).[3] In this respect, we might also expect some of the remaining disputes to be resolved at COP8 (23 October-1 November 2002, New Delhi).[4] In sum, whereas international negotiations on climate change remain heated and rather slow in providing ready-to-implement outcomes, we might still hope that forthcoming COPs would eventually result in an overall international agreement on the interpretation and the scope of application of the whole set of relevant UNFCCC and KP provisions and on the design of the tools to be adopted by Annex I parties to achieve their commitments.

In addition to issues regarding the framing of those rules that would eventually make the KP an applicable instrument, concerns may also be raised regarding impacts on international trade that might result from national and Community climate-based measures adopted pursuant to the UNFCCC, the KP and COP decisions. Indeed, as specified in Article 2(3) of the KP, the latter is not to apply in a legal vacuum, but rather in compliance with international trade law. Accordingly, parties have the duty to strike a balance between GHG-related provisions and World Trade Organisation (WTO) law that govern international trade.[5] In this respect, the examination of the interplay between multilateral and national climate change measures and international trade law may also depend on whether the KP enters into force. Indeed, as a consequence of its rejection by the new US administration led by President G.W. Bush, it is still, at the time of writing, far from clear whether sufficient ratifications by Annex I parties, other than the US, are made in the medium term so as to convert the KP into a legally binding international instrument. Accordingly, the examination of the treatment that may be provided to national and EC climate change regulation that might affect international trade law should take account of three different potential scenarios, including the unlikely case of full ratification (entry into force of the KP following ratification by all Annex I parties), the case of partial ratification (entry into force of the KP following ratification by sufficient Annex I parties) and the case of non-entry into force of the KP as a result of insufficient ratifications.

Against this background, this chapter provides an overview of the potential international trade law conflicts that may occur in the course of the implementation of national climate change policies and measures. Indeed, whereas climate change policy-makers are still engaged in an ongoing international negotiation process that may affect the content and the entry into force of the KP, there are countries and regional economic organizations, including Member States and the EC that are already developing climate change policies and legislation. In this regard, questions are thus raised as to the legality of such measures in case of resulting international trade restrictions in related goods and services, including in the light of each of the three above-mentioned ratification options.

Accordingly, Section 2 will look at the background upon which the relationship between international trade and climate change should be examined. More particularly, attention will be paid to the US and EC regulatory developments as well as to the current ratification status of the KP, which may then alter the scope of potential international trade disputes. Within this framework, Section 3 will study the case of full and partial ratification. In doing so, we shall examine how potential conflicts between multilateral environmental agreements (MEAs) such as the KP that may contain trade-restrictive measures and international trade law might be addressed. In this regard, attention will be devoted to those legal provisions included in the KP whose implementation may result in international trade impacts and disputes among ratifying parties (full ratification) and between ratifying and non-ratifying countries (partial ratification). Section 4 will look at the case of unilateral domestic climate change policies that may be adopted by individual States in the absence of ratification of the KP and the conditions under which they may be held (in)compatible with international trade law. More specifically, we shall examine the extent to which individual countries may be able to provide subsidies to certain companies or sectors for promoting, for instance, energy-efficient devices, to adopt taxation and technical regulations that would aim at, *inter alia*, reducing the marketing and use of high GHG-emitting products and to establish a national emissions trading scheme (ETS). Section 5 will draw relevant conclusions.

2 CLIMATE CHANGE NEGOTIATIONS AND INITIATIVES: LAYING DOWN THE GROUNDS FOR SEVERAL RATIFICATION SCENARIOS AND INTERNATIONAL TRADE DISPUTES

As mentioned above, the occurrence of potential international trade conflicts as a result of the implementation of climate change measures may depend on whether national (and EC) GHG-related regulations are adopted and, if so, on their nature and scope as well as on the status of ratification of the KP.

Regarding climate change regulatory frameworks, differences in approach, from which trade barriers may further derive, are perhaps best visible when we compare relevant US and EC measures. In fact, at the time where COP8 is about to resume, common grounds of understanding between those two parties on the significance of climate change is still a far prospect.

The examination of the current US policy may first be based upon a declaration of Christy Whitman, 27 March 2001, who, as head of the US Environmental Protection Agency, stated that the USA had no interest in implementing the KP.[6] This statement made official the position previously announced by President Bush himself who confirmed that there would be no US binding law adopted to limit carbon dioxide emissions (CO_2) from power stations.[7] He argued that CO_2 did not qualify as a 'pollutant' under the US Clean Air Act and, more importantly, that he opposed to the KP on the ground that it would exempt 80 per cent of the world from compliance (namely, developing country parties), that it would cause significant harm to the US economy at a time of rising energy prices and national energy shortages, that there was still incomplete scientific knowledge of the causes of, and solutions to, global climate change and a lack of commercially viable technologies for removing and storing CO_2.[8] Additionally, he affirmed on 11 June 2001 that the KP was 'fatally flawed in fundamental ways' and that US action of global warming would consequently merely focus on developing climate change science and technology to the exclusion of any concrete measures that would aim at curbing emissions of CO_2 and other GHGs.[9] In sum, the US are clearly unwilling to adopt any ambitious national climate change policies and measures, which is likely to reflect internal economic and political factors, including the fear of affecting the US energy-related industry and citizens' energy-consuming patterns.[10] In practice, not only has the review of US climate change policy concluded that national measures are, for now, only to focus on developing climate change science and modeling, but the May 2001 draft US *'National Energy Policy'* goes even further by mainly calling for the strengthening of US energy supply and sources without addressing the need for measures on the control of the energy-demand and on energy saving and efficiency.[11]

In contrast, the EU has asserted, at the highest political level, its strong commitment to the KP and its intention to continue international negotiations with a view to reaching agreement on modalities for implementing it.[12] In the same vein, the Swedish Environment Minister Kjell Larsson, the then president of the EU Environment Council, declared on 31 March 2001 that the KP was still alive, that no individual country had the right to declare a multilateral agreement as dead and that the EU clearly intended to ratify the KP, with or without the US, by 2002 at the latest.[13] Similarly, the 7 June 2001 Luxembourg Environment Council Conclusions, which set out the general EU position for forthcoming climate change negotiations, made clear that the US arguments for opposing the KP were ill founded. In particular, they recalled that developed country parties must take the lead in combating global warming on the basis of the principle of common but differentiated responsibilities, that the January 2001 Intergovernmental Panel on Climate Change (IPPC) Third Assessment Report on evidence of existing and growing climate change had laid a sound scientific basis for international actions, that the core elements of the KP on binding GHG emission reduction targets, environmental integrity and compliance were essential and that any attempts to replace the KP by another instrument would cause unnecessary delays. In addition, contrarily to the US argument on the present lack of cost-effective measures, attention drew to the current and potential technological and economic abilities to implement low-cost and win-win climate change policies that would, in turn, address potential major internal and international political and competitive concerns.[14] In sum, following official statements that whereas the EC agrees with the US on the need to continue working on relevant science and technology, it has committed to more concrete actions, which clearly differentiate the US and the European approach to the KP and, more generally, to climate change.

In this respect, the EU has not only recently ratified the KP,[15] but it has also developed a European Climate Change Programme (ECCP), which aims at a twin-track approach. In particular, it calls for the establishment of a EU-wide emissions trading system (ETS) and for the adoption of priority common and coordinated policies and measures (addressing the energy, industry, transport, waste and agriculture sectors) that will form the EC Commission's forthcoming legislative proposals.[16] The latter would include directives on, *inter alia*, the promotion of renewable energies, energy performance of buildings or on fluorinated gases. Other measures for which more work shall be done are also expected such as a (framework) directive on minimum efficiency standards for electrical equipment and on energy services, the enacting of fiscal measures for passenger cars and the conclusion of long term agreements with energy intensive industries.

In addition to EC measures, supplementary regulatory means will also be adopted by Member States at the national level to ensure that they realize at the domestic level their emission reductions imposed under the KP and the EU burden-sharing agreement.[17] As an illustration, the June 1999 Dutch Climate Policy Implementation Plan draws a distinction between measures falling under a basic package (including voluntary agreements with all major sectors completed by regulation, subsidies and positive tax incentives aiming at energy conservation, reduction of carbon intensity of fuel use and promotion of renewable energy) and a reserve package.[18] The latter, which would include measures on the raising of the regulatory energy tax and of the excise duties on motor fuels, would eventually be implemented when the basic package would not be sufficient in achieving sufficient carbon dioxide (CO_2) reductions.[19] Simultaneously, an innovation package would also be adopted, which would seek to foster climate-friendly technology developments such as climate-neutral energy carriers and to establish a national emissions trading scheme.[20] In the same vein, the January 2000 French national program against climate change also provides for the signing of CO_2 voluntary agreements, energy taxation, public aid, eco-labelling and information campaigns and covers, *inter alia*, the industry, transport, forestry, waste and energy production sectors.[21] Regarding emissions trading, reference should also be made to those national schemes that have already been implemented in Denmark and in the UK and which may have to be adjusted once the Community ETS would have entered into force.[22/23]

The comparison between the US and the EC regulatory approach towards climate change shows the extent to which developed countries may or may not be willing to embrace relevant international commitments and to reduce their GHG emissions. While the US administration has rejected the KP on the ground that such an instrument is not science-driven but biased and has announced that no ambitious domestic climate-related policies would be adopted at least in the short term, the EC and Member States have, conversely, ratified it and are setting up a relevant regulatory framework. In this respect, beyond the mere EC-US opposition, concerns may thus be raised regarding the potential impacts that may affect trade between countries which would take climate initiatives that may differ in nature and in degree, or between a State that would adopt GHG-related legislation and one which would not enact any of such provisions.

In fact, as examined below, the scope and acceptability of trade conflicts and the parties that may be involved may depend on whether we deal with national and Community measures adopted under the KP as a legally binding MEA [which would have been ratified by all Annex I parties or by a sufficient number of such parties] or pursuant to unilateral climate legislation enacted in the absence of entry into force of such an instrument.

At the time of writing, criteria laid down in Article 25 KP that set the conditions under which the Kyoto Protocol would become binding are not yet fulfilled.[24] More specifically, whereas more than 55 parties to the UNFCCC have now ratified, approved, accepted or acceded to the KP, ratifying Annex I parties do not yet account for 55 per cent of the total Annex I parties' emissions as required, but only for 37.1 per cent.[25] Thus, for the KP to enter into force, further ratifications by concerned countries are needed, including, for instance, by Russia (17.4 per cent), Canada (3.3 per cent), Australia (2.1 per cent) and by the US (36.1 per cent). However, whereas recent announcements by Poland (3 per cent) and Russia at the September 2002 Johannesburg World Summit on Sustainable Development (WSSD) of their intention to ratify it have raised expectations on the achievement of a partial ratification of the KP in the medium term, serious doubts still exist on the occurrence of a full ratification given the rejection of the Kyoto Protocol by the US administration. Therefore, the occurrence and nature of international trade disputes may, to a large degree, depend on the ability of Annex I parties, including the EU to convince a sufficient number of other concerned countries to ratify the KP so as to ensure, at a minimum, its entry into force, via partial ratification.

In sum, general questions may be raised as to what impacts on international trade may occur in case of implementation of climate change policies and measures. If the KP is ratified by the EU, its Member States and most Annex I parties, it should then result in the enacting of Community and national legislation and regulations on energy-related and GHG-emitting production that may impede trade in relevant goods and services among ratifying countries and, to an even greater extent, between the latter and non-ratifying States that would not have undertaken any international environmental obligations on climate change.

3 NATIONAL CLIMATE CHANGE MEASURES ADOPTED IN THE COURSE OF THE RATIFICATION IMPLEMENTATION OF THE KP

As mentioned earlier, Section 3 will attempt to provide an overview of the treatment that may be given to national climate change measures, which would be adopted under the KP either in the course of full or partial ratification and that may affect the trade in energy and energy-related products among ratifying countries and between ratifying and non-ratifying States. In this regard, we shall first provide a survey of the current debate on the relationship between MEAs and WTO law, in particular where international environmental instruments contain trade-restrictive provisions. The question is to determine what treatment *vis-à-vis* international trade law should be given to national measures that would be adopted pursuant to trade-related provisions contained in MEAs. On this basis, we shall then look at whether Article 2 KP on policies and measures (PAMs) and Article 17 KP on international emissions trading call expressively for trade restrictions or whether international trade conflicts would eventually arise indirectly depending on how they are implemented by each individual party in national law.

A. Overview on the Interaction between MEAs Containing Trade-Restrictive Measures and International Trade Law

When we examine, under an international trade law perspective, the compatibility of national measures adopted pursuant to an MEA such as the KP and which may affect two ratifying parties or, conversely, a ratifying and a non-ratifying party, we should first study the state of the existing debate on the interaction between international environmental and trade instruments.

In fact, attention has been drawn to the MEA-WTO relationship following the 1991 US Tuna/Dolphin case. In particular, as the panel decision held that Article XX(b) and (g) GATT could not be relied upon to justify the protection of resources that would be found outside the jurisdiction of the invoking country,[26] many authors then argued that such a ruling would prevent WTO members that are also parties to MEAs from taking trade-restrictive measures on the imports of certain products, resources or species in accordance with international environmental agreement provisions or in the light of their objectives.

Within this framework, a broad distinction may be drawn between those MEAs that contain trade-restrictive measures, which apply to both parties and non-parties and, on the other hand, those that expressly provide for trade barriers against non-parties.

Concerning the former, we can refer, for instance, to the 1973 Convention on International Trade in Endangered Species of Wild Fauna and Flora (CITES).²⁷ In particular, whereas Articles III, IV and V require the enacting of a strict regulation for controlling the exporting and importing of species listed in Appendix I, II and III (trade in species may only be allowed on the basis of an export and import permit system), Article X stipulates that trade may also take place with non-parties, but provided that documentation comparable to the Convention's requirements is produced. Thus, not only is the trade in species between parties regulated but also the one that may occur between a non-party and a party to the Convention. In sum, whether a country is party or not to CITES, it is not able to trade freely with another State that is a party to the Convention even though it has not committed itself to any international fauna and flora protection requirements. However, provided that all States are treated equally, we may assume that such provisions would not be considered as discriminatory. Yet, we should recall that Article XIV also allows parties to take stricter domestic measures, including the prohibiting of trade in endangered species. In that respect, such bans may clearly amount to a violation of Article XI GATT on the prohibition of quantitative restrictions on imports and exports.²⁸ In the same vein, we may also mention the 1989 Basel Convention on Transboundary Movements of Hazardous Wastes (the Basel Convention).²⁹ While the Convention seeks to regulate the transport and disposal of hazardous and other waste, it provides parties with right to ban the entry or disposal of foreign hazardous wastes in its territory (Article 4(1)(a)). In such a case, any other parties would have the obligation to prohibit the export of wastes to those countries (Article 4(1)(b)). In addition, parties are also to prohibit the import and export of wastes if they have reason to believe that they will not be managed in an environmentally sound manner (Article 4(2)(e)). Article 4(6) also provides that no hazardous waste will be disposed of within the area below the 60° south latitude. Regarding the case of non-parties, Article 4(5) stipulates that parties to the Convention will prohibit the import from them and the export to them of hazardous wastes, except if it takes place in the framework of bilateral, multilateral or regional agreements that would not derogate from the obligation to ensure an environmentally sound management (Article 11). Thus, contrary to Article XI GATT, the Basel Convention permits and, in some cases, requires the banning of the importation and exportation of hazardous wastes. This applies not only to parties but also to non-parties. Where transboundary movements of waste are not prohibited by domestic regulations, they will still be subject to a set of requirements, including notification, packaging, authorization and prior documentation (Article 4(7)). Furthermore, Article 4(9) specifies that such movements are only to take place when the exporting State does not have the technical capacity or suitable sites for disposal. Finally, for a transboundary waste movement to occur, the exporting State would first have to notify the importing and transit State who would then be free to give their consent in writing or to reject it or to impose further conditions (Article 6). Therefore, a set of strict requirements has been imposed on all countries that limit and control imports or exports of hazardous wastes. Thus, there may be room for violation of the GATT/WTO principles on international trade, in particular of the above-mentioned Articles I (the most-favored-nations principle) and III GATT (the national treatment principle) where a differentiated treatment is made between imported hazardous wastes and between imports and domestic wastes.[30]

As far as the second set of MEAs is concerned, we may cite the 1987 Montreal Protocol on Substances that Deplete the Ozone Layer (the Montreal Protocol).[31] Whereas Articles 2 and 3 establish timetables for quantified reductions of controlled substances that deplete the ozone layer, Article 4(1-4) prohibits parties from importing such substances from and exporting them to countries that are not parties to the Protocol. Similarly, Article 4(5) requires parties to discourage the export of technology that could allow the production and use of ozone depleting substances to non-parties.[32] Therefore, the Montreal Protocol provides expressly for trade restrictions that apply to non-parties. Whereas the objective was clearly to create an incentive for all countries to sign and ratify the Protocol, which may, in turn, ensure its effectiveness, we may also argue that such provisions do violate the GATT/WTO principles of non-discrimination as referred to in Articles I and III and XI.

Therefore, the above-mentioned environmental instruments show that there can be conflicts between MEAs' trade-restrictive provisions and the principles and rules of the GATT/WTO, including Articles I, III and XI.[33] However, since there have never been until now any disputes brought before the WTO following the implementation of a national measure taken pursuant to an MEA, some authors have argued that this issue does not deserve *de facto* much attention. Yet, would such an argument remain the same in the event of the ratification and implementation of the KP? Indeed, the political context, the content and outcomes of the climate change negotiations show that great economic and competitiveness concerns have emerged as a barrier to its entry into force. This relates, in particular, to the fear that the consumption of and the international trade in energy and energy-related products would be seriously affected. As examined below, the question would then be to determine whether trade-restrictive national measures adopted pursuant to Articles 2 and/or 17 KP would derive directly from the content of those provisions or from the way each individual party will implement them at the national level.

In any case, we should also note that the interaction between MEAs and WTO law is still subject to on-going debates and work by the WTO

Committee on Trade and Environment (CTE).[34] Indeed, there has as yet been no agreement on the treatment that should be given to MEA trade-restrictive provisions or on the choice of the dispute settlement mechanism (WTO panels or MEA dispute settlement forum) that should be used in the case of a conflict.[35]

More particularly, regarding disputes that may arise between two countries that are parties to both an MEA that contains trade-restrictive provisions and to the WTO (the case of full ratification of the KP), several proposals have developed with a view to reconciling them. Some have claimed, for instance, that an MEA should be considered as an *inter se* agreement based upon the principle of mutual consent that would supersede GATT obligations. Other authors have argued that Article 30(4)(a) of the Vienna Convention on Law of Treaties should apply, which would imply that, where two treaties deal with the same subject matter, the more recent in time should prevail. However, whereas the KP is undoubtedly later in time *vis-à-vis* the WTO, it remains to be seen whether the two instruments would be considered as addressing the same issues. In this regard, many would claim that the WTO is not an environmental treaty but is concerned with environmental issues only when they affect trade liberalization. Third, we may refer to the MEA, including the KP, as a more specific instrument (*ex specialis derogat generali*) than the WTO under customary international law.[36] In practice, in the event of a trade-related conflict between the KP and the WTO, much may depend on which dispute settlement body would eventually be in charge of addressing relevant issues. In this context, it has been argued that conflicts arising between two WTO members that are also parties to the MEA over trade measures taken pursuant to the MEA should first be dealt with under the MEA dispute settlement mechanism.[37]

Alongside full ratification, a more complex situation may arise in the event of partial ratification resulting in trade conflicts between parties and non-parties. Indeed, whereas full ratification would imply that all parties have mutually agreed to the same commitments, therefore reducing the likelihood of conflicts, we cannot reasonably expect a party that has not committed itself to specific environmental obligations, to waive its rights under international trade law. In the event of a trade-restrictive national measure taken pursuant to a specific MEA provision or to achieve one of the instrument's objectives, the affected non-party may claim a breach of Articles XI, I or III GATT. Regarding the forum for dispute settlement, we can assume that recourse to the WTO would be the appropriate means as a non-party to an MEA could not reasonably be brought before a mechanism set by an instrument to which it has not agreed.

In this framework, we can then examine whether relevant provisions of the Kyoto Protocol do fall within the range of those MEAs that contain trade-restrictive measures, and, if so, whether this would affect both ratifying and non-ratifying countries. Indeed, if this were the case, would ratifying parties be entitled, *vis-à-vis* international trade law, to implement national measures in the light of such provisions even if they were to violate GATT/WTO rules and principles?

B. KP Provisions: Background for Country-Specific and Trade-Restrictive National Climate Change Measures

The following paragraphs will examine whether relevant KP provisions allow its implementation at the national level in such a way as to minimize international trade impacts and conflicts.

As the core KP provision, Article 3(1) stipulates that Annex I parties will ensure that their aggregate anthropogenic carbon dioxide equivalent GHG emissions do not exceed their assigned amounts (AAs) as specified in Annex B.[38] Thus, contrary to Articles 4(2)(a) and (b) UNFCCC, which only provide for emission commitments in broad terms, the Protocol imposes legally, clear and quantified targets on developed-country parties. More specifically, while all Annex I parties, taken together, will collectively reduce their overall GHG emissions by at least 5 per cent below 1990 levels by the first 2008-2012 commitment period, each of them is to achieve individually or jointly their assigned level of emission reductions. In this regard, we can recall that a degree of differentiation has been agreed upon so as to take account of each party's specific circumstances. Thus, while the EU and all Member States have to reduce their aggregate emissions by 8 per cent, Japan and the US have a 6 and 7 per cent quantified emission reduction commitment imposed on them. Conversely, whereas the Eastern and Central European countries must achieve a minus 8 to minus 5 per cent target, Russia and Ukraine are entitled to simply stabilize their GHG emissions by 2008-2012 *vis-à-vis* 1990 levels while Norway, Australia and Iceland are allowed to increase them by respectively 1, 8 and 10 per cent.[39] Concerning in particular the EU, note that a burden-sharing agreement has been concluded whereby each Member State has been allocated differentiated QELRCs.[40] More specifically, whereas Denmark, Germany and Austria have to reduce their emissions by 25 per cent, Belgium, the Netherlands and the UK have to comply with a minus 10 per cent target. France only has to stabilize its GHG emissions whereas the four 'cohesion countries' (Ireland, Spain, Greece and Portugal) are allowed to increase their emissions by, respectively, 15, 17, 30 and 40 per cent.

Thus, on the one hand, the KP imposes compulsory targets, which require the adoption of concrete actions by ratifying States. Climate change measures may include, *inter alia*, the provision of subsidies for low-energy-intensive industries, the enacting of taxation schemes on fossil fuel production or use, the setting of regulatory standards on GHG-emitting products and the establishment of national (or EC) emissions trading systems. Yet, on the other hand, the scale of efforts that would have to be undertaken by those parties may well vary among ratifying States. Indeed, not only have QELRCs been set at country-specific levels, but also the volume of emission reductions they require will depend on the amount of emissions each individual party reported for the year 1990.[41] Whether such a differentiation in emission reduction commitments would affect the scope of potential trade impacts and trade-related national measures remains to be seen. Indeed, considering their 1990 emission level and the fact that GHG emissions have continued to increase in the last decade, some Annex I parties might have to cut their emissions considerably so as to achieve their QELRCs by 2008–2012. The scope of such reductions may then entail the adoption of comprehensive trade-restrictive measures, including import restrictions on certain energy products such as fossil fuel.

Within this framework, the relationship between international trade law and the achievement of the above-mentioned emission-related objective of the Protocol shall also be apprehended by examining the instruments that Annex I parties may be using to achieve their commitments and, in particular, whether the KP provides expressively for trade restrictions and for their harmonization. Indeed, we can assume that, if the Kyoto Protocol were to contain detailed and clear provisions on the type of measures that would have to be adopted by all ratifying States to curb emissions, there would then be less room for trade disputes as a level playing field would be established.[42] In this respect, distinctions can be made between provisions that address emission reductions to be realized at the national and international level through the use of emissions trading.[43]

With regard to the national policies and programmes that Annex I parties would enact to reduce GHG emissions at the domestic level, Article 2(1) KP refers explicitly to measures that would seek, *inter alia*, to improve energy efficiency in relevant national sectors, promote and develop renewable forms of energy, progressively reduce market imperfections, fiscal incentives and subsidies and limit emissions from transportation. Yet, Article 2(1) KP provides only a non-exhaustive list of measures that are not compulsory. Indeed, Annex I parties will retain the right to decide what type of policies and measures they will adopt and to what degree on the basis of their national circumstances, namely the structure their economies and development priorities. Against this background, Articles 2(1)(b) and 2(4) KP do nevertheless provide for cooperation and possibly coordination. More particularly, whereas Article 2(1)(b) KP stipulates that concerned parties will cooperate throughout experience and information-sharing so as 'to enhance the individual and combined effectiveness of their policies and measures'[44],

Article 2(4) KP provides that the Conference of the Parties serving as the meeting of the parties to the Kyoto Protocol (COP/MOP) may adopt decisions on the coordination of national PAMs.[45] It would do so if it felt that this would be beneficial and by taking national circumstances into consideration. In this context, the term 'beneficial' may imply that PAM coordination may become necessary if, in violation of Article 2(3) and 3(14) KP, the implementation by Annex I parties of their own of national climate change measures would affect developing country parties referred to in Article 4(8) and (9) UNFCCC and would lead to impacts on international trade. However, even if coordination occurs, to what extent will it take place since national circumstances, namely economic, industrial, social and political, will still be a priority?

Accordingly, we can argue that the lack of harmonization of national PAMs may increase the likelihood of international trade disputes. Indeed, the mere fact that Article 2 KP does not require all ratifying parties to adopt similar domestic climate change measures would result in different policies on energy, energy-related and emitting products. In turn, the absence of international standards is likely to affect international trade and competition. Country A may opt, for instance, for the implementation of regulatory standards or taxation regimes on energy efficiency, which may force manufacturers to produce more climate-friendly goods while country B may give priority to a domestic emissions trading scheme that may lead to lower consumption of high intensive energy sources such as coal.[46] Thus, the enacting of country-specific climate change measures under Article 2 KP may give rise to trade conflicts not only between ratifying and non-ratifying States (partial ratification) but also among ratifying parties (partial and full ratification). In this context, we should note, however, that any conflicts would derive directly from Article 2 KP, which does not expressly require national trade barriers or quantitative restrictions. They would rather indirectly result from the achievement of an objective, namely the realization of GHG emission reductions at the domestic level. Thus, in the event of a conflict, a ratifying State may not be able to argue that its national GATT/WTO inconsistent measures stemmed from a specific international environmental obligation taken pursuant to an MEA, but would rather be considered as a unilateral trade-restrictive provision. Indeed, since Article 2 KP does not state what specific trade-restrictive measures would eventually have to be enacted, this provision would not be questioned in the light of international trade law principles. In fact, in the absence of international standards, the parties themselves would have to make sure that the enacting of PAMs would not result in trade barriers inconsistent with GATT/WTO rules.

Yet, we can also recall that both the UNFCCC and the KP contain provisions that may limit the scope of climate change measures that may be adopted under it to ensure compliance with WTO law. First, Article 3(5)(2nd sentence) UNFCCC on principles holds that measures, including unilateral ones 'should' not constitute a means of arbitrary or unjustifiable discrimination or a disguised restriction on international trade. Therefore, we can argue, on the one hand, that for climate policies enacted under the UNFCCC, governments are to be careful not to breach Article XX GATT head-provision on general exemptions. National authorities may enact climate change policies and legislation that might result in trade barriers in violation of Article I, III and XI GATT, but the same measures would have to be saved or justified in line with the conditions set out in Article XX GATT. Yet, in addition to being a mere principle, note, on the other hand, that Article 3(5)(2nd sentence) UNFCCC uses the term 'should' (as opposed to 'shall') to qualify such an obligation. Thus, does such an expression impose a clear-cut duty on parties not to violate Article XX GATT or is it only a principle that will guide national climate actions? However, it should also be recalled that such a principle is contained in Article 3 UNFCCC, which falls under the operational part of the Convention (as opposed to being mentioned in its preamble) and that its head-provision stipulates that parties 'shall' be guided by, *inter alia*, Article 3(5) UNFCCC principle. Regarding the KP, first, the fourth indent of its preamble stipulates that the parties are 'guided by Article 3 of the Convention', including Article 3(5). Second, Article 2(3) KP holds that Annex I parties will 'strive' to implement their policies and measures in such a way as to minimize the effects on international trade. How should the term 'strive' be interpreted? We can assume that it would imply a duty to take all reasonable measures to try to reduce as much as possible violations of GATT rules. In sum, from a theoretical point of view, we could question whether Articles 3(5) UNFCCC and 2(3) KP would affect the examination of the compatibility of the climate change regime *vis-à-vis* the WTO. More particularly, to what extent do relevant provisions contained in both the UNFCCC and the KP impose a well-defined obligation on Annex I parties not to breach international trade law rules and principles when adopting national climate measures?

As to GHG emission reductions that would be achieved internationally, the KP provides for the possibility of using flexible instruments, namely joint implementation, the clean development mechanism and emissions trading. In this regard, this chapter will address only the case of the establishment of international emissions trading as referred to in Article 17 KP. In fact, when we examine the consistency of such a regime *vis-à-vis* international trade law, we should draw a distinction between the relevant provisions that would eventually be adopted by COPs and those that would be enacted by each individual party at the domestic level. In particular, similarly to Article 2 KP on PAMs, it seems as if the setting-up of an international emissions trading scheme will be subject to a division of responsibilities. On the one hand, COPs will define, in accordance with Article 17 KP, the modalities, principles and guidelines on verification, reporting and accountability. On the other hand, parties will be in charge of organizing the participation of national legal entities (regulated companies, industries or sectors) into such a regime, which would include the enacting of national rules on emission allowance allocation and trading. In this context, concerns may thus be raised under which terms both relevant COP and national provisions may or may not result in international trade law conflicts.

Regarding draft emissions trading guidelines that were formulated at COP7, we should recall that they propose, among others, to impose on Annex I parties the duty to comply with a set of eligibility criteria so as to enable them to participate into the international trade in parts of assigned amounts (PAAs).[47] In particular, concerned countries would have to demonstrate that they realize the requirements set in Articles 5 and 7 KP on the establishment of a national system of estimation on GHG emissions by sources and removals by sinks and on the production and submission of an annual inventory. Accordingly, a developed-country party might not be able to trade PAAs with another Annex I party who would not have fulfilled its eligibility criteria. In other words, Country A would be prohibited from selling or acquiring PAAs from Country B if the latter has not complied with some of its other KP requirements. In the same vein, the draft guidelines refer to the possibility for Annex I parties to conclude bilateral, regional or multilateral arrangements, which would organize trading in PAAs among themselves. Thus, whether such an agreement is concluded between Countries A, B and D, it may imply that other parties would then be barred from trading PAAs with those three States. In this context, would such provisions violate, in particular, above-mentioned Articles XI and I GATT? Indeed, whether those proposals are adopted, it would imply, first, that non-ratifying countries would not be able to participate at all in an international emissions trading scheme. Yet, by not committing themselves to the KP, they would obviously not have AAs or eligibility criteria imposed on them. However, serious international trade law concerns may be raised regarding the development of bilateral, regional or multilateral trading agreements between ratifying countries that would result in the exclusion of other ratifying parties who could not trade with them.

In fact, when we examine those preconditions and limits imposed on international trading in PAAs, we notice that most authors have concluded that emissions allowances as government-issued permits are not, as such, a product under the GATT/WTO, which covers only tangible and manufactured goods, including natural gas and petroleum products. Consequently, a party to the KP would be free to decide with whom it would do trading in PAAs. It could, for instance, refuse to sell AAs to non-parties or, in the light of COP and COP/MOP guidelines, to parties which had not fulfilled their commitments.[48]

In other words, such trading amounts to a State-by-State exchange that is not affected by international trade law disciplines. Thus, we might argue that the provisions that would eventually be adopted by COPs and COP/MOPs which would restrict the trade in PAAs among ratifying parties and between ratifying and non-ratifying countries would not be questioned under GATT/WTO law.[49]

However, the same analysis may not hold regarding national rules that individual parties would eventually adopt to undertake emissions trading whether in the framework of an international system or at the domestic level. Indeed, as noticed earlier, it seems as if each party would have the responsibility to organize its participation and those of legal entities to such an international scheme. Thus, while ratifying countries would remain free to organize emissions trading at the State level in accordance with COP and COP/MOP guidelines, they would have to be cautious not to violate Articles XI, I or III GATT when they determine in the national laws who would be obliged to hold emission permits and, more importantly, how those permits would be allocated among regulated companies, industries or sectors and how emissions trading between regulated actors would be organized. In this regard, they would have to make sure that their national provisions on emissions trading do not affect in a discriminatory manner international trade in energy and energy-related products to which emission allowances will be attached. More specifically, both domestic and foreign extractors, importers or sellers of fossil fuels for instance would certainly have to be treated equally in accordance with Articles I and III GATT. This would especially be the case for non-ratifying countries who would not accept that their companies be denied access to the domestic markets of ratifying parties on the ground that their governments would not have ratified the KP.

In sum, similarly to Article 2 KP, the implementation of emissions trading will result in a reduction of energy and energy-related products. Trade restrictions will arise as ratifying parties diminish their own consumption and emissions. However, we can argue that the KP, in itself, does not include specific trade-restrictive measures, either under Article 2 or under Article 17. In fact, any trade effects will largely depend on the way ratifying parties implement their commitments. In particular, they would have to make sure that their national PAMs do not discriminate against foreign industries and imports and do not amount to Article XI GATT quantitative restrictions. They would have to be cautious when they organize the participation of their national regulated industries or sectors in emissions trading so as not to deny access to their domestic markets in energy and energy-related products to imports. Yet, in this context, we can argue that international trade disputes would be less likely to arise among ratifying parties as they would have committed themselves to similar international environmental objectives. However, the lack of specification on policies and measures and on national emissions trading provisions, which will remain country specific, does not exclude the prospect of conflicts. Regarding non-ratifying countries, we cannot expect them to waive the rights they enjoy under international trade law in the sense that they would not have to be discriminated against on the basis of implementing national measures *vis-à-vis* ratifying States. Accordingly, section 4 will address the case of unilateral measures and the compatibility with GATT/WTO law.

4 INTERNATIONAL TRADE LAW AND CLIMATE CHANGE POLICIES: UNILATERAL NATIONAL MEASURES IN THE ABSENCE OF RATIFICATION

In the event of the non-ratification of the KP, the following will address the case of the enacting and implementation of two broad categories of national measures that may have both a positive effect on domestic GHG emissions and a trade or competition-restrictive impact at the international level. In particular, we shall first examine the extent to which an Annex I party may be entitled, under international trade law, to subsidize national climate-friendly products or methods of production so as to foster technology research and improvements. Second, we shall look at degree of discretion that such a party may or may not enjoy when it plans to adopt national fiscal and technical regulations that would aim at curbing GHG emissions in its territory by creating incentives for consumers to purchase the most energy-efficient products and by imposing emission or efficiency product standards. Finally, we shall study the issue of domestic emission-trading schemes and the conditions under which they may be held incompatible with WTO rules and principles.

A. Climate-Friendly Subsidies

To determine whether an Annex I party is entitled to establish a national subsidy program to support the production and sale of the most-climate-friendly products, we refer to the 1994 Agreement on Subsidies and Countervailing Measures (SCM Agreement), which sets new disciplines on the use of subsidies by WTO members.[50]

Regarding the definition of a subsidy, Article 1 of the SCM Agreement stipulates that it refers to a financial contribution from the government or from a public body which may take the form of, *inter alia*, direct transfers of funds, the non-collection of government revenue (tax incentives) or the public provision of goods and services other than general infrastructure. Therefore, a national subsidy program that would intend, for instance, to promote the use by companies of the most efficient cars as company cars by providing relevant funds would amount to a subsidy as such. The same would apply to those industries that use large quantities of fossil fuels in their production processes and which would benefit from tax reductions by switching to least emitting energy sources.

Concerning the examination of the legality of a national subsidy program, Article 2 of the SCM Agreement first draws a distinction between specific and non-specific subsidies. Regarding the former, Article 2(1)(a) specifies that they are the ones that are provided *in law* or *in fact* only to a certain company or industry or to a certain group of companies or industries. As to non-specific subsidies, Article 2(1)(b) holds that they are allocated on the basis of objective criteria which do not favor certain enterprises over others. The objective criteria are economic in nature and apply horizontally, for example, according to the number of employees or the size of the company. Within this framework, the SCM Agreement then refers to three categories of subsidies, including prohibited subsidies (Articles 3 and 4), actionable subsidies (Articles 5, 6 and 7) and non-actionable subsidies (Articles 8 and 9).

First, prohibited subsidies refer to any subsidy that would seek to promote expressly the use of domestic goods to the detriment of imports (Article 3(1)(b)). It may relate, for instance, to a national subsidy program that would encourage the acquisition of given domestic climate-friendly products to the exclusion of similar imports. In such a case, whether a WTO member claims that a prohibited subsidy has been granted or maintained, it may then ask for consultations to take place (Article 4). If no mutually acceptable solution has then been reached, any member to such consultations may refer the matter to the WTO Dispute Settlement Body (DSB), which may decide, in turn, to establish a panel. If the latter finds that the concerned measure is indeed a prohibited subsidy in the sense of Article 3(1)(b) of the SCM Agreement, it will then recommend in its final report its immediate withdrawal. The DSB will adopt this report unless one of the parties to the dispute decides to appeal against it. Where the recommendation of the DSB is not followed on time, it will then authorize the complainant to impose countermeasures.

Second, actionable subsidies are any subsidy (specific and non-specific) that would adversely affect the interests of another WTO member. More specifically, an adverse effect, as defined in Article 5, refers to a subsidy which would cause injury to the domestic industry of another member or which would lead to serious prejudice to the interests of another member.[51] Similarly to prohibited specific subsidies, claims may be brought by a complaining member who may ask for consultations, or, in case of failure to reach an agreement on compensation, may refer the matter to the DSB for the establishment of a panel who could then recommend the withdrawal of the measure by a fixed deadline. If the recommendation is not followed, the complainant can then be authorized to take proportionate countermeasures. Accordingly, an Annex I party may not be able to provide a subsidy for a climate change purpose that would greatly affect foreign industries and competitors, for example, a national subsidy program that would result in the whole domestic energy-intensive industry switching to natural gas to the detriment of oil or coal.

Third, Article 8(1) of the SCM Agreement specifies that non-actionable subsidies relate to non-specific subsidies and specific subsidies that, under Article 8(2)(c), seek to provide assistance to promote the adaptation of existing facilities to new environmental requirements imposed by law which result in greater constraints and financial burdens. Yet, the subsidy concerned will only benefit those facilities that have been in operation for at least two years before the new requirements are imposed. In addition, it will be granted only once, be limited to 20 per cent of the costs of adaptation, will not cover the costs of replacing and operating the new investments, be proportionate to the company's expected reduction on pollution, be available to all firms that can adopt the new equipment and will not cause any adverse effects to another WTO member. Therefore, an Annex I party that has ratified the KP and adopted relevant national regulations on GHG emissions may be entitled to provide support to affected existing industries to assist them in complying with the new laws and requirements. Yet, limits would be imposed regarding which ones would benefit from assistance, the scope and duration of such support and the fact that no WTO member should suffer from any adverse effects as defined in Article 5 of the SCM Agreement.

In sum, in the light of the achievement, at the domestic level, of GHG emission reductions and energy efficiency, an Annex I party may be entitled to provide relevant subsidies, but under strict conditions. It should not aim at supporting expressly the use of domestic products against foreign goods, even if the formers are more environmentally friendly. It should not result in injury to a domestic industry of another member nor should it seriously affect the interests of that member. It can be assumed that non-discriminatory subsidy programs that seek to protect the environment be more likely to be saved as it may be difficult to prove the negative impacts on a foreign industry. Examples are, for instance, subsidies provided to industries to encourage the use by their employees of public transport or the use of energy-saving devices in offices. In addition, Annex II of the SCM Agreement provides that tax refunds may be given regarding the use of energy, fuel and oil used in the production process of exported goods. Would such an exemption to subsidy disciplines conflict with climate change policies that are intended to reduce GHG emissions, including the consumption of energy-related inputs? In particular, we can refer to Article 2 KP that provides that PAMs shall, among others, aim at improving energy efficiency.

B. National Taxation and/or Technical Regulations for Climate Change Objectives

Introduction
In addition to subsidies for achieving domestic GHG emission reductions, an Annex I party may, depending on its specific national circumstances, opt for a range of instruments, including, *inter alia*, taxation schemes on high emitting products such as cars or electrical goods, the enacting of product standards imposing maximum emission or energy-efficiency levels and the setting up of domestic emissions trading systems. We shall discuss how such tax-based and regulatory measures should be framed so as to ensure compliance with principles of international trade law. The latter have expressly been laid down to guarantee, within the context of a global economy, that no arbitrary obstacles to international trade or distortions to competition arise. Therefore, we will examine the legal framework within which countries that have committed themselves to climate change objectives either in the course of the ratification of the KP or on the basis of a voluntary and unilateral policy of global warming, may adopt relevant taxation and regulatory measures while maintaining compliance with their international obligations under WTO law.

Let us take, as an analytical basis, a theoretical case study. We can assume, for instance, that an Annex I party would intend to address the issue of CO_2 emission and energy efficiency of certain electrical products such as refrigerators, dish-washers, washing machines, dryers and air-conditioners.[52] In particular, the competent national authorities would plan to adopt an internal taxation regime whereby a charge would be imposed on the relevant electrical products, the rate of which would depend on the level of energy efficiency and/or of CO_2 emissions. The objective would be to create an incentive for consumers to acquire the least polluting and most energy efficient goods and to foster technological innovations. In addition, they may also think of adopting emission and/or energy-efficiency standards for such products. Such a regulatory regime will be either a substitute for or a complementary provision to the tax-based system. It remains to be seen under what conditions such measures would comply either with the rules and principles of the GATT/WTO or with Article 2 of the Agreement on Technical Barriers to Trade (TBT).[53]

Principles of international trade law
The study of the compatibility of national environmental measures with WTO law, requires first an examination of the GATT principles that govern

international trade and set the limits on which national measures should rely. As mentioned earlier, they include the most-favored-nations principle (Article I GATT), the national treatment principle (Article III GATT) and the principle of the prohibition of quantitative restrictions to imports and exports (Article XI GATT). Such measures may alternatively fall under the TBT if they qualify as a technical regulation. Second, we should also consider the key concepts of 'like' products as the scope of application of the GATT and TBT principles varies whether we are dealing with products that are considered as like or unlike. In addition, an overview of the environmental exemption as provided in Article XX GATT and Article 2 TBT will also be made so as to determine under which conditions a national measure that would violate one of the above-mentioned principles may nevertheless be saved.

GATT principles
First, the most-favoured-nation principle (MFN), Article 1 GATT, a non-discrimination principle, rules that any WTO contracting party or member is to provide to like or similar imported products the same treatment.[54] Thus, whereas the Annex I party would remain free to fix the level of environmental protection or public health it hopes to achieve domestically, it would not be entitled to impose a different fiscal or regulatory treatment on a like imported product on the basis of its origin. A similar product imported from the UK, Belgium or the US would have to benefit from the same tax-based or product-standard provisions.

Second, the national treatment principle (NT) as referred to in Article III(1) GATT, including in Article III(1) GATT, stipulates that internal taxes, charges and regulations, which may affect the sale, purchase, transportation, distribution or use of products should not be applied to imported and domestic products so as to afford protection to national production.[55] As far as like products are concerned, whereas Article III(2) GATT rules that imported products will not be subject to internal taxes or charges of any kind in excess to those applied to like domestic products,[56] Article III(4) GATT states that imported products will be accorded treatment no less favorable than the one which is accorded to the like domestic product in respect of laws, regulations and requirements affecting its internal sale, purchase, transportation or use.[57] Accordingly, while, on the one hand, an Annex I party would not be entitled to impose different taxation and regulatory provisions on like domestic and imported electrical products, it also derives from Article III GATT that a degree of differentiation would be allowed between two electrical products that would not be considered as being similar or like.

In addition to Article I and III GATT principles of non-discrimination, any internal national taxation and regulation would have to make sure that no restriction to international trade (on importation and exportation) would arise under Article XI GATT.⁵⁸ In particular, the latter provides that no quantitative prohibitions or restrictions in the form of bans, quotas or import/export licenses should be adopted. This relates to those national measures that would set an entire prohibition of imports or a quantified limit on imports. However, import bans may alternatively fall under Article III GATT, such as a national measure that would prohibit the use of certain products containing specific hazardous substances. As an illustrative example, the WTO panel and Appellate Reports ruled that the prohibition imposed by France on the use of asbestos products in its territory fell under Article III GATT.

TBT principles

As mentioned above, there may be national environmental measures that would qualify as technical regulation under Annex I of the TBT.⁵⁹ In such a case, the legality of the measure would be examined under the TBT alone to the exclusion of the GATT.

More specifically, as a general principle, Article 2(1) TBT stipulates that national technical regulations will ensure that imports be accorded no less favorable treatment than the one provided to like domestic products (NT principle) and to their imported like products (MFN principle).⁶⁰ In this framework, Article 2(2)(3rd sentence) TBT rules that technical regulations may be adopted for achieving a 'legitimate objective', including the protection of human health or safety, animal or plant life or health, or of the environment.⁶¹ Yet, if WTO members are entitled to adopt national environmental technical measures, Article 2(2)(1st and 2nd sentences) TBT specifies that they should not create unnecessary obstacles to international trade in the sense that they should not be more trade restrictive than necessary.⁶²

Therefore, any WTO member would be entitled to adopt a technical regulation such as an emission or energy-efficiency standard on electrical products even if this were to result in barriers to trade. Indeed, foreign companies may then have difficulties in exporting their product unless they comply with the national product standard concerned. However, for such national measures to be valid, the national authorities would have to make sure that those trade restrictions only result from the application of a non-discriminatory and least trade restrictive technical regulation.⁶³

Like products

As seen above in the course of the examination of the application of the GATT and TBT principles, the concept of like products is crucial. Indeed, whether, for instance, two electrical products would be considered like or unlike, a similar or a differentiated treatment may be allowed or prohibited. Therefore, a key question is then to determine what is meant by like products and what are the consequences for those national authorities that propose regulating their energy-efficiency or emission levels such a way so as to favor the most climate-friendly ones.

Provided that two products are never entirely similar, criteria have thus been developed to draw distinctions.[64] As referred to in the WTO jurisprudence, those criteria are concerned with the tariff classification of the products, its end use on the relevant market (what is it for, how is it to be used?), with the taste and habits of concerned consumers (whether consumers prefer low-emitting and energy-efficient products) and the properties, nature and qualities of the products (i.e. whether a given electrical product is energy-efficient or noisy).

In this context, would two products that have different energy efficiency or CO_2 emission levels be considered as like or unlike under the WTO? As illustrative cases, we can refer to the 1990 WTO panel report on the Thailand Cigarette dispute, which ruled that the different level of content of hazardous substances in cigarettes does not differentiate them.[65] In the same vein, in the 1996 US Gasoline, the WTO Appellate Body report held that the different level of content of specific hazardous substances in imported gasoline marketed in the US does not make it dissimilar to domestic gasoline.[66] Conversely, we can refer to the 2001 Asbestos dispute and in particular to the Appellate Body report, which rules that asbestos fibers and substitutes fibers should not have been considered as like products by the panel report since they had different physical characteristics. Perhaps, more importantly, it also declared that health risks that may be posed by asbestos fibers and products may be considered in the course of the examination of the likeness of two products under the existing criteria that relate to physical properties and of consumers' tastes and habits. Thus, it remains to be seen whether the environmental risks that may result from a specific product, such as high GHG-emitting electrical equipment, could also be considered under existing criteria for differentiating it from another electrical device that would be more energy efficient or less polluting.[67]

Regarding the application of the concept of like products to production and processes methods (PPMs), we should also recall that the 1991 and 1994 Dolphin/Tuna cases have confirmed PPMs could not be used for products as they were not like products. Thus, an imported electrical product will be considered as a like domestic electrical one even if the imported good has been manufactured in a less environmentally or climate-friendly way.[68]

Application of the fundamental principles of international trade law to the case study on electrical equipment and reliance on environmental exemptions

With a view to examining the degree to which an Annex I party would be entitled to adopt internal taxation or regulatory provisions on imported and domestic electrical products on the basis of their energy-efficiency or CO_2 emission level so as to achieve domestically GHG emission reduction targets, we shall examine whether the national measure would fall under the GATT and under what conditions it may be justified under Article XX. Then, we shall look at the case where such a measure would conversely qualify as a technical regulation under the TBT.

Does the national measure fall under Articles XI or III GATT?

Does a national measure that would impose a different taxation or regulatory treatment on electrical products marketed in its territory on the basis of energy efficiency or emission levels resulting in trade barriers fall under Article XI or under Article III GATT? We could argue that such a measure would be likely to fall under Article III as it would be considered as an internal tax or regulation affecting, among other things, the sale or use of the product on the relevant market. The trade restrictions would be deemed to derive indirectly from the electrical product standards laid down in the measure. Conversely, violation of Article XI GATT would arise in the case of clear quantitative limits imposed on imports or exports via quotas or licenses that would not be based on specific product characteristics.[69] However, should the national measure fall under Article XI GATT, the defending WTO member would then have to rely on Article XX GATT to try to have its measure saved.

To determine if the national measure falls under Article III GATT, we should then examine whether electrical products having different energy efficiency or emission levels would be considered as like or unlike products.[70]

If we assume that those products are not like products, the competent national authorities would then be entitled to impose either a differentiated taxation treatment (under Article III(2)) or regulatory provisions (under Article III(4)) in the light of the energy and emission performance. However, for such a treatment to take place legally, the concerned WTO member would have to make sure that the measure is applied in a non-discriminatory manner *in law*, which would imply that the differentiated taxation or regulatory treatment would apply under the same conditions both to imports and to domestic products (MFN and NT principles). In addition, the national measures should not seek to provide protection for domestic production (Article III(1)). If those conditions are not realized, a complaining WTO member may claim a violation of Article XI GATT. In such a case, the defending country would then have to rely on Article XX GATT for justification of its measures.

Conversely, if those products are considered as like products, the WTO member concerned would have to make sure that the same taxation rate is imposed on all products whatever the level of energy efficiency or emission performance (Article III(2)). The same applies to a regulatory provision whereby imports should not be subject a less favorable treatment than the one accorded to like domestic products (Article III(4)). In addition, the national measure will not aim at providing protection for national production (Article III(l)). This would imply that the national measure would have to be non-discriminatory not only *in law*, but also *de facto*. The latter relates to a situation where a similar treatment could still benefit domestic production to the detriment of imports, which would be denied trade opportunities on the national market. If Article III(1) were found to be violated, the defending State would then have to rely on Article XX GATT.

The GATT (environmental) exemptions – Article XX GATT

As examined above, there may be cases where a national environmental measure would violate either Article III or Article XI GATT. However, Article XX GATT provides for general exceptions, including environmental ones as in Article XX(b) and(g).[71] In this context, note that it is up to the party that invokes the exception to demonstrate that its measure meets the requirements laid down in Article XX.[72] Regarding the sequencing of verification, a party would have to show first that its measure falls under one of the exceptions mentioned in Article XX(b) or (g) GATT and second, that it fulfils the conditions laid down in the head-provision of Article XX.[73]

Regarding reliance on Article XX(b) GATT, the defending party would have to demonstrate that its policy in respect of which measures were adopted fell within the range of those that are designed to protect human, animal or plant life or health. For example, the panel in the 1996 US Gasoline dispute agreed that a policy to reduce air pollution from the consumption of gasoline was a policy falling within the range of those concerning protection of human, animal and plant life or health.[74] Note that the very same panel also found that such a policy aimed at conserving a natural resource (air) within the meaning of Article XX(g) GATT.[75] In this context, it remains to be seen whether national measures that aim at reducing CO_2 emissions, including those from electrical products would also eventually fall under Article XX(b) GATT. In other words, it would have to be determined whether reducing emissions of GHGs at the domestic level would benefit human, animal and plant life or health under the same terms as in the US gasoline case. In addition, the invoking party would have to prove that the measures, although inconsistent with GATT principles, were necessary to achieve the relevant policy objective. Thus, whereas the policy objective in itself (its scope) is not subject to scrutiny, the measures that are employed to realize it will be the least-restrictive ones that could also allow accomplishing it.[76] In the Thailand Cigarette case, the import restrictions on certain types of cigarettes was found not necessary as alternative less trade-restrictive measures could have been adopted such as a regulation on the quality of the cigarettes through strict and non-discriminatory labeling accompanied by a ban on unhealthy substances. Thus, it remains to be seen whether a taxation scheme on electrical products based on their energy efficiency would be considered as a necessary measure under Article XX(b) GATT or whether other types of regulation would have to be preferred, such as labeling and information disclosure.

When resorting to Article XX(g) GATT, the invoking party would have to demonstrate, as in Article XX(b) GATT, that the policy that it pursues and under which its measures, found to be inconsistent with GATT/WTO law, were adopted, fell within the range of those that relate to the conservation of exhaustible natural resources. In this regard, the above-mentioned US Gasoline case panel report ruled that clean air was an exhaustible natural resource since it could be exhausted by the emission of pollutants.[77] Thus, as for Article XX(b), we can argue that a national policy on climate change, which may include measures to curb CO_2 emissions and improve energy efficiency would fall under Article XX(g) GATT. Second, the concerned member would have to prove that its measure relates to the conservation of exhaustible natural resources. The term relates has been interpreted as primarily aimed at in the sense that a measure would have to primarily aim at achieving the concerned policy objective as opposed to being necessary or essential to the conservation of the exhaustible natural resources.[78] Accordingly, in the case of national measures on emitting electrical pollutants, they would only have to aim primarily at conserving the atmosphere without being essential to its protection. Third, the invoking State would have to demonstrate that the measures taken to conserve exhaustible natural resources are made effective in conjunction with restrictions on domestic production or consumption. In fact, such a requirement has been interpreted as an obligation under which domestic products or activities that may exhaust a natural resource are also made subject, in addition to imports, to restrictions that primarily aim at conserving the concerned natural resources.[79]

Alongside the realization of criteria laid down in Articles XX(b) and/or (g) GATT, it remains to be seen whether WTO members can enact national measures that would aim at protecting human, animal or plant life or health (under Article XX(b) GATT) or at conserving exhaustible natural resources located outside national jurisdictions (under Article XX(g) GATT). In other words, we can wonder how a WTO panel may assess an environmental national trade-restrictive measure that would also apply in an extraterritorial manner. First, we can refer to the Tuna-Dolphin case II whereby the Appellate Body hold that both Articles XX(b) and (g) GATT would not prevent a member from enacting national regulations aiming at protecting the life of health of living things or which would relate to the conservation of exhaustible natural resources located outside national jurisdiction.[80] However, on the ability of a State to adopt a national environmental measure with external effect, panels in the Tuna-Dolphin cases I and II hold, on the other hand, that the relevant US measures could not be considered as being necessary to protect the life or health of concerned resources (under Article XX(b)) nor as being primarily aiming at their conservation (under Article XX(g)). In addition, they added, *inter alia*, that a national environmental protection policy could not include measures that would force other countries to alter their policies with respect to persons or things within their own jurisdiction and where such changes would be required so as to make the relevant national policy effective. Worded differently, a national environmental measure would have to be sufficient in its own right without bearing any coercive effect. Yet, the Appellate Body in the Shrimp-Turtle dispute has departed from above-mentioned reasoning in holding that it is not necessary to assume that requiring from exporting countries compliance with, or adoption of, certain policies (...) prescribed by the importing country, renders a measure incapable of justification under Article XX.[81] More particularly, whereas it acknowledged, on the one hand, that the concerned US measure felt under Article XX(g) (as they related to the conservation of exhaustible natural resources, namely sea turtles), it ruled, on the other hand, that the way they had been implemented amounted to an unjustifiable and arbitrary discrimination under Article XX.[82] Thus, the rejection of US measures was not founded on the fact that they may force other countries to alter their conservation policies, but merely because they had been applied in violation of conditions set in Article XX head-provision. In other words, a WTO member may be entitled to adopt a national measure that would aim at protecting the life and health of human, animals or plants or at conserving exhaustible natural resources located out side national jurisdiction, provided that no unjustifiable or arbitrary discrimination arises. In addition, we can recall that all concerned panels referred to the need to try to reach international or regional agreements with potentially affected parties before trade-restrictive national environmental measures are enacted.[83] If no such attempts are made, panels are then likely to consider the national measure as 'unilateralism' and posing unjustifiable trade restrictions.

Once a WTO member has demonstrated that its measures, found to be inconsistent with GATT/WTO law, met the requirements contained in Articles XX(b) or (g) GATT, it would have to prove that they realize the conditions spelt out in the head-provision of Article XX. In particular, the concerned national measure would have to have been applied in a manner that did not amount to an unjustifiable or arbitrary discrimination between countries where similar conditions prevail or to a disguised restriction on international trade. In fact, as specified by the Appellate Body Report in the US Gasoline case, whereas the terms arbitrary discrimination, unjustifiable discrimination or disguised restriction have not been subject to precise definitions, their object and purpose is the avoidance of abuse of illegitimate use of the exceptions to the substantive rules available in Article XX GATT.[84] In the same vein, the Appellate Body Report in the Shrimp Turtle case recalled that the analysis of the application of the head-provision of Article XX GATT consisted in maintaining a balance (...) between the right of a member to invoke an exception under Article XX and the duty of that same State to respect the treaty rights of the other Members. In fact, the Shrimp Turtle case may provide an illustrative example of what could amount to an arbitrary and unjustifiable discrimination. The dispute questioned the GATT consistency of Section 609 (1989) of the 1973 US Endangered Species Act, which provided that to export shrimp to the US, all countries in whose waters shrimp and sea-turtles co-existed were to be certified by US authorities as having enforced a turtle-excluder-device (TED) legislation for commercial shrimp trawlers and comparable in effectiveness to those used in the US. The Appellate Body found that such legislation amounted to an unjustifiable discrimination as it had an unjustifiably coercive effect on policy decisions of foreign countries. In practice, no serious attempts had been made by the US to consider any other conservation programs that may have been in place in other countries nor the different conditions that may prevail and to engage in prior international cooperation and negotiations. In addition, various countries were discriminated against since they were subject to different phasing-in periods and were not treated equally regarding US transfers of TED technology. Furthermore, arbitrary discrimination was also recognized on the ground that the US did not ensure that its policies were appropriate to the specific local conditions prevailing in other countries and that the certification process lacked transparency and predictability and did not provide for a formal opportunity for an applicant country to be heard or to respond to any arguments that may be made against it.[85] Regarding the application of the terms disguised restriction on international trade, the panel report hold in the Asbestos case that a national measure would qualify as such if compliance with it were only a disguise to conceal the achievement of trade-restrictive objectives.[86] The panel added that the mere fact that a national measure benefits domestic production did not constitute a protectionist aim as such, provided that it remains within certain limits.

Therefore, whereas a national fiscal or regulatory measure on energy efficiency or emission level of electrical products, which would be inconsistent with the GATT/WTO rules because of resulting trade restrictions, may well be justified under Articles XX (b) or (g) GATT, it would have to meet the conditions set out in the head-provision of Article XX GATT. In particular, the WTO member would have to make sure that such measures would not be applied in a manner that would constitute an arbitrary or unjustifiable discrimination or a disguised restriction on international trade.

Does the national measure fall under Article 2 TBT?

As mentioned earlier, a national measure may be considered as a technical regulation in the meaning of Annex I TBT and thus fall under Article 2. It might certainly be the case of a national measure that imposed a product characteristic on electrical products in terms of energy efficiency performance. Although no WTO cases have so far dealt with the application of this provision, it would be interesting to see how a panel would interpret it and whether it would do so in a similar manner as has been reported regarding measures falling under Article XX GATT.

In fact, Article 2(2) TBT implies that a national technical regulation that resulted in trade restrictions, whether it applies to like or unlike electrical products would be justified, provided that it did not lead to unnecessary obstacles to international trade (the least trade restrictive measures available for achieving the policy objective pursued) and that it would be necessary to fulfil the concerned legitimate objective. Note, as under Article XX GATT, that policy objective itself is not considered in the sense that a WTO member would remain free to set any environmental protection objectives at the domestic level. Rather, the concerned State would have to demonstrate that the means that had been employed to achieve such an objective were necessary and the least TBT inconsistent. Note also that the preamble (6th indent) provides, in the same terms as the head-provision of Article XX GATT, that such technical regulation will not be applied in a manner that would constitute a means of arbitrary or unjustifiable discrimination between countries where the same conditions prevail or a disguised restriction on international trade. However, what would be the impact of such a provision, provided that it is found in the non-operational part of the TBT? It remains to be seen how and to what extent future panels will rely on it.

Finally, we should refer to Article 2(5) TBT, which stipulates that any national technical regulation that would be taken in conformity with international norms, recommendations or directives would presumably not create unnecessary obstacles to international trade. In the same vein, Article 2(2)(4) TBT would even entitle Annex I parties to set higher norms if they demonstrated that the international norms were inappropriate to achieve the environmental objective concerned. In this context, note that the international climate change regime would set only quantified emission reduction targets but not specific objectives in terms, for instance, of energy efficiency. As seen earlier, Article 2 KP provides simply a non-exhaustive list of measures, which would be implemented on the basis of national circumstances. Thus, unless COP/MOP or another international body set specific norms for products, the above-mentioned provisions may not be relied upon to justify obstacles to trade.

C. International Trade Law and Climate Change Policies: The Case of National Rules on the Implementation of an Emission Trading System

In addition to the adoption of national subsidies, taxation and regulatory measures, countries may also opt for the setting of a national emission trading scheme that would aim at ensuring that domestic CO_2 emissions are reduced. The question is then to determine under what conditions such a regime may be established so as not to violate above-mentioned GATT/WTO principles.

In fact, whether we examine the case of full, partial or non-ratification of the KP, national rules, which will either determine the participation of each individual party in an international trading scheme (full or partial ratification) or set a purely domestic ETS (an ETS being established as a unilateral measure in the absence of ratification), will ensure that the operationalization of such schemes will not amount to discriminatory practices affecting the trade in energy and energy-related products upon which emission allowances will be imposed.[87]

We should note that a domestic ETS might operate at different levels. First, the holding of emission permits may be imposed on those who extract and import fossil fuels in a given national market. In this case, all concerned domestic and foreign companies who extract and import, for instance, oil or coal would have to have an emission permit to be entitled to put those products on the domestic market of the importing country. Alternatively, emission allowances may be imposed on those who sell fossil fuels and other energy-related products. Finally, such an emission permit precondition may be of concern to those who use or burn fossil fuels in producing goods. Companies concerned would then have to possess sufficient permits that would correspond to their annual emissions.

In addition to the identification of the different levels where emission permit trading could take place, we should also make a preliminary remark. More specifically, a party may not be entitled to prevent as such the entry into its domestic market of energy and energy-related products on the ground that only a limited volume of such products is to be combusted so as to reduce GHG emissions. Indeed, a party that intends to restrict such emissions at the national level, either in the course of the realization of its commitments under the KP or in the framework of a unilateral national climate change policy, will limit the number of permits available in accordance with the amount of emissions that would be allowed. Yet, the mere fact that allowances are of a limited number may not authorize a party to block the imports of energy and energy-related products at its border. Should it be the case, it would then be considered as a violation of Article XI GATT. The party would then have to have recourse to Article XX GATT to demonstrate that such quantitative restrictions are justified. However, it may be difficult to demonstrate that such a national measure could not be replaced by alternative means that would be less trade restrictive, such as an emission permit system. In sum, a party may regulate the use of energy and energy-related products in its territory but may not prevent imports as a whole.

If we assume that no border restrictions on the imports of fossil fuels are set, a party would then have to make sure whether emissions permits were imposed on either those who extract and import energy and energy-related products or those who sell them on domestic markets, that all regulated companies or industries enjoy a non-discriminatory access to available national allowances. In this context, the party would have first to establish an allocation scheme, which would allow both domestic and foreign extractors, importers or sellers to apply, on an equal basis, for emission permits.

If grandfathering is the option that is retained, it should not be held incompatible with WTO law as this would result in maintaining the *status quo*. However, if the domestic market concerned had not been liberalized before the initial allocation took place, such an allocation system may conversely result in a violation of Article III GATT as it would not accord to imported energy and energy-related products a similar treatment to that accorded to like domestic products and would clearly benefit domestic production. If auctioning is the preferred allocation method, then the party would have to make sure that both domestic and foreign regulated actors are treated equally in an open and transparent manner.

As well as initial allocation, the party would also have to address the case of new entrants, whether domestic or foreign. It would have to ensure that new entrants are able to acquire permits from those who would have initially been allocated allowances. Foreign suppliers should not be discriminated against. A solution may be to provide a stock of emission permits, which would not be allocated in the first place but reserved for the newcomers. Another alternative would be to limit in time the validity of emission permits so as to organize new allocations at regular intervals.

Note that Article III GATT, as mentioned earlier, allows a differentiated treatment to be imposed on products that are not like. Thus, it remains to be seen whether natural gas, coal and petroleum products for instance are like or unlike products. If not, we can then assume that a different allocation method could be imposed whether we are dealing with an extractor, an importer or a seller of natural gas or of coal. However, as seen above, such a differentiated treatment should not result in discriminatory practice against foreign products and/or protection of national production.

Finally, in the event where emission permits are imposed on those companies who use energy and energy-related products, international trade disputes would be less likely as all industries located in country A would be

To design and implement climate change measures 99

considered as national industry. However, if the party intends to favor those companies that use the least-polluting energy sources such as natural gas, we should then examine whether this would amount to a violation of the 1994 Agreement on Subsidies and Countervailing Measures. Indeed, there are cases where the allocation of permits to specific companies may distort international competition to such an extent that the interest of another WTO member may be affected.

5 CONCLUSION

This chapter has provided an overview of the potential impacts on international trade that may result from the implementation of national measures on climate change. Indeed, whether such measures are adopted pursuant to the KP or unilaterally, it is possible to be in violation of the GATT/WTO rules and principles, unless they are appropriately formulated.

In the event of the ratification of the KP, ratifying parties would have to make sure that they implement their obligations in a non-discriminatory manner. This would be all the more important as the current political context shows that not all major economic players will commit themselves to CO_2 emission reductions. In this context, whereas we can assume that international trade disputes would be less likely among parties, we cannot expect non-ratifying States to waive their rights that they hold under GATT/WTO law. For this reason, countries would have to ensure that no quantitative restrictions on importing and exporting and no violation of Articles I and III GATT arise from implementing national climate change policies and measures. The same may be argued regarding the enacting of unilateral measures in the absence of ratification of the KP. In such a case, caution would have to be exercised *vis-à-vis* international trade law. In any event, Article XX GATT lays down the possibility for national environmental policies and trade restrictions. The bottom line will remain the avoidance of arbitrary or unjustifiable discrimination and of disguised restrictions to international trade.

NOTES

1. *United Nations Framework Convention on Climate Change*, New York, 9 May 1992, *31 International Legal Material* (1992) at 849; *Kyoto Protocol to the United Nations Framework Convention on Climate Change*, Decision 1/CP.3 (1997), posted at: http://www.unfccc.int/resource/convkp.html.
2. For a summary of COP6(I), see *Earth Negotiations Bulletin*, Vol. 12, No. 163, Monday, 27 November 2000, posted at: http://www.iisd.ca/climate/cop6/index.html. See also homepage of COP6: http://cop6.unfccc.int/.
3. For further information on COP6 (II), see COP6 (II) homepage at: http://unfccc.int/cop6_2/.

See also *Earth Negotiations Bulletin*, Vol. 12 No. 176, Monday, 30 July 2001, posted at: http://www.iisd.ca/climate/cop6bis/. For further information on COP7, see COP7 homepage, posted at: http://unfccc.int/cop7/. See also *Earth Negotiations Bulletin*, Vol. 12, No. 189, Monday, 12 November 2001, posted at: http://www.iisd.ca/climate/cop7/.
4. See COP8 homepage at: http://unfccc.int/cop8/. Coverage of COP8 by the *Earth Negotiations Bulletin* will be available at: http://www.iisd.ca/climate/cop8/.
5. The 1994 World Trade Organisation (WTO) encompasses several WTO agreements, including, *inter alia*, the *Agreement establishing the World Trade Organisation* (WTO Agreement), the *General Agreement on Tariffs and Trade* (GATT), the *General Agreement on Trade in Services* (GATS), the *Agreement on the Application of Sanitary and Phytosanitary Measures* [SPS Agreement], the Agreement on Technical Barriers to Trade (TBT Agreement) and the *Agreement on Subsidies and Countervailing Measures* (SCM Agreement), posted at: http://www.wto.org/english/docs_e/legal_e/final_e.htm.
6. See *ENDS, Environment Daily*, 28 March 2001.
7. White House Press Release, 13 March 2001, posted at: http://www.whitehouse.gov/news/releases/2001/03/20010314.html.
8. For more details on the interactions between science and international climate change, see Grimeaud (2001a).
9. '*Remarks by the President on Global Climate Change*', 11 June 2001, posted at: http://www.whitehouse.gov/news/releases/2001/06/20010611-2.html. President Bush also reiterated that the targets on GHG emissions as contained in the KP were unrealistic, not science driven, and would lead to negative economic impacts and that climate change actions should result in commitments, imposed on all parties, including developing countries. See also conclusions of the EU-US Göteborg Summit, 14 June 2001, posted at: http://europa.eu.int/comm/gothenburg_council/eu_us en.htm.
10. Recall that the US has the highest CO_2 emission per capita. They are responsible for more than 20 per cent of global CO_2 emissions while their population accounts for only 4 per cent of worldwide population.
11. May 2001 Report of the National Energy Policy Development Group '*Reliable, Affordable and Environmentally Sound Energy for America's Future*': posted at http://www.whitehouse.gov/energy/. See *ENDS Environment Daily*, 18 May 2001 for details on related reactions by J. Pronk [chairman of CDP6] and EC officials.
12. Stockholm European Council Presidency Conclusions, Annex II '*European Council Declaration on Climate Change*', 23-24 March 2001: posted at http://ue.eu.int/presid/conclusions.html. See also the Göteborg European Council Presidency Conclusions (15-16 June 2001) whereby the EC Commission was required to prepare a proposal for ratification of the KP before the end of 2001: posted at http://europa.eu.int/comm/gothenburg_council/index_en.htm.
13. Statement on Climate Change by Mr Kjell Larsson, Minister for the Environment in Sweden, 31 March 2001, held in conjunction with the Informal Meeting of Ministers for the Environment in Kiruna (30 March-1 April 2001): posted at http://www.eu2001.se/eu2001/news/.
14. 2235th Environment Council Meeting, Luxembourg, 7 June 2001, Press Release No: 9116/01: posted at http://ue.eu.int/newsroom/.
15. *Council Decision (2002/358/EC) of 25 April 2002 concerning the approval, on behalf of the European Community, of the Kyoto Protocol to the United Nations Framework Convention on Climate Change and the joint fulfilment of commitments thereunder*, Official Journal L.130, 15/05/2002, p. 1-3.
16. See EC Commission's *Proposal for a Directive of the European Parliament and of the Council establishing a scheme for greenhouse gas emission allowance trading within the Community and amending Council Directive 96/61/EC* (COM (2001)581) adopted on 23 October 2001, Official Journal C.75E, 26/03/2002 p. 33-44. See also *Communication from the Commission to the Council and the European Parliament on EU policies and measures to reduce greenhouse gas emissions: towards a European Climate Change Programme (ECCP)* (COM/2000/88), adopted on 8 March 2000 and *Communication from the Commission on the implementation of the first phase of the European Climate Change*

Programme (COM/2001/580) adopted on 23 October 2001. Both documents and further information on the European Climate Change Programme are available at: http://www.europa.eu.int/comm/environment/climat/eccp.htm.
17. See Council Conclusions of 16 June 1998. To note that the EU's burden sharing agreement will become legally binding under EC law once the Kyoto Protocol will have entered into force as a consequence of its inclusion in the April 2002 EC Council's decision on its ratification. The adoption of such an agreement is based upon Article 4 of the Kyoto Protocol.
18. *The Netherlands' Climate Policy Implementation Plan*, June 1999, available at: www.minvrom.nl.
19. The Dutch government announced that The Netherlands would achieve 50 per cent of its emission reductions at the domestic level and 50 per cent abroad throughout the use of the KP flexible instruments (joint implementation, clean development mechanism and international emissions trading).
20. Note that such a national emissions trading scheme would eventually have to comply with the one that would be set at the EU level. For inforrnation on national climate change programmes, see also Germany's *National Climate Protection Programme*, 18 October 2000, which provides for the continuing of existing measures and the developing of new climate actions. They include, *inter alia*, a reform of the ecological tax that would result in a gradual increase of energy prices, the promotion of renewable energy sources, the expansion of combined heat and power generation and the enacting of voluntary agreements on energy efficiency, available at: www.bmu.de/english/fsetl024.htm.
21. See the *Programme national de lutte contre le changement climatique*, available at: www.environnement.gouv.org See also *Law No. 2001-153 tendant à conférer à la lutte contre l'effet de serre et à la prévention des risques liés au réchauffement climatique la qualité de priorité nationale* (19 February 2001), Journal Officiel 43, 20 February 2001, p. 2783, available at: www.legifrance.gouv.fr. For further details on the French action plan on climate change, see the Mission Interministérielle de l'Effet de Serre *'Programme National de lutte contre le changement climatique (2000)'*: posted at http://www.effet-de-serre.gouv.fr/.
22. A Danish CO_2 emissions trading scheme entered into force on January 2001. See *Act No. 376 of 2 June 1999 on $C0_2$ Quotas for Electricity Production*, available at: www.ens.dk/uk/energy_reform/index.htm For further information, see Pedersen (2000).
23. See the *UK Greenhouse Gas Emissions Trading Scheme* 2002, available at: http://www.defra.gov.uk/environment/climatechange/trading/index.htm. See Rees and Evers (2000) and Hobbey (2001).
24. Article 25(1): *'This Protocol shall enter into force on the ninetieth day after the date on which not less than 55 Parties to the Convention, incorporating Parties included in Annex I which accounted in total for at least 55 per cent of the total carbon dioxide emissions for 1990 of the Parties included in Annex 1, have deposited their instruments of ratification, acceptance, approval or accession'.*
Article 25(3): *'For each State or regional economic integration organization that ratifies, accepts or approves this Protocol or accedes thereto after the conditions set out in paragraph 1 above for entry into force have been fulfilled, this Protocol shall enter into force on the ninetieth day following the date of deposit of its instrument of ratification, acceptance, approval or accession'.*
Article 25(4): *'For the purposes of this Article, any instrument deposited by a regional economic integration organization shall not be counted as additional to those deposited by States members of the organization'.*
25. As of 27 September 2002, 95 parties to the UNFCCC, including 25 Annex I Parties had ratified the KP. The latter comprise the EU, the 15 Member States, Iceland, Norway, Slovakia, Slovenia, Romania,.Latvia, Hungary, the Czech Republic and Japan. For up-to-date data on the status of ratification of the KP, see http://unfccc.int/resource/kpthermo.html.
26. See below for further details.
27. Convention on International Trade in Endangered Species of Wild Fauna and Flora (1973),

12 International Legal Materials 1085 (1973). For more details on CITES, see Sands (1995, pp. 375-381).
28. See below for further details on Article XI GATT. Also, see the 1998 Rotterdam Convention on the Prior Informed Consent (PIC) Procedure for Certain Hazardous Chemicals and Pesticides in International Trade (not yet into force), which requires an import and export permit system to be in place for certain chemicals and allows parties to ban or to severely restrict the importation of the most hazardous ones. The text of the convention is posted at: http://www.pic.int/index.html.
29. Basel Convention on Transboundary Movements of Hazardous Wastes and their Disposal (1989), 28 International Legal Materials 649 (1989).
30. See below for further details on the most-favoured-nations and the national treatment principles.
31. Montreal Protocol on Substances that Deplete the Ozone Layer [1987], 26 *International Legal Materials* 1550(1987).
32. Note that article 4(8) provides that import bans against non-parties may not apply if the importing State demonstrates that it complies with article 2 of the Protocol on reduction of the production and consumption of controlled substances.
33. The Committee on Trade and Environment identified 20 MEAs containing trade provisions. See, *inter alia*, the 1966 International Commission for the Conservation of Atlantic Tuna, the 1980 Convention for the Conservation of Antarctic Marine Living Resources, the 1992 UN Convention on Biological Diversity, the 2000 Cartagena Protocol on Biosafety, the 1995 UN Fish Stocks Agreement and the 2001 Stockholm Convention on Persistent Organic Pollutants.
34. The CTE was established in 1995 with the objectives, *inter alia*, of studying the 'relationship between the provisions of the multilateral trading system and trade measures for environmental purposes, including those pursuant to multilateral environmental agreements' and initiating 'recommendations on whether any modifications of the provisions of the multilateral trading system re required'. See Marrakesh Ministerial Decision on Trade and Environment of 15 April 1994 (Annex I),which sets the CTE's mandate and terms of reference: posted at www.wto.org.
35. See 1996 Report of the Committee on Trade and Environment, WT/CTE/1 (96-4808): posted at www.wto.org. See also Doha WTO Ministerial Declaration (adopted on 14 November 2001), WT/MIN(01)/DEC/1, 20 November 2001: posted at www.wto.org. See in particular para. 31 and 32 (Work programme). Paragraph 31 stipulates that negotiations will be launched on (i) the relationship between existing WTO rules and specific trade obligations set out in multilateral environmental agreements (MEAs), provided that the negotiations shall be limited in scope to the applicability of such existing WTO rules as among parties to the MEA in question. The negotiations shall not prejudice the WTO rights of any Member that is not a party to the MEA in question; (ii) procedures for regular information exchange between MEA Secretariats and the relevant WTO committees, and the criteria for the granting of observer status; (iii) the reduction or, as appropriate, elimination of tariff and non-tariff barriers to environmental goods and services. Paragraph 32 requires the CTE to give attention to (i) the effect of environmental measures on market access, especially in relation to developing countries, in particular the least-developed among them, and those situations in which the elimination or reduction of trade restrictions and distortions would benefit trade, the environment and development; (iii) labeling requirements for environmental purposes. Work on these issues should include the identification of any need to clarify relevant WTO rules. The Committee shall (…) make recommendations, where appropriate, with respect to future action, including the desirability of negotiations. The outcome of this work as well as the negotiations carried out under paragraph 31(i) and (ii) shall be compatible with the open and non-discriminatory nature of the multilateral trading system, shall not add to or diminish the rights and obligations of members under existing WTO agreements (…) nor alter the balance of these rights and obligations (…).
36. For more details see, for instance, S. Charnovitz, *Multilateral Environmental Agreements and Trade Rules, Environmental Policy and Law*, 26/4 (1996), pp. 163-169.

37. Article 14 UNFCCC provides that, in the event of a dispute between parties over the implementation or application of the Convention, they may decide to have recourse to the International Court of Justice and/or to arbitration. Note that Article 19 KP holds that Article 14 UNFCCC will apply *mutatis mutandis* to the protocol. Conversely, note that Annex Section 6 of the 1994 Agreement relating to the implementation of part XI of the 1982 United Nations Convention on the Law of the Sea provides that trade-related disputes which occur between two parties that are also members of the GATT/WTO will with by the WTO panel. On the contrary, where such a dispute affects two parties who are not both GATT/WTO members, recourse wlll be made to the Convention dispute settlement mechanism: posted at: http://www.un.org/Depts/los/unclos/closindx.htm.
38. 'Carbon dioxide equivalent emissions' refers to a basket of six greenhouse gases that include carbon dioxide (CO_2), Methane (MH_4), Nitrous oxide (N_2O), Hydrofluocarbons (HFCs), Perfluorocarbons (PFCs) and Sulphur hexafluoride (SF_6). In accordance with article 5(3) KP, 'global warming potentials' has been adopted as a common measure.
39. See Grimeaud (2001b) for more details on the background of article 3(1) KP and on Annex I parties' AAs.
40. See Council Conclusions of 16 June 1998. Note that the EU's burden sharing agreement will become legally binding under EC law once the Kyoto Protocol will have entered into force as a consequence of its inclusion in the EC Council's Kyoto Protocol ratification decision. The adoption of this agreement is based upon article 4 of the Kyoto Protocol. Note also that Article 3(1) KP provides that QERLCs may be achieved individually or jointly.
41. Note that, whereas article 3(1) and 7 KP refer to 1990 as the emission base year, article 3(5) allows economies in transition to use a different base year for the six GHGs. In this regard, Bulgaria, Poland and Romania chose to use 1989 as the base year whereas Hungary has been entitled to use the average emission levels of the year 1985-1987.
42. This would especially be true for trade impacts that might arise between ratifying States.
43. Note that parties have not yet agreed on the degree to which GHG emission reductions should be achieved domestically via relevant programmes and measures (PAMs) and/or internationally through the use of flexible instruments, including joint implementation (JI), clean development mechanism (CDM) and emissions trading (ET). Indeed, whereas articles 5, 6(1)(d), 12(3)(a) and 17 KP stipulate that the emission reduction units (ERUs), the certified emission reductions (CERs) and the parts of assigned amounts (PAAs) that will be held, acquired or transferred from the use of JI, CDM and ET will only be supplemental to national emission reductions, no quantitative limits or caps have yet been set. In this context, note also that the EU has always argued for giving priority to domestic emission reduction as opposed to the US who claim that no cap should be imposed in that matter and that Annex I parties should feel free to achieve most of their QELRCs internationally by relying upon the flexible instruments. See also the Dutch climate policy, which provides that the Netherlands would achieve half of its emission reduction commitments at home and the other half at the international level by using JI, DM and ET.
44. Article 2(1)(b) KP provides that COP/MOP should decide on the ways to facilitate such a cooperation.
45. See FCCC/CP/2001/13/add.1, decision 13/CP.7 *'Good Practices' in Policies and Measures among Parties included in Annex I of the Convention*, posted at: http://unfccc.int/resource/docs/cop7/13a01.pdf.
46. As examined below, the absence of international standards may prevent WTO members from relying upon articles 2(4) and 5 of the Agreement on Technical Barriers to Trade (TBT agreement) to justify national technical regulation that may lead to trade restrictions.
47. FCCC/CP/2001/13/add.2, decision 18/CP.7 *Modalities, rules and guidelines for emissions trading under article 17 of the Kyoto Protocol*, posted at: http://unfccc.int/resource/docs/cop7/13a02.pdf. Eligibility criteria include the obligation for each Annex I party to, *inter alia*, have in place a national system for the estimation of anthropogenic GHG emissions by sources and anthropogenic removals by sinks (in accordance with article 5(1) KP and relevant guidelines and a national registry (in accordance with article 7(4) KP and relevant guidelines; report annually on emissions and

removals (in accordance with articles 5(2) and 7(1) KP and relevant guidelines) and to supply any supplementary information (in accordance with article 7(1) KP and relevant guidelines).
48. For more details, see Werksman (1999).
49. Note that it may be claimed that emission allowances are negotiable instruments under Agreement on Financial Services (General Agreement on Trade in Services (GATS)). In such a case, those WTO members who have adhered to this agreement would not be entitled to prohibit the trade in emission allowances from their territory.
50. See *Supra* No. 5.
51. Serious prejudice is defined in Article 6 of the SCM Agreement and concerns, *inter alia*, subsidies that cover operating losses sustained by an industry or the direct forgiveness of government-held debt.
52. We can also assume that no relevant international norms have so far been enacted.
53. See *Supra* No. 5.
54. Article I (I) GATT: With respect to customs duties and charges of any kind imposed on or in connection with importation or exportation or imposed on the international transfer of payments for imports or exports, and with respect to the method of levying such duties and charges, and with respect to all rules and formalities in connection with importation and exportation, and with respect to all matters referred to in paragraphs 2 and 4 of Article III, any advantage, favour, privilege or immunity granted by any contracting party to any product originating in or destined for any other country shall be accorded immediately and unconditionally to the like product originating in or destined for the territories of all other contracting parties.
55. Article III (I) GATT: The contracting parties recognize that internal taxes and other internal charges, and laws, regulations and requirements affecting the internal sale, offering for sale, purchase, transportation, distribution or use of products, and internal quantitative regulations requiring the mixture, processing or use of products in specified amounts or proportions, should not be applied to imported or domestic products so as to afford protection to domestic production.
56. Article III (2) GATT: The products of the territory of any contracting party imported into the territory of any other contracting party shall not be subject, directly or indirectly, to internal taxes or other internal charges of any kind in excess of those applied, directly or indirectly, to like domestic products. Moreover, no contracting party shall otherwise apply internal taxes or other internal charges to imported or domestic products in a manner contrary to the principles set forth in paragraph 1.
57. Article III (4) GATT: The products of the territory of any contracting party imported into the territory of any other contracting party shall be accorded treatment no less favourable than that accorded to like products of national origin in respect of all laws, regulations and requirements affecting their internal sale, offering for sale, purchase, transportation, distribution or use. The provisions of this paragraph shall not prevent the application of differential internal transportation charges which are based exclusively on the economic operation of the means of transport and not on the nationality of the product.
58. Article XI GATT: No prohibitions or restrictions other than duties, taxes or other charges, whether made effective through quotas, import or export licenses or other measures, shall be instituted or maintained by any contracting party on the importation of any product of the territory of any other contracting party or on the exportation or sale for export of any product destined for the territory of any other contractingparty.
59. Technical regulation is defined in Annex I of the TBT as a document, which lays down product characteristics or their related processes and production methods, including the applicable administrative provisions, with which compliance is mandatory. It may also include or deal exclusively with terminology, symbols, packaging, marking or labeling requirements as they apply to a product, process or production method.
60. Article 2(1) TBT: 'Members shall ensure that in respect of technical regulations, products imported from the territory of any Member shall be accorded treatment no less favourable than that accorded to like.products of national origin and to like products originating in any other country'.

To design and implement climate change measures 105

61. Article 2(2) TBT: 'Members shall ensure that technical regulations are not prepared, adopted or pplied with a view to or with the effect of creating unnecessary obstacles to international trade. For this purpose, technical regulations shall not be more trade-restrictive than necessary to fulfil a legitimate objective, taking account of the risks non-fulfillment would create. Such legitimate objectives are, *inter alia*, national security requirements; the prevention of deceptive practices; protection of human health or safety, animal or plant life or health, or the environment. (…).
62. In this context, article 2(8) TBT rules that members should specify, whenever appropriate, technical regulations based on product requirements in term of performance rather than design or descriptive characteristics.
63. Note also that article 2(4) TBT provides that, where international standards exist and require national measures, domestic technical regulations will use those standards, except when they would be ineffective in achieving the legitimate objectives pursued. In such a case, article 2(5) TBT also specifies that those technical regulations will be presumed not to create an unnecessary obstacle to international trade.
64. See Report of the 1970 Working Party on Border Tax Adjustments, BSID I 8/S/97 (1972), which first identified criteria to distinguish between like and unlike products.
65. Thailand – Restrictions on the Importation of Cigarettes, BSID 37S/200, adopted on 7 November 1990.
66. United States – Standards for Reformulated and Conventional Gasoline, Appellate Body Report, WT/DS./AB/R adopted on 20 May 1996.
67. European Communities – Measures Affecting Asbestos and Asbestos-containing Products, Appellate Body Report, WT/DS135/AB/R, adopted on 21 March 2001. For a detailed analysis of the Panel and Appellate Body reports, see van Calster (2001a,b).
68. United States – Restrictions on Imports of Tuna (Tuna I), circulated on 3 September 1991, unadopted, BSID 39S/155 and United States – Restrictions on Imports of Tuna (Tuna II), circulated on 16 June 1994, unadopted, DS29/R.
69. See *Supra* No. 67, European Communities – Measures Affecting Asbestos and Asbestos-containing Products.
70. See above discussion for an explanation of likeness.
71. Art. XX *General Exceptions*: Subject to the requirement that such measures are not applied in a manner which would constitute a means of arbitrary or unjustifiable discrimination between countries where the same conditions prevail, or a disguised restriction on international trade, nothing in this Agreement shall be construed to prevent the adoption or enforcement by any contracting party of measures:
    (b) necessary to protect human, animal or plant life or health;
    (g) relating to the conservation of exhaustible natural resources if such measures are made effective in conjunction with restrictions on domestic production or consumption.
72. See *Supra* No. 66, United States – Standards for Reformulated and Conventional Gasoline, Appellate Body Report, paras. 6.20, 6.31 and 6.35.
73. See United States – Import Prohibition of Certain Shrimp and Shrimp Products, Appellate Body report WT/DS58/AB/R, circulated on 12 October 1998.
74. See *Supra* No. 66, United States – Standards for Reformulated and Conventional Gasoline, Appellate Body Report, para. 6.21.
75. See *Supra* No. 66, United States – Standards for Reformulated and Conventional Gasoline, Appellate Body Report, para. 6.37. Note that the question of the applicability of article XX(b) GATT to the US policy on clean air was not dealt with by the Appellate Body as it was not a matter that had been appealed by the US. Thus, only the application of art. XX(g) was then considered.
76. See *Supra* No. 65, Thailand – Restrictions on importations and internal taxation on cigarettes, para. 75: The import restrictions imposed by Thailand could be considered to be necessary in terms of art. XX(b) only if there were no alternative measure consistent with the [GATT] or less inconsistent with, which Thailand could reasonably be expected to employ to achieve its health policy objectives.
77. See *Supra* No. 66.
78. See, Canada – Measures Affecting Exports of Unprocessed Herring and Salmon, adopted on

22 March 1988, BSID 35S/98, para. 4.5. See also *Supra* No. 66, United States – Standards for Reformulated and Conventional Gasoline, Appellate Body Report, paras. 17–18.
79. See *Supra* No. 66, United States – Standards for Reformulated and Conventional Gasoline, Appellate Body Report, paras. 20–22. See also *Supra* No. 73, United States – Import Prohibition of Certain Shrimp and Shrimp Products, Appellate Body Report, paras. 143–145.
80. See *Supra* No. 68 United States – Restrictions on Imports of Tuna (1994), para. 5.15.
81. See also *Supra* No. 73, United States – Import Prohibition of Certain Shrimp and Shrimp Products, Apellate Body Report, para. 121.
82. Section 609 (1989) of the 1973 US Endangered Species Act provided that shrimp harvested with technology that may adversely affect sea turtles may not be imported into the US, unless the harvesting exporting country has a regulatory program that is comparable to that of the US and that governs the incidental taking of sea-turtles and that the rate of vessels' incidental taking is comparable to the US average. In this context, 1996 Guidelines were enacted whereby certification for allowing countries to export shrimp to the US depended on the establishment in the exporting country of a program containing a requirement that all vessels use turtle excluder devices [TEDs] comparable in effectiveness to those used in the US. In this regard, the Appellate Body found that this amounted to unjustifiable discrimination as no attempt had been made by US authorities to consider any other conservation programs that may have been in place in other countries nor the different conditions that may prevail.
83. See also *Supra* No. 73, United States – Import Prohibition of Certain Shrimp and Shrimp Products, Appellate Body Report, para 61: (...) do not imply that recourse to unilateral measures is always excluded, particularly after serious attempts have been made to negotiate (...).
84. See *Supra* No. 66, United States – Standards for Reformulated and Conventional Gasoline, Appellate Body Report, para. 25.
85. See *Supra* No. 82.
86. See *Supra* No. 67, European Communities – Measures Affecting Asbestos and Asbestos-containing Products, Appellate Body Report, paras. 8.238–8.239.
87. As mentioned earlier, the mere fact that the proposed draft COP guidelines might result, if adopted, in the establishment of limits on the trade in PAAs by requiring the realization of eligibility criteria (a ratifying country may be prohibited to trade parts of its AAs with another party if the latter has not complied with its obligations under, for instance, articles 5 and 7 KP) and/or by allowing the setting up of bilateral or multilateral arrangements between parties (two or more ratifying parties may set up joint trading arrangements in PAAs to the exclusion of other parties) may not be considered as violations of the GATT/WTO disciplines.

REFERENCES

van Calster, G. (2001a), 'Getting there slowly: international trade law and public health in the WTO Asbestos Case', *European Environmental Law Review*, **10** (4), April, 113-19.

van Calster, G. (2001b), 'Health protection and international trade: back on the right track after appellate body intervention in asbestos', *European Environmental Law Review*, **10** (5), May, 163-4.

Grimeaud, D. (2001a), 'Overview of the policy and legal aspects of the international climate change regime (Part 1)', *Environmental Liability*, **9** (2), April, Lawtext Publishing.

Grimeaud, D. (2001b), 'An overview of the policy and legal aspects of the international climate change regime (Part 2)', *Environmental Liability*, **9** (3), June, Lawtext Publishing.

Hobley, A. (2001), 'Emissions trading in the United Kingdom: an overview', *Environmental Liability*, **9** (1), 3-10.
Pedersen, S.L. (2000), 'The Danish CO_2 emissions trading system', *Review of European Community and International Environmental Law*, **9** (3), 223-31.
Rees, M. and R. Evers (2000), 'Proposals for emissions trading in the United Kingdom', *Review of European Community and International Environmental Law*, **9** (3), 232-8.
Sands, P. (1995), *Principles of International Environmental Law. Volume I: Frameworks, Standards and Implementation*, Manchester: Manchester University Press.
Werksman, J. (1999), 'Greenhouse gas emissions and the WTO', *Review of European Community and International Environmental Law*, **8** (3), 1999, 251-64.

5. Developing carbon trading in Europe: does grandfathering distort competition and lead to state aid?

Edwin Woerdman

1 INTRODUCTION

In March 2000 the European Commission presented a Green Paper on greenhouse gas (GHG) emissions trading within the European Union (EU) (COM, 2000a). The purpose of that document is to find out which design of a European carbon trading market is desirable and/or acceptable by stimulating a discussion among stakeholders, scientists and politicians. In this Green Paper, the Commission refers to permit trading among private entities as a 'unique opportunity' (COM, 2000a: 9) and aims to establish an experimental EU-wide permit trading scheme by 2005. In October 2001 the Commission presented a proposal for a directive on such a scheme that incorporates large emitters (COM, 2001a).

If member states link their domestic permit trading schemes, firms can trade across national borders within the EU. This could reduce EU compliance costs by more than 30 per cent (Capros et al., 2000). If the parties to the Kyoto Protocol also allow private emissions trading under Article 17, European firms may also trade permits with firms in industrialized countries outside the EU, such as Norway or perhaps the United States. Because a larger market widens the scope for efficiency gains, the EU could then achieve cost savings of almost 50 per cent (Capros et al., 2000).

To reap the economic benefits of permit trading within (and outside) the EU, several political barriers must be broken down. One of these is the issue of permit allocation which has both political, economic and legal aspects. In general, there are two ways to allocate permits: private entities have to buy the permits (auctioning) or they get them for free (grandfathering). It is possible that one member state conducts an auction (for instance, to generate revenues), while another uses grandfathering (for instance, to generate support from the energy-intensive industry). The Commission fears that such a difference in the way member states allocate permits to their private entities, especially by means of grandfathering, may distort competition and could lead to state aid (for example, COM, 2000a: 5; COM, 2001a: II).

Consequently, the objective of this chapter is to find out under which conditions European differences in domestic permit allocation procedures in general and grandfathering in particular (a) will lead to competitive distortions according to economic theory and (b) will lead to state aid according to European Community (EC) law. The following (and second) section contains the economic analysis which partly builds upon an article by Woerdman (2000a). The third section contains the legal analysis which explicitly considers the recently revised Community guidelines on state aid for environmental protection (OJ, 2001). The fourth section approaches the issues of competitive distortions and state aid from a political science perspective deemed necessary (but not performed) by Van der Laan and Nentjes (2001) to supplement a law and economics analysis of competitive distortions. The outlines of a potential legal solution are discussed by considering the harmonization of permit allocation. Finally, in the fifth section a conclusion is presented.

2 ECONOMIC ANALYSIS OF PERMIT ALLOCATION AND COMPETITIVE DISTORTIONS

Van der Laan and Nentjes (2001) note that there are two interpretations of the competitive distortion concept: as an inefficiency in allocation of resources and as an inequity of firms' starting conditions. The issue of permit allocation will be analysed below in relation to competitive distortions by using and specifying both the efficiency approach and the equity approach.

Competitive Distortions, Efficiency and Opportunity Costs

Following neoclassical welfare economics, the concept of competitive distortion can be interpreted as an inefficiency or price distortion. A competitive distortion is then defined as a measure which entails a price deviation from the welfare optimum under perfect competition, thereby reducing the efficiency of (inter)national trade. According to economic theory, international differences in domestic permit allocation procedures will not lead to competitive distortions. To follow the argument, it is crucial to understand the concept and effect of opportunity costs.

Not only auctioning, but also grandfathering entails costs for firms: grandfathered permits have an opportunity cost when they are used for covering the emissions of the permit owner (Nentjes et al., 1995). The opportunity cost, which is equal to the price for which the permit can be sold, must be included in the product price. Instead of using the permits, the firm could have sold the permits. The revenue forgone is a cost to the firm, comparable with the 'interest forgone' on own capital. Hence, grandfathered firms do not have a cost advantage over auctioned firms abroad (or over domestic newcomers), just because they received permits for free.

Although there are no price distortions when comparing grandfathering with auctioning, grandfathering does imply a transfer of wealth to firms, since they receive an input which has a certain market value. Therefore, grandfathering permits could be viewed as granting a subsidy to the firm. However, this subsidy is a capital gift to the firm with the character of a lump-sum subsidy (for example, Jensen and Rasmussen, 1998). In efficiency terms, a lump sum subsidy is not distorting in the product market, since it does not affect marginal emission reduction costs and it does not alter the output and price decisions of firms.

The opportunity cost aspect is no more (and no less) than an economic characteristic of grandfathering and should not be seen as some sort of 'policy advantage' in favour of grandfathering. In fact, a disadvantage of grandfathering arises if there are other distortionary taxes in the economy. In that situation, auctioning is more efficient than grandfathering if the auction revenues are recycled to lower these taxes (for example, Goulder et al., 1999). Another potential disadvantage is that firms could have an incentive to strategically increase their emissions in the run-up to the allocation of permits in order to receive more permits if the permits are grandfathered based on historical emissions. Nevertheless, from an economic perspective, this incentive will only exist if the benefits of receiving more permits in the future are expected to outweigh the costs of increasing emissions now. Moreover, even if this incentive exists, it can be (and is usually) neutralized by choosing the emission reduction target on the basis of a reference year sufficiently back in time, so that polluters 'burn their own fingers' if they increase their emissions (more than they would in a business-as-usual scenario) prior to the allocation (Lyon, 1982).

The economic analysis presented above assumes perfectly competitive markets. Imperfect competition is unlikely, because the direct participation of private entities in a European (and finally international) GHG emissions trading system is expected to create a thick market with many small traders. There are only a few exceptional cases of imperfect competition where a competitive distortion could arise. An example is a situation where a grandfathered firm starts a price war with an auctioned competitor abroad according to the so-called 'deep purse theory' or theory of predatory pricing (Nentjes et al., 1995). The grandfathered firm can outlast the auctioned firm (or entrant) in a price war because of its larger capital reserve. However, predatory pricing is unlikely to occur in practice, not only because it is a risky and expensive strategy, but also because energy-intensive firms usually do not compete on monopolistic markets and the additional capital requirements to buy emission permits are a small percentage of the total capital requirements. Furthermore, a dominant firm which starts a price war to push aside its competitor abroad could be prosecuted by the EU authorities that enforce antitrust policy.

Competitive Distortions, Equity and Level Playing Field

Although the opportunity costs are the same under grandfathering and auctioning, European permit allocation differences will affect the financial position of firms. This aspect becomes relevant if the concept of competitive distortion is interpreted in terms of equity. Unlike an auctioned competitor, a grandfathered firm does not have to buy its permits. A grandfathered firm only has to pay for its emission reductions and not for its emissions, so that it has a lower cash outflow than an identical firm which has to buy its permits. In other words, a grandfathered firm initially buys the permits from itself (opportunity costs), while an auctioned firm buys the permits from the government or the public (cash outflow). If a grandfathered firm receives its permits for free, it obtains a non-distortionary windfall profit (see Bohm, 1999). Since grandfathering implies a capital gift to the firm, a grandfathered firm has more financial resources, or own capital, than an auctioned firm, which (*ceteris paribus*) gives the former a stronger financial position than the latter as a result of permit allocation alone. This can be seen as a distortion of fair competition.

There is no single interpretation of the concept of equity or fairness, either in philosophical theory or in political practice. Nevertheless, it appears that an unfair competitive distortion in the context of permit allocation usually refers to a distortion of the 'level playing field' for firms, where the associated inequity is primarily defined or perceived in terms of an inequality or asymmetry (for example, Jepma and van der Gaast, 1999; Yamin and Lefevere, 2000). Our definition extends the analysis of Van der Laan and Nentjes (2001) from target to instrumental level and elaborates the equity view by emphasizing that change matters. Woerdman (2000a) argues that the problem of international differences in permit allocation procedures is the inequality of the changes in firms' starting conditions resulting from the mere process of permit allocation. Furthermore, the equity view does not so much think of competition being distorted because firms face different laws, but rather because these different laws have different financial consequences for firms and their competitive relations. The level playing field then refers to the competitive relations between firms. This implies that an unfair competitive distortion would arise if the allocation of permits leads to unequal changes of firms' relative financial positions.

The level playing field approach does not object to the fact that the competitive positions of firms can be unequal because they have different market shares and that their relations may change, both before and after permit allocation, because of their economic activities and strategies. Rather, the level playing field or financial position approach contends that the competitive positions of firms are not allowed to change because of the political process of permit allocation itself. Consequently, the level playing field is maintained if permit allocation leaves the financial positions of firms and their competitive relations unaltered. Moreover, our analysis does not imply that an efficient permit trading scheme cannot be equitable: this depends on what one perceives to be a fair distribution of permits. Our point is rather that grandfathering is not distorting from an efficiency perspective, but that it could be seen as distorting from an equity perspective. In this respect, there is in fact a trade-off between efficiency and equity.

Again, there are only a few exceptional cases in which imperfect competition could play a role. In an imperfect capital market, the findings of Koutstaal (1997) imply that a grandfathered firm has a competitive advantage if the auctioned competitor abroad needs to borrow money to buy the permits. The interest to borrow money could exceed the interest on own capital due to the imperfect capital market. The permit expenditures of the auctioned firm are then higher than the opportunity costs of the grandfathered firm due to the interest it has to pay for its loans. However, the practical relevance of this argument is negligible, not only because the interest difference under consideration is small compared with total production costs, but also because the loans will be short term in a liquid permit market and (contrary to the interest on own capital) the interest charges for loans are tax deductible.

3 LEGAL ANALYSIS OF PERMIT ALLOCATION AND STATE AID

The member states of the EC face the question whether grandfathered permits could be interpreted as a form of state aid under EC Article 87 or not. To find an answer to this question, we shall analyse EC Article 87 on state aid as well as Commission (and Court) decisions and reports on state aid, including the revised Community guidelines on state aid for environmental protection (OJ, 2001). After discussing the criteria for state aid, the conditions will be analysed under which state aid is exempted from its prohibitory status.

Permit Allocation and State Aid Criteria

Article 87(1) on state aid as formulated in the EC Treaty determines that 'any aid granted by a Member State or through State resources in any form whatsoever which distorts or threatens to distort competition by favouring certain undertakings or the production of certain goods shall, in so far as it affects trade between Member States, be incompatible with the common market'. Two lines of reasoning can be applied following this legal text. According to the opportunity cost argument, as was shown before, grandfathered firms have no cost advantage over auctioned firms, which would imply that grandfathering is no aid at all or constitutes aid which does not distort efficiency, trade and competition between member states. However, according to the financial position or level playing field argument, grandfathering should be regarded as state aid as well as a competitive distortion. The reason for this is that grandfathered firms receive a capital gift from a member state, thereby financially advantaging or favouring those firms relative to their auctioned competitors abroad. The definition of state aid is thus insufficient to decide whether grandfathering should be seen as state aid or not.

Although the EC Treaty does not give a full definition of state aid, the concept has been elaborated by the European Commission (for example, COM, 1999: 84) and the European Court of Justice (for example, Cases E/1/98 and E/2/98 of the Flemish region versus the Commission), who both describe state aid in terms of an 'advantage'. This could suggest that grandfathering should be regarded as state aid, since grandfathered permits are a capital gift which implies a financial advantage for firms. However, it could also be argued that grandfathered firms have no cost advantage over auctioned firms by following the opportunity cost argument. The European Commission (COM, 2000b) uses four criteria to determine whether or not a measure is to be regarded as state aid which is incompatible with the common market. A measure is considered to be state aid if it satisfies the criteria of (a) state origin, (b) firm advantage, (c) specificity and (d) trade effect.

First, with respect to the first criterion of state origin, the aid must be granted by the state or through state resources. It could perhaps be claimed that grandfathering (although it is a transfer of permits) is not a genuine or direct transfer of resources, since the permits are allocated for free by the state. Nevertheless, the opposite can well be defended by stressing that these permits have market value and that the capital gift induced by grandfathering is an (in)direct transfer of state resources. Furthermore, Jepma et al. (1999) indicate, on the basis of COM (1998), that the state origin criterion requires a transfer of resources from the state (or in the state) receiving, actually or potentially, less revenue in order for state aid to exist. It could be argued that the state will receive less revenue in the case of grandfathering compared to either (pre-existing) taxation or auctioning, because grandfathering can be interpreted as giving the (hypothetical) auction revenue to the polluters (Welch, 1983: 168).

Second, with respect to the other three criteria, grandfathering should be seen as an advantage which affects trade by favouring specific firms and thus distorts competition according to the level playing field approach, but not according to the reasoning of the opportunity cost approach. Interestingly, the Commission not only mentions the desirability of a level playing field, where firms are treated on an equal footing, in the context of state aid (for example, COM, 1998: 79; OJ, 2001: 13), but it also describes the firm advantage criterion as a financial advantage which improves a firm's market position (for example, COM, 1999: 84). Furthermore, the Commission emphasizes that when the state confers even a limited advantage on an undertaking which is active in a sector characterized by competition, there is a distortion or risk of distortion of competition (COM, 2000b: 75), which could run counter to the Commission's goal to ensure the competitive functioning of markets (OJ, 2001: 5). However, it can also be argued that grandfathering does not distort efficiency or trade because its opportunity costs will be reflected in the product price. From this perspective, grandfathered permits internalize costs as much as auctioned permits do. This aspect would then imply that grandfathering is allowed, because cost internalization is a priority objective in the Commission's policy on the control of state aid (OJ, 2001: 5).

Permit Allocation and State Aid Exemptions

It was shown above that neither Article 87(1) on state aid nor its elaboration by the European Commission or the European Court of Justice seems to provide a decisive answer whether grandfathering should be regarded in terms of efficiency (opportunity costs) or equity (financial advantage). However, not all state aid is prohibited under European competition law. Article 87(3) as well as the Community guidelines on state aid for environmental protection (OJ, 1994) which have been recently revised (OJ, 2001) provide the basis for the exceptions under which to regard state aid as compatible with the common market. Briefly, state aid can be allowed if:

- the aid promotes the execution of an important project of common European interest;
- the aid remedies a serious disturbance in the economy of a member state;
- the aid facilitates the development of certain economic activities or areas;
- the European Council decides that the aid is compatible with the common market.

First, even if grandfathering should be seen as state aid, it can nevertheless be allowed on the basis of Article 87(3)(b) if the aid is used to promote the execution of an important project of common European interest. In 1987 the Court recognized (*Glaverbel* Case 62/87) that concerted action by a number of member states to combat environmental pollution is an example of an important project of common European interest for the purposes of Article 87(3)(b). Opinions differ whether grandfathering literally 'promotes' climate change mitigation under the Kyoto Protocol as this section of the law would require. On the one hand, emissions trading reduces costs and thus makes emission reductions easier to achieve, while its political acceptance is increased by the wealth transfer grandfathering induces. On the other hand, grandfathering may be insufficient or even counterproductive to generate political support for permit trading, because it induces time-consuming allocation problems with conflicting interests between emitters about the basis of grandfathering (for example, energy-efficiency versus historical emissions) and it does not raise revenues for the government(s) involved (Woerdman, 2000b). Moreover, the Community guidelines require that the aid, in this case considering grandfathering, must be necessary for the adoption or continuation of the project (OJ, 2001: 6, 13), which may be difficult to defend unless the political acceptance of emissions trading exclusively hinges on grandfathering.

Second, Article 87(3)(b) also allows state aid if the aid is used to remedy a serious disturbance in the economy of a member state. It is clear that auctioning (as well as taxation) entails a financial burden for polluters, but it is not evident that auctioning would thus create a 'disturbance' in the economy of a member state which is 'serious' enough to allow for grandfathering as the remedy against it. Environmental economists rather defend the opposite by indicating that auctioning is more efficient than grandfathering, because the auction revenues can be recycled to lower distortionary taxes (for example, Goulder et al., 1999). But if grandfathering should be seen as aid, it may still be allowed provided that it is treated as a temporary second-best solution (OJ, 2001: 5).

Third, grandfathering may be exempted from the state aid prohibition on the basis of Article 87(3)(c), which provides that aid may be considered compatible with the common market if it facilitates the development of certain economic activities or of certain economic areas, where such aid does not adversely affect trading conditions to an extent contrary to the common interest. On the one hand, if the opportunity costs of using grandfathered permits imply that they do not affect efficiency and hence trading conditions, it could also be argued that they have the same effect as auctioned permits on the development of certain economic activities or areas precisely because they have no cost advantage. On the other hand, it could be defended that grandfathered permits do facilitate the development of certain economic activities or areas, because they are a capital gift giving the firm a stronger financial position than under auctioning. In practice, Article 87(3)(c) has already been used several times to allow for state aid, for instance in the case of the energy-intensive industries in the Netherlands and Denmark as well as Norway, Sweden and Finland, which were exempted from a tax on CO_2 emissions. Although some would make a comparison by claiming that grandfathering will be exempted from state aid because it resembles a tax exemption, it should be noted that tax exemptions distort efficiency because they induce different prices per unit of GHG for different firms, whereas grandfathered permits do not distort efficiency because they have opportunity costs and therefore entail a price. Consequently, the opportunity cost argument would imply that if grandfathering is to be regarded as state aid, it will be exempted from the state aid prohibition rules even more easily than the aforementioned tax. Nevertheless, in the political arena grandfathering could perhaps be perceived as being even 'worse' than a tax exemption, because the former necessitates a new framework which formally gives an emitter the right to produce a limited amount of pollution, which is more visible and definitive than exempting an emitter (temporarily) from an arrangement in an existing scheme.

Fourth, if grandfathering were state aid, it could be allowed, in principle, on the basis of Article 87(3)(e), which refers to the discretionary power of the European Council to decide by qualified majority – on the basis of a proposal by the Commission – that an aid measure is compatible with the common market. In itself, this provision does not help to judge the relevance of the opportunity cost argument and the financial position or level playing field argument in EC law and politics. However, it underlines that the issue whether or not permit allocation differences between member states are desirable and whether or not grandfathering should be seen as state aid could well be (and probably will be) decided upon, not so much solely on the basis of legal considerations, but rather on the basis of a political decision.

An additional issue is whether the polluter-pays principle should be defined in terms of either opportunity costs or financial expenses. The polluter-pays principle requires that the costs of measures to deal with pollution should be borne by the polluter who causes the pollution (OJ, 2001: 3). The European Commission (COM, 2000a) explicitly stated that auctioning applies the polluter-pays principle, simply because polluters literally pay for their pollution by means of purchasing the permits from the government. Whether grandfathering will also be seen as compatible with the polluter-pays principle depends on the perspective taken. From an effiency perspective, although grandfathered polluters do not pay for their emissions, they do pay for their emission reductions (assuming they did not receive 'hot air' permits) and the essential point is that every unit of GHG emissions has a price because each unit has the opportunity of being sold. From an equity perspective, however, grandfathering is a wealth transfer from the public to the polluters, which implies that the 'public pays' by giving the polluters the revenues it might have received by auctioning the permits (see Welch, 1983). Which of these perspectives will finally dominate is not clear yet and will be discussed in the next section on politics and perceptions.

The Community guidelines on state aid for environmental protection could suggest that firms can only receive a part of their permits for free and that grandfathering is only allowed temporarily (OJ, 2001). We assume that emission permits are a form of operating licence, so that only the provisions for operating aid are considered. This assumption seems reasonable, because without such permits a firm is not allowed to operate. Emission permits (grandfathered or auctioned) are legally required in an emission-trading scheme if firms want to produce a certain (limited) amount of pollution.

In the context of operating aid, which is only relevant if grandfathering should be viewed as aid in the first place, firms may receive the aid no longer than five years. If the level of aid decreases each year ('degressive aid'), which should be the general rule, the intensity may amount to 100 per cent of the extra costs in the first year but must have fallen in a linear fashion to zero by the end of the fifth year. If the level of aid is the same during these years ('non-degressive aid'), a firm may receive no more than 50 per cent of the extra costs necessary to meet the environmental objectives. Tax exemptions are also seen as operating aids and may only be granted, among other things, for a limited period of time with a maximum of ten years (or five years in the case of energy-efficiency improvements). A temporary relief from environmental taxes may be authorized by the Commission with a view to the risks of a loss of international competitiveness. This could imply that grandfathering, which could be introduced to accommodate similar competitiveness concerns, may also be deemed compatible with European law as long as it is a temporary transition (of possibly five or ten years) to an auctioned scheme.

The operating provisions above could imply that firms may receive no more than a certain percentage (for instance 50 per cent) of their permits for free during a limited period after which they have to buy their permits via an auction. Alternatively, if governments were to agree upon a five-year transition period, the implication could be that the annually (re)allocated permits are allowed to be grandfathered for 100 per cent in the first year, 80 per cent in the second year, 60 per cent in the third year, 40 per cent in the fourth year and 20 per cent in the fifth year, so that all permits are auctioned (and thus 0 per cent is grandfathered) in the sixth year when the transition period is over.

However, the Community guidelines only allow firms to receive aid over their extra investment costs necessary to reduce emissions, so that it could be argued that they may not receive aid over their entire emissions. In efficiency terms, the implication could be that firms may not only be grandfathered, but that they may also receive some aid (under the conditions specified above) for the costs they spend to reduce their emissions. In equity terms, grandfathering is seen as aid (as demonstrated in a previous section), but auctioned firms then still seem to be allowed under the EC rules to receive some aid for their expenses on emission reductions.

4 POLITICAL ANALYSIS OF PERMIT ALLOCATION AND PERCEPTIONS

From the law and economics perspective applied above, it is ambiguous whether grandfathering, in the case of dissimilarities in permit allocation between EC member states, satisfies each of the aforementioned four criteria for state aid to exist, and if it does, whether or not it will be exempted from the state aid injunction. Therefore, this section extends and supplements the law and economics perspective with a political science perspective on competitive distortion and state aid issues, first, by explaining the impact of perceptions on political behaviour in the context of competitiveness and permit allocation, and second, by analysing the decisions of the European Commission in the case of domestic permit trading in Denmark and in the United Kingdom (UK) to approve the grandfathering of permits to legal entities. Moreover, the outlines of a potential legal solution are discussed by considering the harmonization of permit allocation.

Perceptions in Political Decisions on Permit Allocation

From a political science perspective, it does not so much matter in the political process whether international differences in domestic permit allocation can 'objectively' lead to competitive distortions according to some economic theory or equity principle. Rather, the agents involved act 'subjectively' on the basis of their perceptions of such issues (for example, Woerdman, 2002). The discretionary power of the European Council to decide – on the basis of a proposal by the Commission – that grandfathering is exempted from the state aid provisions means that a political decision will be pivotal to the issue of grandfathering, state aid and permit allocation differences between EC member states. In principal, financial and equity arguments are likely to play a role in such a decision, not only with a view to the historical relevance of the level playing field argument in European environmental legislation (Hargrave et al., 1999: 11) and state aid policy (Cini and McGowan, 1998: 158), but also considering the continuously recurring reference made by the Commission in its recent Green Paper on GHG emissions trading in the EC to 'fair competition' (pages 7 and 12), 'conditions for equal competition' (page 14) and a 'level playing field' (page 15) as well as to the relation between competitive distortions and financial (dis)advantages (page 19) for firms (COM, 2000a). However, the Green Paper does mention the concept of opportunity costs once (page 20) when discussing the issue of new entrants to the European carbon trading market (COM, 2000a). In its proposal for a Directive on GHG emissions trading, the Commision writes:

> [I]t is feared that if allowances were allocated on the basis of auctioning in one Member State but allocated free in another, competition may be distorted. ... The proposal does not spell out what would be consistent or inconsistent forms of allocation with regards to State aid as each situation will have to be examined on its merits, [but] ... State aid scrutiny examines the possible distortions of competition ... (COM, 2001a: 11-12).

Cini and McGowan (1998) have also drawn the conclusion that politics plays a role in the state aid decisions of the Commission, both in terms of political values and perceived national interests. The general impact of values (see van Deth and Scarbrough, 1995) suggests that not only efficiency, but also equity is likely to be considered in a political decision, for instance on permit allocation, which – in itself – increases the chance that grandfathering will be seen as state aid. However, a member state's perception of its national interests in relation to permit allocation and state aid will lie somewhere between two extremes, that is (a) the desire to protect the national sovereignty of being free to allocate the permits as domestically preferred and (b) the desire to protect the national economy and industry against member states who are free to choose any permit allocation they like which could result in a competitive advantage for the competitors abroad.

Which (mixture of) desire(s) will finally be dominant in the Commission is uncertain. Although the Commission leaves open in its proposal for a directive whether to grandfather or auction during the first commitment period, it does indicate its preference for a harmonized method, since combinations of grandfathering and auctioning may distort competition and could lead to state aid (as will be ascertained on a case-by-case basis) (COM, 2001a: 11). Another political indication is provided by the report of the European Climate Change Programme (ECCP, 2001). This programme, in which many stakeholders were invited to participate, is initiated and intended by the Commission to prepare a proposal on climate change policy to be made to the Council and the European Parliament in the course of 2001. The report not only warns against competitive distortions due to permit allocation and underlines the desirability of a level playing field, but also states that permit allocation differences do not necessarily give rise to distortions within the internal market and claims that distortions are likely to be temporary (ECCP, 2001: 8-10). It concludes that 'Member States should be allowed to choose their own initial method of allocation, subject to obtaining any appropriate State Aid approvals' (ECCP, 2001: 9). From this it is not clear whether the equity interpretation or the efficiency interpretation prevails, not only because the report does not mention any reasons for these views, but also because it prefers a case-by-case assessment of potential state aid issues arising from permit allocation. However, again without providing a foundation for the argument, the report expects a progressive evolution towards auctioning in the longer term (ECCP, 2001: 8). A potential indication of which desires and perceptions will finally be dominant in the Commission is provided by the Commission's decisions on state aid in the domestic permit trading schemes in Denmark and the UK, which will be treated in the following subsection.

The Political Precedents of Permit Trading in Denmark and the UK

Important albeit limited 'test cases' for the competitive distortion and state aid issues in an EU-wide carbon trading market are the political precedents of domestic permit trading for the power sector in Denmark and for various companies in the UK. Denmark set up the first domestic and obligatory permit trading scheme in Europe which became operational in January 2001 (when the emission ceiling for electricity producers came into effect). The domestic and voluntary UK scheme became operational in April 2002. The European Commission reached a decision on the allocation of permits in the context of state aid, both for Denmark (COM, 2000c) and for the UK (COM, 2001b). These decisions are bound to set a political precedent for future thinking about – and decisions on – (differences in) permit allocation in a European carbon trading market. However, the European Commission clearly indicated that its decisions in the cases of Denmark and the UK do not necessarily set a legal precedent for future decisions on emission-trading schemes. Below we shall analyse first the Commission decision on state aid in the Danish case and then its decision in the UK case.

According to Act 376 of 2 June 1999 (originally Bill 235) of the Danish parliament, tradable permits are grandfathered to electricity producers in Denmark – irrespective of whether they are Danish or foreign owned – based on their historical CO_2 emissions during the 1994-98 period (Folketinget, 1999; COM, 2000c). The scheme runs from 2001 to 2003, but if any new entrants were to arrive during this period (which is not expected), they will be allocated quotas 'on the same terms' as incumbents following objective and non-discriminatory criteria. The aforementioned Act will then be amended which will be notified again to the Commission. According to Haites et al. (2000), the reallocation of permits in the case of new entrants before 2003 implies that both newcomers and incumbents will receive grandfathered instead of auctioned permits. The emission ceilings for electricity producers have the effect that they – as a group – must reduce their emissions, but the combined heat and power plants have received less stringent (business-as-usual) emission ceilings because the latter have contributed more strongly to CO_2 savings in the past. The Danish parliament acknowledged that the bill had to be notified to the European Commission, among other things, on the basis of EC Article 88(3) concerning state aid. It also stipulated that the bill would not come into effect before receipt of approval by the Commission. The Commission approved the scheme by means of a letter to the Danish government dated 12 April 2000 and reached two basic decisions (COM, 2000c).

First, the Commission considers grandfathering in the Danish scheme to be state aid, because the tradable permits have a market value and because the state forgoes revenue which could derive from auctioning the permits. Albeit limited to the Danish case, it does provide some support for one of our conjectures, namely that the Commission would see grandfathering as state aid based on the state origin criterion, translating grandfathering in terms of giving the (hypothetical) auction revenue to the polluters. The Commission has interpreted grandfathering as a wealth transfer without considering its opportunity costs and indicated that a company may use its profits from permit sales to improve its competitive position.

Second, the Commission nevertheless decided to allow grandfathering following EC Article 87(3)(c) which exempts state aid to develop certain economic activities or areas. The fact that they see grandfathering as actually developing economic activities or areas implies that they acknowledge the financial advantage of grandfathering over auctioning as in the equity interpretation. If they had used an efficiency perspective, grandfathering would not be seen as having a cost advantage over auctioning because of its opportunity costs, in which view it does not develop economic activities or areas more or less than auctioning would. In short, the eight reasons for the Commission's exemption are that the Danish scheme (1) contributes to environmental protection and generates experience with emissions trading, (2) incorporates large emitters, (3) intends to participate in future international carbon trading, (4) represents emission reductions, (5) is limited to 2003, (6) does not restrict electricity imports or exports, (7) provides annual reports for transparency and (8) treats incumbents and newcomers equally (COM, 2000c: 6, 7). These reasons to accept grandfathering despite its state aid character are not only environmental, legal and economic, but also political in nature. The Commission approved the state aid in Denmark, among other things, because of its desire to gain experience with and prepare for emissions trading (reasons (1) and (3)). This provides some support for our conjecture that a

Commission's state aid decision is at least partly based on political considerations and trade-offs. It also shows that there is broad room for interpreting the exemption rules which goes beyond a strict law and economics approach. Several reasons for the exemption of grandfathering have nothing to do with the allocation *per se* (grandfathering versus auctioning), but rather relate to the (other and more general) characteristics of the scheme itself.

In its decision of 28 November 2001 on state aid in the UK case, the Commission has reached similar conclusions and largely used the same type of arguments as in the Danish case. Unlike the Danish scheme, the UK scheme runs from 2002 to 2007, combines permit trading with credit trading, excludes the power sector and contains absolute targets that are voluntary. Although this broader scheme has a more complex design than the smaller Danish scheme, the bottom line is that companies in the UK that voluntarily wish to take up absolute targets (the so-called 'direct participants') receive grandfathered tradable permits.

The Commission approved the scheme by means of a letter to the UK government (COM, 2001b). Explicitly referring to the Danish case and without mentioning the opportunity costs of free allocation, the Commission also considers grandfathering in the UK scheme to be state aid, among other things because the state forgoes revenue which could derive from auctioning the permits. Grandfathering is seen as an advantage that distorts competition with companies not having access to the scheme. Similar to the Danish case, the Commission nevertheless decided to allow grandfathering following EC Article 87(3)(c) which exempts state aid that is used to develop certain economic activities or areas. And again, the Commission also used political arguments, as well as environmental, economic and legal ones, to defend the exemption.

In short, the thirteen reasons given by the Commission for allowing the exemption are that the UK scheme: (1) is in line with the idea that each member state may choose the policy it wishes to comply with the Kyoto targets as long as Community provisions on emissions trading are absent, (2) uses a competition-oriented instrument, (3) precedes Community regulation, (4) will provide valuable learning insight for the benefit of any later initiatives, (5) is the first multilateral trading scheme in the EU, (6) achieves a net environmental benefit, (7) uses an incentive necessary to ensure voluntary participation, (8) requires companies to reduce emissions below their targets to capitalize the potential aid from free allowances, (9) recovers the incentive in the case of non-compliance, (10) is limited in time and will adapt to the requirements of an EU-wide emissions trading scheme foreseen in 2005, (11) will produce detailed annual reports, (12) undertakes to accept emissions trading based on mutual agreements with other states and (13) will elaborate on and inform the Commission of non-discriminatory ways to include new entrants (COM, 2001b: 11-12).

As we have seen before, these reasons demonstrate that grandfathering is allowed as state aid, among other things, by stating a political desire to gain experience with emissions trading (reason (4)) and by mentioning characteristics of the scheme that have nothing to do with the allocation of permits *per se* (such as reasons (5), (11) and (12)). The Commission emphasizes that the UK might have to adapt its scheme if EU-wide emissions trading starts in 2005 'in order to avoid distortion of competition between allowances issued through different systems' (COM, 2001b: 12).

The presence of political arguments and the absence of the opportunity cost argument in the economic analysis of grandfathering by the Commission can be interpreted in two ways. A pessimist might see them as another example of sometimes imperfect and incomplete case-by-case decisions by the state aid directorate (see Cini and McGowan, 1998: 143) or as another example of 'infant' economic analysis in the state aid directorate, which is more legally oriented than the environment directorate, or in EC competition law itself (see Hildebrand, 1998: 413). An optimist might see them as an example of a Commission that is blind neither to international political developments (regarding the emerging international carbon trading market under the Kyoto Protocol) nor to national political preferences (of Denmark and the UK) and that is able to find a balance between costs and benefits and between risks and opportunities, indicating that: 'The Danish CO_2 quota system has to be assessed in the light of its merits' (COM, 2000c: 6).

It is important to observe that other criteria than efficiency were used by the Commission. The equality principle played a role in the Commission's decision on the Danish scheme, which is also part of the (unnoticed) level playing field argument. The Commission indicated its support of the equal treatment of newcomers and incumbents in the Danish case in order to avoid competitive distortions (COM, 2000c: 5, 7), which apparently arise in the case of unequal treatment, meaning that they interpret such distortions not in terms of efficiency, but in terms of fairness. The political (albeit not legal) precedents created by the Commission's decisions in the Danish and UK cases could suggest that grandfathering in a European carbon trading scheme will not only be seen as state aid, but might also be exempted. However, this remains uncertain because of the difference of assessing domestic permit allocation in two particular EU member states versus the assessment of international permit allocation dissimilarities between several member states.

Political Support in Favour of Harmonizing Permit Allocation

Both the Commission and the European Parliament are in favour of a harmonized method for permit allocation to avoid competitive distortions in an EU-wide permit trading scheme (compare COM, 2001a; EP, 2002). If auctioning is the harmonized rule, the potential problem is that a member state that wishes to deviate by allocating permits for free causes a distortion of fair competition in favour of its own industry, leading to state aid if level-playing-field considerations are considered to be relevant. A potential legal solution, which needs to be explored further, is to make grandfathering the harmonized rule, so that an EU country that wishes to deviate by auctioning permits only causes a distortion of fair competition that is detrimental to itself, which is allowed according to Article 97 of the EC Treaty.

A possible disadvantage is that it could still be seen as conflicting with the polluter-pays principe laid down in Article 174 of the EC Treaty. The reason for this is that if member states acknowledge that permit allocation differences among them distort competition, it means that they emphasize the financial differences between grandfathering and auctioning (from an equity perspective) and not their similarities in terms of opportunity costs (from an efficiency perspective). In terms of opportunity costs, the polluter also pays when permits are handed out for free, but the point is that in terms of financial effects, grandfathered polluters do not have to pay for their emissions (in contrast to auctioned firms).

At the end of 2002 the co-decision procedure on the design of an EU emission trading scheme was not yet finished. The European Council seems to lean in the direction of the aforementioned potential solution as the Ministers want to make 90 per cent grandfathering the harmonized rule during 2008-2012 (and 100 per cent grandfathering during 2005-2007), whereas the European Parliament, which initially insisted on using auctioning, later advocated the use of 85 per cent grandfathering in both periods (see, for instance, Worsley and Freedman, 2002).

5 CONCLUSION

On the basis of a law and economics analysis supplemented with a political science approach, we have indicated under which conditions differences between the domestic permit allocation procedures of the member states of the European Union in general and grandfathering in particular will distort competition and lead to state aid in a European carbon trading market. This chapter shows that it depends on whether one takes an efficiency or equity perspective. A grandfathered firm is not advantaged from an efficiency perspective, because both grandfathered and auctioned firms have to include the opportunity costs of holding the permits in the product price. However, a grandfathered firm is advantaged from an equity perspective, because it has more financial resources (*ceteris paribus*) than its auctioned competitor in another member state which has to buy its permits.

A political (albeit not legal) precedent for a European permit-trading market is set by the decisions of the European Commission on the permit trading schemes in Denmark and the UK. The Commission characterized grandfathering as state aid, but nevertheless exempted it by using not only legal and economic, but also political arguments. If the Commission were to continue to use this interpretation in an EU-wide scheme, the provisions for operating aid in the revised Community guidelines on state aid for environmental protection (2001) could suggest that firms may receive no more than a certain percentage (for instance 50 per cent) of their permits for free during a limited period of time (for instance five years) as a transition phase towards an auctioned scheme. However, grandfathering could, again, be exempted from the aid provisions in the first place.

A normative implication of our analysis is that scientists and policy makers (also those from the Commission) should make explicit in their papers and decisions whether they use an efficiency or an equity interpretation when they think about whether or not grandfathering is state aid, because – on a positive-theoretical level – the issue stands or falls with this distinction. In particular, they should become more aware of the role of opportunity costs when discussing the economic, legal and political implications of grandfathering.

A potential legal solution is to make grandfathering the harmonized rule, so that an EU country that wishes to deviate by auctioning permits causes a distortion of fair competition that is only detrimental to itself, which is allowed according to Article 97 of the EC Treaty. A possible disadvantage, however, is that grandfathering, when not its opportunity cost (efficiency) but its financial advantage (equity) is stressed, could be seen as conflicting with the polluter pays principle laid down in Article 174 of the EC Treaty, because grandfathered polluters, despite the opportunity costs of their permits, do not have to pay for their emissions. This must be studied further.

REFERENCES

Bohm, P. (1999), 'International greenhouse gas emission trading – with special reference to the Kyoto Protocol', *TemaNord* 1999:506, Stockholm: University of Stockholm.
Capros, P., L. Mantzos, M. Vainio and P. Zapfel, (2000), 'Economic efficiency of cross-sectoral emission trading in CO_2 in the European Union', paper presented at the Conference on 'Instruments for Climate Policy: Limited versus Unlimited Flexibility?', 19–20 October, University of Ghent, Belgium.
Cini, M. and L. McGowan, (1998), *Competition Policy in the European Union*, London: Macmillan.
COM (1998), *XXVIIth Report on Competition Policy 1997*, Luxembourg: Office for

Official Publications of the European Communities.
COM (1999), *XXVIIIth Report on Competition Policy 1998*, Luxembourg: Office for Official Publications of the European Communities.
COM (2000a), *Green Paper on Greenhouse Gas Emissions Trading Within the European Union*, Green Paper presented by the Commission, 8 March.
COM (2000b), *XXIXth Report on Competition Policy 1999*, Brussels/Luxembourg: European Commission, document SEC(2000)720 final, 5 May.
COM (2000c), 'State Aid No. N 653/99 – CO_2 Quotas (Statsstøttesag nr. N 653/99 – CO_2 kvoter)', Letter by Mario Monti to the Danish Government, Brussels/ Luxembourg: European Commission, English version (draft), 12 April.
COM (2001a), *Proposal for a Directive of the European Parliament and of the Council Establishing a Framework for Greenhouse Gas Emissions Trading within the European Community and Amending Council Directive 96/61/EC (Presented by the Commission)*, document COM(2001)581, 23 October, Brussels/Luxembourg: European Commission.
COM (2001b), *State Aid No. N 416/2001 - United Kingdom Emission Trading Scheme*, Letter by Mario Monti to the UK Government, C(2001)3739fin, 28 April, Brussels/Luxembourg: European Commision.
ECCP (2001), *European Climate Change Programme - Report June 2001*, Brussels/Luxembourg: European Commission.
European Parliament (EP) (2002), *Draft Report on the Proposal for a European Parliament and Council Directive on Establishing a Scheme for Greenhouse Gas Emission Allowance Trading within the Community and Amending Council Directive 96/61/EC*, provisional document 2001/0245(COD), PR/385503EN.doc, Committee on the Environment, Public Health and Consumer Policy (Rapporteur Jorge Moreira da Silva), 8 April, Brussels: European Parliament.
Folketinget (1999), *Bill on CO_2 Quotas for Electricity Production*, Act no. 376 of 2 June 1999, unauthorized translation of 30 June 1999, Folketinget (Danish Parliament).
Goulder, L.H., I.W.H. Parry, R.C. Williams III and D. Burtraw (1999), 'The cost-effectiveness of alternative instruments for environmental protection in a second-best setting', *Journal of Public Economics*, **72** (3), 329-60
Haites, E., J. Skjelvik, D. Harrison, M. Ward, N. Zarganis and K. Yamaguchi (2000), 'Emissions reduction policies and measures in Annex I countries', in: R.J. Kopp and J.B. Thatcher (eds), *The Weathervane Guide to Climate Policy: An RFF Reader*, Washington, DC: Resources for the Future (RFF), pp. 83-101.
Hargrave, T., N. Helme, T. Denne, S. Kerr and J. Lefevere (1999), *Design of a Practical Approach to Greenhouse Gas Emissions Trading Combined with Policies and Measures in the EC*, Washington, DC: Center for Clean Air Policy (CCAP).
Hildebrand, D. (1998), *The Role of Economic Analysis in the EC Competition Rules*, The Hague: Kluwer Law International.
Jensen, J. and T.N. Rasmussen (1998), *Allocation of CO_2 Emission Permits: A General Equilibrium Analysis of Policy Instruments*, Copenhagen: Danish Ministry of Business and Industry.
Jepma, C.J. and W.P. van der Gaast (1999), 'Flexible instruments' carbon credits after Kyoto', in C.J. Jepma and W.P. van der Gaast (eds), *On the Compatibility of Flexible Instruments*, Dordrecht: Kluwer Academic Publishers, pp. 3-15.
Jepma, C.J., W.P. van der Gaast, N. Schrijver and D. Giesberger (1999), *The Role of the Private Sector in Joint Implementation with Central and Eastern European Countries: An Interpretation of Article 6 of the Kyoto Protocol*, Joint

Implementation Network, Groningen: JIN.
Koutstaal, P. (1997), *Economic Policy and Climate Change: Tradable Permits for Reducing Carbon Emissions*, Cheltenham, UK and Northampton, MA, USA: Edward Elgar.
Lyon, R.M. (1982), 'Auctions and alternative procedures for allocating pollution rights', *Land Economics*, **58** (1), 16-32.
Nentjes, A., P. Koutstaal and G. Klaassen (1995), *Tradeable Carbon Permits: Feasibility, Experiences, Bottlenecks*, Dutch National Research Programme on Global Air Pollution and Climate Change (NRP), Report No. 410 100 114, Groningen/Bilthoven: NRP/RuG.
OJ (1994), 'Community guidelines on state aid for environmental protection', *Official Journal of the European Communities C 72*, Vol. 37, 10 March, 3-9.
OJ (2001), 'Community guidelines on state aid for environmental protection', *Official Journal of the European Communities C 37*, Vol. 03, 3 February, 3-16.
Van der Laan, R. and A. Nentjes (2001), 'Competitive distortions in EU environmental legislation: inefficiency versus inequity', *European Journal of Law and Economics* **11** (2), 131-52.
Van Deth, J.W. and E. Scarbrough (eds) (1995), *The Impact of Values*, Oxford: Oxford University Press.
Welch, W.P. (1983), 'The political feasibility of full ownership property rights: the cases of pollution and fisheries', *Policy Sciences*, **16**, 165-80.
Woerdman, E. (2002), *Implementing the Kyoto Mechanisms: Political Barriers and Path Dependence*, Dissertation, Groningen: University of Groningen.
Woerdman, E. (2000a), 'Competitive distortions in an international emissions trading market', *Mitigation and Adaptation Strategies for Global Change*, **5** (4), 337-60.
Woerdman, E. (2000b), 'Organizing emissions trading: the barrier of domestic permit allocation', *Energy Policy*, **28** (9), 613-23.
Worsley, R. and R. Freedman (eds) (2002), *'Europarl' Daily Notebook 10-10-2002*, Brussels: European Parliament.
Yamin, F. and J. Lefevere (2000), *Designing Options for Implementing an Emissions Trading Regime for Greenhouse Gases in the EC*, Final Report to the European Commission DG Environment, 22 February, London: FIELD.

6. Legal aspects of the Dutch approach to CO_2 reduction

Chris Backes and Reinske Teuben

1 INTRODUCTION

The aim of this chapter is to examine the possible contribution of legal science to finding suitable and effective instruments to control climate change. More specifically, we shall discuss some instruments of climate policy from a legal point of view. The instruments that are discussed may also be of interest to other countries and the European Union.

According to the United Nations Framework Convention on Climate Change (UNFCCC) and the Kyoto Protocol, the so-called Annex I parties (that is, the industrialized countries) have committed themselves to legally binding reduction targets for the emission of greenhouse gases between 2008 and 2012. The European Union is treated as a 'bubble' which means that there is a reduction target for the Union as a whole. In the EU's internal 'burden-sharing agreement' every member state has been assigned its own reduction commitment. The reduction targets vary from −28 to +27 per cent. The Netherlands will have to reduce its greenhouse gas emissions by 6 per cent between 2008 and 2012, compared to the emissions level for the year 1990.

In the last decade several efforts have been made to cut back greenhouse gas emissions, such as various energy-saving measures, improving the energy efficiency of, for example, industrial processes and household appliances, the introduction of a regulating energy tax for smaller energy users, higher fuel taxes for motor vehicles and stimulating sustainable energy.[1] However, despite those efforts, over the past ten years Dutch CO_2 emissions have not decreased or even stabilized at all. On the contrary, by 1998 they had increased by about 11 per cent, and they are expected to rise even further in forthcoming years. By 2010, the difference between the reduction targets and the actual emissions would, without additional policy measures, be about 19 per cent.[2] In no other EU member state is there such an enormous 'policy shortfall'. One important reason for the increasing CO_2 emissions is the strong economic growth that took place in the Netherlands during the 1990s. The Netherlands has not succeeded in separating the CO_2 emission levels from the economic growth rates sufficiently. In this respect CO_2 is different from most other pollutants, like SO_2, CO, lead and NO_x.

Under the Kyoto Protocol the parties are free to choose their domestic instruments and measures in order to comply with their reduction commitments. EU measures are regarded as domestic measures. In 1999 the Dutch government presented its national Climate Policy Implementation Plan, the *Uitvoeringsnota Klimaatbeleid*. This document sums up the additional measures for dealing with the 'policy shortfall'. Along with the existing policy, these additional measures should guarantee that the Dutch reduction commitments will be achieved.

The Dutch government distinguishes between measures to be taken within the Netherlands and measures to be taken abroad, for example in developing countries. Both kinds of measures are to account for 50 per cent of the additional emission reductions. This distinction is reflected in the *Uitvoeringsnota*, which consists of two parts. The first part describes the domestic measures, while the second part focuses on the measures that will be taken outside the Netherlands. The latter include the use of the so-called Kyoto mechanisms, such as the clean development mechanism (CDM) and joint implementation (JI). The idea behind taking measures abroad is the difference in the cost of emission reductions. In other countries, and especially in the developing countries, a reduction of, for instance, 1 ton of CO_2 can often be achieved at lower cost than in the Netherlands, where the easiest and most profitable measures have already been taken. Hereafter, we shall not discuss these 'international' measures any further. Instead we shall restrict ourselves to the domestic measures as mentioned in the first part of the Dutch climate change policy implementation plan.[3]

The first part of the *Uitvoeringsnota* contains a wide range of possible domestic measures. They are divided into three packages of measures with different priorities, namely the 'basic package', the 'reserve package' and the 'innovation package'. The instruments mentioned in the 'basic package' of measures will be used first to cut back the Dutch CO_2 emissions. They are relatively cheap and include, for example, a further 'greening' of the tax system, increasing the tyre pressure of motor vehicles, stimulating the use of renewable energy sources[4] and planting new woodlands that can absorb CO_2 from the atmosphere. In the government's opinion, the 'basic package' will be sufficient to achieve an emission reduction of 25 Mton CO_2.[5] If not, the more expensive measures in the 'reserve package', for example, underground storage of CO_2 and an increase of energy and fuel taxes, will be applied. The innovation package is only meant for emission reductions in the longer term, namely after 2012. This package includes more fundamental technological and instrumental innovations.

The first evaluation of the policy contained in the *Uitvoeringsnota* will take place in the first half of 2002. At that moment the government can decide whether it is necessary to use the reserve package of measures. Recently the government announced that it is planning to introduce two additional measures, namely combining heat and power, which is generally referred to as the total energy principle, and further energy saving in houses and other buildings.

To summarize, the present climate policy already uses many different measures and instruments, and other possible instruments are presently being discussed. The facts show that Dutch climate policy in the past has failed to achieve substantial reductions of greenhouse gas emissions. The effects of many of the new additional policy measures again depend on the development of the price for energy, which is very difficult to predict. It is therefore necessary to make every possible effort to fulfil the Kyoto targets and to ensure, as far as possible, that the proposed additional measures will be successful. This implies – among other things – that the legal problems that sometimes obstruct an optimal functioning of possibly useful instruments must be solved. In this respect, two instruments are of particular interest: first, voluntary agreements with the energy-intensive industry and second, CO_2 emissions trading. These two instruments are not only interesting in the context of Dutch national climate policy, but can be of interest to other countries and the European Union too. However, from a legal point of view, there are some questions that need to be solved before these promising instruments can be used effectively. By doing this, legal science could make its contribution to a more successful climate policy. Of course it is possible that some of these problems cannot be solved nationally, so that perhaps voluntary agreements or emissions trading can only be successfully applied at the EU level.

2 VOLUNTARY ENVIRONMENTAL AGREEMENTS

In the Netherlands there is a long history of voluntary agreements (*convenanten*; that is, legally binding agreements under private law) between the government, on the one hand, and trade and industry on the other. Since the mid-1980s the government has intensively promoted such agreements covering various environmental aspects. The government is convinced that the necessary high reduction rates for all kinds of emissions, as far as industry and trade are concerned, can hardly be realized by revising environmental permits and enforcing them or by the use of other command-and-control instruments only, but that these targets are more likely to be met by a cooperative approach and by using the self-regulating powers of the industrial sector. The idea is that the government lays down the necessary reduction targets and shows the consequences if those targets are not met, for example, in the form of taxes or emission limit values in (binding) general rules. The business sector will then come up with its own ideas as to how to realize these reduction rates in order to avoid such obligatory measures.

At the time of writing, there are more than 100 environmental voluntary agreements between the government and Dutch industry and trade,[6] including important agreements on the use of energy.

Multiannual Agreements on Energy Saving

Since 1992, the government has concluded 'multiannual agreements' (*meerjarenafspraken*) on saving energy with a number of industrial branches. For each branch, sometimes far-reaching reduction targets have been laid down. These targets were derived from the general reduction targets in the Dutch national environmental policy plan (*Nationaal milieubeleidsplan*, NMP). About 1000 enterprises, which together consume about 90 per cent of the energy used by the industrial sector, signed up to the agreements. The enterprises committed themselves to drawing up an energy saving plan, all of which was evaluated by a neutral agency, called NOVEM.[7] The central question of this evaluation was whether the contribution of the concrete enterprise to the overall reduction target of the whole branch was large enough. Only those energy-saving measures were prescribed that could be recouped within three years. It is interesting to see that the measures which successfully passed this evaluation were subsequently included in the environmental permits and thus became binding and enforceable. Thus the use of energy was not excluded from the command-and-control instruments and was not merely regulated by the use of voluntary agreements, but rather the voluntary agreements and the environmental permits as command-and-control instruments were linked together to form a consistent and effective mix of instruments. This seems to be a successful combination of licensing and voluntary agreements, which overall meets the preconditions of the European Directive on Integrated Pollution Prevention and Control (IPPC Directive).[8] By adopting this combination of instruments, energy use can in fact be reduced further than by only using the command-and-control instruments, which would mean that thousands of environmental permits would have to be amended.

However, one question arises. When evaluating the energy-saving plans of each individual enterprise, an individual cost–benefit analysis is made. Only measures that can be recouped by the individual enterprise within three years can be prescribed.[9] From an economic point of view, this may make a great deal of sense, but it is probably not in accordance with the so-called 'best available techniques' (BAT) criterion in the IPPC Directive. The directive concerns larger industrial (and agricultural) enterprises and requires the member states to request individual environmental permits for those enterprises. The maximum emissions allowed by such a permit must be based on the best available techniques. The directive defines more exactly what BAT means. The BAT criterion of the IPPC Directive, as well as the corresponding criteria in Article 8.11 (3) of the Dutch Environmental Management Act (EMA, *Wet milieubeheer*),[10] prescribe an objective and general standard, relating to the whole sector. All measures which are available and cost-effective for an average going concern in a certain sector have to be taken. What is cost-effective and available for each individual enterprise, can differ from what may be required of an *average* enterprise in a certain sector. However, this difficulty does not concern the basic principles and the main structure of the multiannual agreements on energy use. It would generally be possible to use a more objective, general criterion when evaluating the energy saving plans.

The Benchmarking Agreement on Energy efficiency

Meanwhile, the next generation of environmental agreements on energy use has been developed. This concerns the so-called 'benchmarking agreement', which was signed in the summer of 1999.[11] About 100 energy-intensive enterprises, which together consume about 80 per cent of the industrial energy, participate in this agreement. The participants have undertaken to be among the world's most efficient users of energy within their sector by the year 2012 at the latest. The agreement contains detailed descriptions as to how this is to be measured objectively. Again, NOVEM plays an important role in the process of certifying the necessary evaluations. Government and industry expect that the measures necessary in order to be among the world leaders in energy efficiency will lead to additional savings of about 2 or 2.5 Mton CO_2.[12] Most of this, some 1.5 or 2 megatons, can be saved by taking measures to renovate coal-fired power stations, which are not yet among the world leaders.

In return the public sector (central government and the provinces) agreed not to introduce ceilings for the overall CO_2 emissions or new energy taxes and to take no additional measures in the environmental permits of the enterprises, which go further than or differ from those that are necessary to achieve top rating in energy efficiency. On an individual scale, as a first step the enterprises have to take all the measures that are necessary to reach the top in energy efficiency, and that can be recouped at an interest rate of 15 per cent. If this is not enough, as a second step, which continues until 2008, more expensive measures will have to be taken, which still have to be cost-effective, but looking at the sector as a whole. If even this is not enough, (international) instruments such as international emissions trading, joint implementation and so on may be used.

This new way of handling the energy use of the industrial sector in principle seems to be in keeping with the IPPC Directive and the corresponding parts of the Dutch EMA. Again an exception has to be made for the individual but not general and objective way of evaluating cost effectiveness. This especially concerns the first step. Taking all measures that for the individual enterprise can be recouped at an interest rate of 15 per cent may, in certain cases, not be enough to meet at least the requirements of the BAT standard related to an averagely functioning enterprise of the sector concerned. Besides, there are some other details that give rise to legal questions. One of these is that a company group is allowed to decide where and when it wants to invest in energy-efficiency measures. Within a company group, energy saving at one plant or installation can compensate for fewer or no measures at another plant or installation. Although economically this seems to make sense, it can be problematic from a legal point of view. On the level of each installation, it is not certain whether all the measures have been taken that are necessary to meet the requirements of both the BAT criterion (IPPC Directive), and the corresponding criteria of Dutch national law (Article 8.11 (3) EMA). Legal experts are currently discussing whether the way of defining the top rating in energy efficiency and the details of the benchmarking process ensure that the measures necessary to be among the world leaders at least meet the BAT standards in every individual case. However, although this point of discussion seems a rather theoretical one, it can be solved quite easily. If in a single case the requirements of the IPPC Directive and Article 8.11 (3) EMA appear to be more stringent than the results of the benchmarking process, then the more stringent requirements can be enforced by using the command-and-control instrument of the environmental permit. In that case, the government or province is legally forced to break the promise provided in the benchmarking agreement, namely not to enforce additional measures over and above the 'benchmarking measures' with the instrument of the environmental permit.

The EC Treaty, more specifically Articles 81 and 87 EC, could give rise to some legal obstacles for the benchmarking agreement. The exchange of information that enterprises must deliver within the benchmarking process, especially information on the use of energy and energy costs and other information concerning the production process and capacity of energy-intensive plants, may cause a competition distortion. However, much depends on the details of the benchmarking process. Rutten[13] has suggested some ideas to make this process 'distortion-proof', for example, to provide only data that are at least one year old. Therefore it seems that the benchmarking agreement can be implemented without violating Article 81 EC.

Whether the benchmarking agreement may be qualified as state aid in the sense of Article 87 EC cannot be judged yet. This will depend on which governmental measures for other enterprises, for example, additional energy taxes, will not be applied to the firms that have joined the benchmarking agreement and, on the other hand, what additional costs the benchmarking enterprises incur. Only if the benchmarkers save more than they have additionally spent can this qualify as state aid.[14]

From a more political perspective, our main concern relates to a different question. Even if the Dutch enterprises really succeed in becoming the world's best in energy saving, it is not at all certain that the consumption of energy by the industrial sector will be cut back or even stabilized. Which and on what scale energy-saving measures will be realized, in the end directly depends on the energy price. Although energy prices are relatively high at the moment, it is not certain whether this will be so in the mid-term future. The experience of the last 15 years has shown that it is very difficult to predict the energy price and economic growth with at least some degree of certainty. The prognosis that the benchmarking agreement will provide an extra energy saving of 2 or 2.5 Mton CO_2, is to a large extent a theoretical one.

By signing this agreement, the Dutch government will, at least until 2012, no longer be able to develop a market for tradable CO_2 emission allowances with a fixed ceiling or a market for tradable CO_2 reductions. The energy-intensive industry, responsible for the overwhelming majority of industrial energy consumption, will have to be left out if a CO_2 trading system is introduced. This will place a very heavy burden on the effectiveness of such an instrument. The question is whether, taking into account the enormous gap between the Dutch reduction targets for CO_2 and the actual growth of the CO_2 emissions, it was wise to renounce the possible introduction of a market for tradable CO_2 emission allowances or tradable reduction credits for the energy-intensive industry. The benchmarking agreement does not ensure that the necessary reductions of industrial CO_2 emissions will be realized. After all, the agreement regulates only the *efficiency* of energy use, not the total amount of CO_2 emissions. This is not only debatable from a national policy point of view, but may also become problematic in the light of the ideas for a European emissions trading scheme, which were proposed in the European Commission's Green Paper.[15] The benchmarking agreement would conflict with such a development, but of course it can in principle be revoked by the government.

3 CO_2 EMISSIONS TRADING

As we have already said, the Dutch climate policy has not been very successful so far. Instead of decreasing, CO_2 emissions have increased considerably since 1990. It is doubtful whether the present policy will be able to attain sufficient emission reductions before 2012. Moreover, far more radical reductions (possibly up to 80 per cent) will be necessary in the long term in order to achieve a situation of real sustainability. The failure of the present policy as well as the long-term perspective make it clear that new instruments are needed in addition to the existing ones.

A possible new instrument to lower CO_2 emissions is a national or a European system of emissions trading.[16] In the Netherlands and the EU, this instrument is often discussed, but to date no Dutch or European trading scheme has actually been used in practice. In the US, various trading systems have been applied, although none of them concerned CO_2 emissions. In Denmark, a domestic CO_2 emission-trading scheme for the electricity production sector became operational in 2001.[17]

The Idea of Emissions Trading

The basic idea of emissions trading is simple. For every amount of CO_2 that is emitted an allowance is needed. The total number of allowances, and therefore the total emitted volume of CO_2, is limited. This is the so-called 'emissions ceiling'. The allowances are transferable between the polluters. If an enterprise cuts back its emissions, it can sell the superfluous allowances. Higher emissions than are covered by the allowances (for example, when an enterprise wants to increase its production) are only permitted when extra allowances are bought. Thus a CO_2 permit market with sellers and buyers of CO_2 emissions develops.

As long as reduction measures are relatively cheap in a certain enterprise, it is worthwhile taking these measures and selling the allowances that are no longer required. If measures are, on the other hand, relatively expensive (that is, when the costs of the measures are higher than the market price of the allowances), it is then more advantageous to buy permits. As a result, measures to reduce CO_2 emissions are taken where this is most profitable. This could lower the total costs of climate policy considerably. Furthermore, the system creates an incentive for technological innovation. Since one could recoup one's investments in 'clean' technology by selling superfluous emission allowances, innovative efforts become more attractive.

Thus, emissions trading can bring about great benefits, as it can be both an effective and an efficient means of reducing CO_2 emissions: effective, because there is an emissions ceiling that limits the total emissions,[18] and efficient, because the use of the market mechanism can lower the overall costs of climate policy. Moreover, CO_2 is highly suitable for inclusion in an emission-trading scheme, because the global warming effects caused by CO_2 are independent of where the gas is emitted. CO_2 emissions trading will therefore not lead to unacceptably high concentrations of pollution at certain locations ('hot spots').

The Present Discussion in the Netherlands and the EU

In the Dutch *Uitvoeringsnota Klimaatbeleid*, CO_2 emissions trading is only mentioned in the so-called 'innovation package'. This means that the introduction of the instrument cannot be expected in the short term, but in principle only after the first commitment period of the Kyoto Protocol, that is, after 2012. Still, experiments and studies are planned in the coming years in order to prepare for a possible emission-trading system in the future. Probably, emissions trading will first be introduced on a small scale (for instance, in one specific industrial sector) and on a more experimental basis. The experiences from these experiments can be useful when designing a more comprehensive trading scheme. Only the so-called 'sheltered' sectors[19] are likely to be involved in any experimental emissions market.

The government's decision not to introduce any comprehensive trading scheme in the short term does not seem to be a very wise stance. One reason is that the present policy has so far failed to achieve any real emission reductions.[20] The Netherlands should use all possible instruments to meet the Kyoto targets. Another reason is the European climate policy. According to the European Commission's Green Paper,[21] the EU is already considering the introduction of emissions trading in 2005. The Green Paper is also meant as a discussion paper, which explains why many design options are still left open, for example, with regard to the number of member states involved, the method of allocating of the allowances, the participating economic sectors and so on. An important aim is to gather experience with emissions trading before the international trading based on Article 17 Kyoto Protocol is introduced.

The proposals for tradable NO_x emission reductions

Although the Dutch government seems to think of tradable CO_2 allowances only in the longer term, a system of emissions trading for NO_x is already planned for introduction in 2003. It may be interesting to look at the proposals for a domestic trading scheme for NO_x as well. These proposals are at a more advanced stage than the plans for tradable CO_2 permits. Of course NO_x emissions are in some ways different from CO_2 emissions (NO_x has local effects, for instance), but as we shall see later, the legal problems related to both trading systems are very similar.

In the NO_x proposals a different type of system is chosen, namely a system of tradable emission reductions. In this type of trading scheme there is no emissions ceiling and there is no need to 'cover' every emission with an allowance. Instead, there is a general emissions standard for all polluters. This standard is a relative one, in which the allowed CO_2 emissions are related to the production levels. Thus, the total CO_2 emissions are not limited, only the emissions per product unit. As long as one complies with the general norm, there is no need to possess emission allowances. If an enterprise emits less CO_2, the additional reduction compared to the standard can be sold. If the enterprise's emissions exceed the norm, it can choose between the following: either it takes reduction measures, or it buys the reductions that have been achieved elsewhere (see Figure 6.1). Like tradable emission allowances, tradable emission reductions can increase the cost efficiency of climate policy, as the system leaves the choice to the emitters whether they take measures by themselves or buy the reductions that result from measures taken by other emitters.

In the proposed NO_X system, a general emissions standard will be set. Additional reductions can be sold. Complying with the norm is possible by taking reduction measures or by buying the tradable reductions that have been attained by other companies.

Yet, this type of trading system has some important drawbacks, at least as far as CO_2 emissions are concerned. First, there is no emissions ceiling. The allowable emissions are expressed in relative, not in absolute terms. In the absence of an emissions ceiling it remains uncertain whether the overall reduction targets will be achieved. As the general emissions standard is a relative one, the total CO_2 emissions are still influenced by economic growth rates. Further, this type of system may be difficult to apply to CO_2 emissions. The general emissions standard would be related to production levels. This

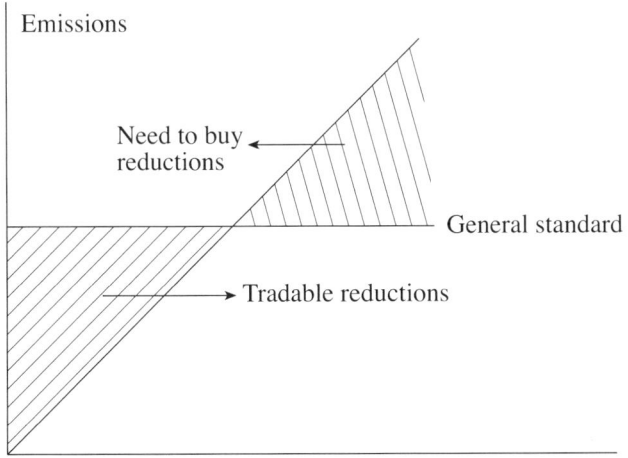

Figure 6.1   Tradable reductions

seems very problematic, because there are enormous numbers of products and production processes that cause CO_2 emissions. Therefore, it would be necessary to set many different emission standards, related to the production levels for one specific product. This problem does not exist for all types of emissions. In the case of NO_X the standard can be related to the energy use, which is much easier to measure. In other words, a system of tradable emission reductions for CO_2 is not very likely to be introduced. Tradable allowances seem to be the more logical option. This is also reflected in the present discussion on CO_2 emissions trading. In their official documents, both the Dutch government and the EU focus on possible systems of tradable allowances, while tradable CO_2 reductions do not seem to be regarded as an alternative.[22]

Legal Aspects of Emissions Trading

Most of the existing studies on emissions trading focus on the economic aspects. The legal aspects are hardly ever discussed. Nevertheless, CO_2 emissions trading causes many legal problems. These problems often seem to be overlooked, but they can be real obstacles to a successful introduction of a permit-trading scheme. This means that the legal questions need to be solved before a national or European emissions market can actually become operational.

The problems related to emissions trading are very diverse and concern different fields of law, such as European and national environmental law, European Community/Union law, administrative law, the protection of property rights and enforcement questions.[23] Sometimes basic principles of environmental law are involved. For example, it is doubtful whether the 'grandfathering' (that is free distribution) of emission allowances is compatible with the polluter-pays principle. It would be impossible to discuss all these problems in this chapter. In the following we shall therefore concentrate on the legal problems in the field of environmental law.

Environmental law
One of the major obstacles that could complicate the introduction of emissions trading is the compatibility of a trading regime with the existing environmental law system. The Dutch environmental law system is not only a national matter, but is strongly influenced by European law, which makes the problem somewhat more complicated. The question of the compatibility of CO_2 emissions trading with the present legal system, especially the Dutch Environmental Management Act (EMA) and the European IPPC Directive, has hardly been discussed so far. Therefore, the discussion concerning NO_X trading can also be of interest.

The Present Dutch Environmental Law System (EMA/IPPC)

The Dutch environmental law system is based on the national Environmental Management Act (EMA) and, for the larger-scale polluters, the European IPPC Directive. The central element of the system is the environmental permit. This permit is restricted to a specific emission source ('installation') at a specific location. Usually, an integrated assessment of all environmental aspects is prescribed. These aspects also include the use of energy, so CO_2 emissions are covered by the environmental permit. At least in theory, all environmental aspects of a certain installation have to be looked at in concert.

A high level of protection is aimed at. When emissions cannot be prevented, they should be reduced as much as possible. In this context, Article 8.11 (3) EMA mentions the ALARA ('as low as reasonably achievable') principle: the highest possible level of protection must be applied, unless this is not reasonably achievable. The ALARA principle is closely related to the BAT principle, which is mentioned in Article 2 (11) of the IPPC Directive. According to this article, the best available techniques are the best technological means to prevent or reduce emissions that can be reasonably required. This means that it should not only be technologically, but also economically feasible to use these techniques. In this context the installation must be compared to other (similar) enterprises in the same branch of industry.

Emissions trading can conflict with this environmental permit system. The problem has two aspects: first, the prescribed integrated approach, and second, the compatibility of emission-trading schemes with the ALARA and BAT principles.

Integrated approach

The first problem is the integrated assessment of all environmental impacts, as required by the EMA and the IPPC Directive. The environmental permit covers almost all environmental 'sectors', such as air pollution, soil pollution and noise, but also the use of energy. When a permit is granted, the authorities should base their decision on an integrated assessment of all possible impacts. This integrated approach is one of the main characteristics of the EMA and the IPPC Directive. The idea behind the principle of an integrated assessment of all environmental effects of an installation is twofold. In the first instance a separated approach could cause the pollution to shift from one medium to another. If, for example, the effects of air pollution are looked at first, then the maximization of the reduction of air pollution could cause more pollution to the water or the production of more waste. Second, when assessing certain environmental effects separately, crossover and cumulative effects are not taken into account.

At first sight CO_2 emissions trading seems to exclude CO_2 emissions and the use of energy from the integrated assessment. This seems incompatible with the present legal system. Basically there are two possible solutions. The first one would be to alter the existing system by excluding this specific aspect from the integrated approach of the environmental permit system. This would require an amendment of both the national and the European legislation. In principle, this is possible. However, one has to be aware of making too many exceptions to the integral approach. Especially in the Netherlands, where there are already such exceptions for the emissions of ammonia and others are being developed for NO_X emissions, there may come a time when there is not much of the integral approach left.

Another possibility is that the emission-trading regime would be set up in such a way that it *is* compatible with the existing legal system. This is not necessarily impossible. If, for example, the integrated permit for installations still covers the use of energy and the trade of CO_2 is only allowed to meet the criteria of that integral permit, an integrated assessment which includes the use of energy and a trading system need not to be a contradiction. The energy target, which was developed in an integrated process of balancing all environmental aspects and the total costs of all environmental measures, can be met by real reductions or by buying reduction or emission rights. More attention should be directed towards how a trading system and a process of integral permitting can be combined. However, there seem to be more possible solutions than are apparent at first sight.

ALARA and BAT principles

The second, more important aspect is the meaning of the ALARA and BAT principles.

In the present legal system, every individual installation should comply with the requirements that follow from the ALARA and BAT principles. These requirements imply a maximum emission level for the installation. Emissions trading has a completely different approach. It focuses on the overall emissions of all installations together. In the case of CO_2 trading, there is a national target that must be met. It is less important whether the emissions of a specific installation are as low as possible. This follows from the economic background of the instrument. When the choice between reducing emissions or buying permits is left to the polluters themselves, considerable cost benefits can be achieved. It is clear that this can only be acceptable as far as the location of the emissions makes no difference. This is the case for CO_2 emissions.

The question is whether these two approaches are compatible. In this context, a distinction must be made between the ALARA principle on the one hand, which is a principle of Dutch national environmental law, and the BAT principle, based on the European IPPC Directive, on the other hand. Although the two principles are intrinsically closely related, they operate differently because of their different backgrounds. Of course legal problems that only occur on the national level can be solved more easily than problems that have their basis in European law.

National environmental law: the ALARA principle Article 8.11 (3) EMA contains the ALARA principle. As mentioned above, each enterprise's emissions should be as low as reasonably possible. 'Reasonably' indicates that the economic feasibility of the measures is also relevant. In this context a comparison with other enterprises in the same sector is made.

The ALARA principle is a minimum standard: the requirements laid down in the environmental permit must at least be this strict. However, at the same time the principle is explained as a maximum standard. The authorities cannot prescribe measures that are more stringent than ALARA. Thus, there is only one possible outcome, namely the ALARA level itself. This makes emissions trading impossible, as it is not compatible with the national legal system. Only if the EMA is changed, would a national emissions trading scheme be possible. One option would be to change the meaning of the ALARA principle in the EMA, by making it only a minimum, like the BAT principle in European law (see later), and no longer a maximum standard. The other possibility is that the national legislation would make an exception to the ALARA principle for the specific purpose of emissions trading.

In its proposals for a system of tradable NO_X reduction credits (see above) the government wants to set a general emissions standard that minimally meets the ALARA level. This denies the fact that ALARA is – at least at the moment – also explained as a maximum requirement. Even if it were possible to apply a general standard based on ALARA or a higher level than that, it would be questionable whether this system could work properly, as it may limit possibilities for trading too much (see next subsection).

As far as the compatibility of the proposed NO_X trading system with the ALARA principle is concerned, interesting developments have taken place recently. In its advice of October 2000 [24] the Dutch Council of State considered the proposed emissions trading system to be structurally incompatible with the system of the EMA and the ALARA principle.

According to the Council of State, moving the pollution to another location is not allowed under the present location-based environmental permit system. This would be incompatible with the idea of the individual installation as the object of environmental regulation. Further, the Council assumes that the objective of the EMA is not directly to reduce emissions, but to protect the environment from negative impacts. Emission reductions can only be imposed when there is a direct relationship between the emissions and the negative impacts on the environment at that specific location. Finally, it is not considered possible to 'comply' with the requirements based on the ALARA principle by buying emission reductions. In the Council's opinion, the ALARA principle must be applied at the installation level. This restrictive interpretation of the ALARA principle is also based on the fact that the permit is restricted to a specific place.

The Council's opinion may be somewhat disappointing, but it still has an element of truth in it. In any case it is clear that there is serious tension between the existing environmental law system with its installation-oriented approach and the ALARA (EMA) and BAT (IPPC) requirements for specific emission sources at specific locations on the one hand, and emission-trading schemes on the other. The most important instrument of the existing system, the environmental permit, is strictly limited to individual installations at specific locations. In many cases this may make a lot of sense, especially when the activities have strong impacts on the local environment. However, being strictly bound to individual installations, the present permit system does not seem to be a very suitable instrument to deal with environmental problems that occur on a national or even – like climate change – an international level.

European environmental law: BAT principle   The BAT principle is based on secondary EU law, namely the IPPC Directive. Unlike the ALARA principle, BAT only represents a minimum requirement. This follows from the legal basis of the IPPC Directive (Article 130S EC, now Article 175). More stringent national measures are allowed, as long as they are compatible with the Treaty, especially with the provisions concerning fair competition, the free movement of goods, non-discrimination and so on.

In a system of tradable emission allowances it is problematic that the IPPC Directive always requires an environmental permit that is in accordance with BAT. This approach is fundamentally different from the idea of tradable emission allowances. In such a system it would be very difficult to guarantee that every single enterprise complies with the BAT requirements. This would only be possible under far-reaching restrictions. An emission-trading scheme based on tradable allowances seems to be structurally incompatible with the IPPC Directive. An amendment of the directive would be necessary to introduce this type of trading system.

In the case of tradable reduction credits it would be possible to set a general emissions standard nationally that is stricter than BAT. As we have said before, this is in principle admissible under the IPPC Directive.

Yet, this seems to be a rather theoretical option. Given the IPPC Directive and its interpretation, BAT would serve as an absolute maximum. This maximum cannot be exceeded: every emitter has to take measures to meet at least the BAT level, otherwise the system would not be compatible with the IPPC Directive. As reduction credits may only be bought for those emissions

Legal aspects of the Dutch approach on CO_2 reduction 143

that do not exceed BAT, the general standard must represent a more ambitious level (BAT+, see Figure 6.2). This is necessary to create a market with both suppliers and demanders.

However, if the general standard requires BAT+, only a few enterprises will be able to achieve reductions that go further than this ('BAT++') and thus earn reduction credits. An emissions market will only operate properly when there is a sufficient number of sellers and buyers. With a very strict emissions standard based on BAT+, the scope for trading is likely to be too restricted. The number of reduction credits that enterprises are willing to sell will probably be relatively small, so that full advantage cannot be taken of the potential economic benefits of emissions trading (for example, efficiency gains). As a result, for practical reasons it may be desirable to consider an amendment of the IPPC Directive in the case of tradable reduction credits as well, even though this type of system would theoretically be possible under the existing directive. Only if European and national environmental law are changed, can a trading regime be as effective as it should be in order to lead to substantial emission reductions. Especially in the case of European law this can be a very long and sometimes difficult process, so action will have to be taken as soon as possible.

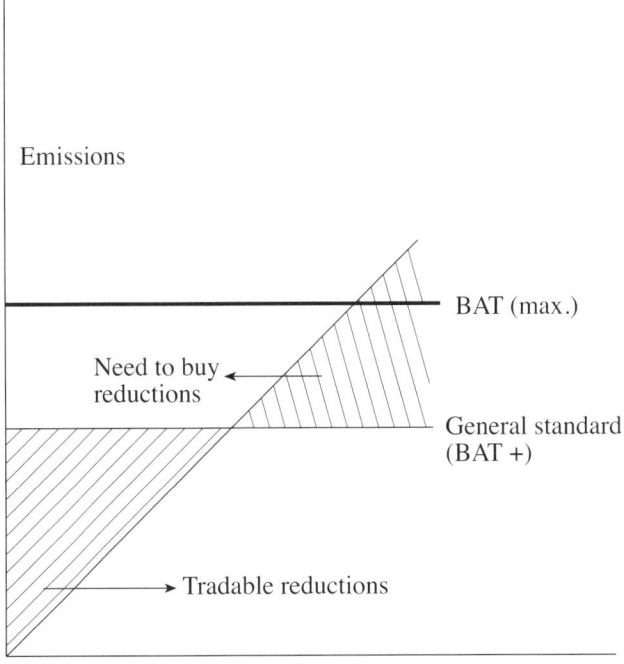

Figure 6.2   Tradable reductions and BAT

4 CONCLUSION

It is astonishing that in the Netherlands the gap between what the government promised during the Kyoto process and the reality is wider than in any other European country. Despite ambitious policy plans and all the measures that have already been taken, the CO_2 emission figures show an increase instead of a decrease. Comparing the Dutch situation with, for example, the German figures, one can say that the strong economic growth in the Netherlands during the last few years cannot be the only reason for that gap. Dutch climate policy seems to have failed in the past.

To reduce the gap between the Dutch reduction targets and the real situation the use of command-and-control instruments only is far from sufficient. These 'traditional' instruments regulate each source of pollution separately and restrict only those emissions that cause direct environmental effects. Therefore the traditional instruments of licensing and general rules do not seem to be very suitable for solving environmental problems that occur on a higher level, for instance – in the case of CO_2 – on a global level.

Voluntary agreements seem to be a good way to complement the traditional methods of regulation. The experience in the Netherlands shows that such voluntary agreements can be combined with and fit into the traditional system of licensing. However, that is only possible if enough attention is paid to the legal aspects of the details of the voluntary agreements. The actual agreement on benchmarking with energy-intensive industry is another interesting example in this field. Legally, there seem to be no major problems. However, the question is whether it was wise for the government to commit itself not to use any additional instruments to reduce the CO_2 emissions until the year 2010. This means that the option to introduce a market with tradable emission allowances or reduction credits is excluded also. Taking into account the Dutch 'policy shortfall' and the developments on the European level and in other EU member states, it seems advisable to ignore the preliminary restrictions of the benchmarking agreement on that point and to think about the legal aspects of the introduction of a national or European system of CO_2 emissions trading, which is hardly being done at the moment.

Emissions trading could help to reach the reduction targets more easily. Moreover it could contribute to achieving greater reductions in the longer term. With regard to environmental law it seems likely that not only the Dutch Environmental Management Act, but also secondary EU law will have to be amended, otherwise the system would probably be incompatible with the existing system with its ALARA and BAT requirements. Theoretically, a system of tradable reductions may be allowable without an amendment to the IPPC Directive, but only under such restrictions that its effectiveness would decrease considerably. Because on the one hand emissions trading seems to be an instrument which can hardly be dispensed with in the future, while on the other hand it differs so much from the structure and content of the existing environmental law, a thorough legal discussion is urgently needed. The introduction of a market with tradable NO_X reduction credits, which is being prepared by the Dutch government, presents a good occasion to promote the discussion on at least an important part of the legal aspects of CO_2 emissions trading. Climate policy may indeed benefit from such a discussion. Only if the urgency of the climate problem is really recognized and if an effective combination of all possible measures and instruments is used in practice, are the Kyoto targets and the long-term aim of sustainability feasible.

NOTES

1. *Uitvoeringsnota Klimaatbeleid, deel 1* TK 1998/1999, 26603, No. 2, p. 17 et seq. (The Netherlands' Climate Policy Implementation Plan, Part I, *Domestic Measures*; an English translation can be downloaded from www.minvrom.nl). VROM-Raad, Transitie naar een koolstofarme energiehuishouding. Advies ten behoeve van de Uitvoeringsnota Klimaatbeleid (Council for Housing, Spatial Planning and the Environment, Transition to a low-carbon energy economy) (Advies 010, 1998), pp. 23–7.
2. That is, 50 Mton CO_2-eq.
3. The fact that these are domestic measures does not, however, mean that these instruments could not be used in other countries or in the European Union as well.
4. By 2010, 5 per cent of all energy must come from renewable sources.
5. This is the required 50 per cent of the Dutch policy shortfall that is to be solved by domestic measures.
6. See van den Broek and Korten (1997).
7. The Dutch Organization for Energy and Environment, Nederlandse Organisatie voor Energie en Milieu (NOVEM).
8. Council Directive 96/61/EC of 24 September 1996 concerning Integrated Pollution Prevention and Control, *Official Journal* 1996, L 257, pp. 26–40.
9. Later on this criterion was changed to five years.
10. See also Section 3, below.
11. For a good description, see van den Broek and Niezen (1999).
12. The total amount of energy that will be saved by the industry in the next ten years is estimated to be a great deal more than this. However, the greatest part of the expected reduction in energy use has already been calculated in previous estimations. The consequence is that the contribution of the benchmarking agreement to the overall target of saving 50 megatons is relatively poor.
13. Rutten (2001).
14. See further Chapter 5 in this volume (Woerdman) on a carbon-trading market and state aid.
15. European Commission, *Green Paper on Greenhouse Gas Emissions Trading within the European Union* (8 March 2000), COM (2000) 87 final.
16. This must be distinguished from *international* emissions trading between Annex I parties, as mentioned in the Kyoto Protocol (Article 17). Both national and EU trading systems are domestic measures under the Protocol.
17. See Pedersen (2000).
18. However, the effectiveness of an emissions ceiling strongly depends on the enforcement regime. Strict enforcement is necessary in order to ensure that all emissions are really covered by the possession of allowances.
19. That is, those economic sectors that are subject to little or no international competition. Most

energy-intensive industries belong to the 'exposed' sectors, which have to compete internationally.
20. See Section 1, above.
21. COM (2000) 87 final.
22. See, for instance, Instellingsbesluit Adviescommissie plafonnering CO_2-emissies, 16 August 2000, Staatscourant, 7 September 2000, No. 173 p. 11 (government decree establishing the Dutch CO_2 Emissions Trading Committee) and the European Commission's Green Paper, COM (2000) 87 final.
23. See also Peeters' contribution to this volume (Chapter 7).
24. Not published. The advice is referred to in Minister Pronk's letter to the Dutch parliament of 19 February, 2001, TK2000/2001 26578 no. 3, p. 6.

REFERENCES

Pedersen, S.L. (2000), 'The Danish CO_2 emissions trading system', *Review of European Community and International and Environmental Law*, **9** (3), 223–31.

Rutten, G.J. (2001), 'Casestudy milieu en mededinging: Energie-efficiency benchmarkingovereenkomsten in het licht van art. 81 EG-Verdrag' (Case study environment and competition: the benchmarking agreements on energy efficiency in the light of Article 81 EC Treaty), in J.W. van de Gronden, and K.J.M. Mortelmans (eds), *Mededinging en niet-economische belangen (Competition and non-economic interests)*, Deventer: Kluwer, pp. 179–214.

van den Broek, J.H.G. and M.P.H. Korten (1997), *Milieu- en energieconvenanten in Nederland. Een overzicht voor de praktijk (Voluntary agreements on energy and the environment in The Netherlands)*, Deventer: Tjeenk Willink.

van den Broek, J.H.G. and J. Niezen (1999), 'Convenant Benchmarking energie-efficiency' (Benchmarking agreement on energy efficiency), *Milieu en Recht*, **26** (6) 152–9.

7. Legal feasibility of emissions trading: learning points from emissions trading for ozone-depleting substances

Marjan Peeters

1 LEGAL COMPATIBILITY, FEASIBILITY AND EFFECTIVENESS[1]

The choice for a particular policy instrument and the specific way of designing and executing it is a political one. Environmental regulatory instruments are meant to deal with polluting behaviour and are instituted to achieve environmental protection. Therefore, effectiveness and efficiency are important considerations. In developing and instituting these instruments a balance must be found between the legal parameters and the instrumental conditions (efficiency and effectiveness). Having an instrument in accordance with the law is also important from the perspective of effectiveness and efficiency: when there are problems with the legal aspects the instrument will be less likely to reach its (environmental) goals and may be very costly because the government is obliged to pay financial compensation.

To achieve the climate change policy goals, effective and efficient regulatory tools for realizing the desired protection against the greenhouse effect are needed. In this context, attention must be paid to legal compatibility and feasibility of the alternative regulatory tools. As the Kyoto Protocol shows, emissions trading is seen as an important policy measure for reducing greenhouse gas emissions. Also joint implementation and the clean development mechanism are flexible mechanisms that both involve the transfer of emission reduction credits from one state to another in relation to concrete emission abatement projects. The aim of the three Kyoto mechanisms is to achieve a cost-effective implementation of the required restrictions on greenhouse gas emissions. According to the Bonn Agreement on climate change, the use of the flexible mechanisms has to be supplemental to domestic action to reduce emissions. This domestic action must constitute a significant element of the emission reduction effort by industrialized countries.[2] In fact, countries are using or considering a domestic emissions trading scheme as a national measure against the greenhouse effect (as in the UK[3] and Holland[4]). And, as the so-called Green Paper on greenhouse gas emissions trading within the European Union shows, emissions trading is also being considered within the EU.[5] Under the European Climate Change Programme (ECCP) an emissions trading scheme is being prepared which is due to start within the EU by 2005.[6] According to the ECCP a pre-Kyoto EU system should serve as a 'learning by doing' process.[7]

Although the possibility of trade in emission rights was already discussed in economic literature a long time ago,[8] in practice not many examples exist. Our knowledge about emissions trading will increase when we examine implemented emissions trading schemes in order to identify principles for instituting and executing them for other or new environmental problems, like greenhouse gases. It is clear that we are in a learning process concerning the pros and cons of emissions trading, and also its legal aspects. In fact, some experience has been gained in the United States, with emissions trading for air pollution.[9] An international example of emissions trading can also be found, since emissions trading is part of international, European and national law for the protection of the ozone layer. It is likely that new learning points might be formulated when this emission-trading scheme is examined in detail. Both environmental threats (the depletion of the ozone layer and the greenhouse effect) are international environmental problems, whereby the environmental threat occurs regardless of where the gases are emitted. The emission-trading instrument is meant to be an effective and efficient regulatory tool, especially for polluting activities where the location of the emissions is not important.

Of course, the problem of ozone-depleting substances (ODSs) is far from fully comparable with greenhouse gases. Therefore, experiences with the ozone-depleting emission-trading scheme (ODS trading) must be translated to the specific characteristics of the greenhouse gas problem. But – as no other global environmental problem has been regulated through emissions trading – only the ODS market can serve as an example of an existing international emission-trading scheme. In this contribution legal aspects of emissions trading for ozone-depleting substances will be examined, in order to obtain a better idea about this means of regulating polluting behaviour, which is considered to be an important regulatory tool for the climate change problem.

Structure and Method

Before a specific emission-trading scheme can be examined, it is important to identify the legal aspects. Section 2 offers a general overview of the specific legal aspects of the emission-trading instrument. Section 3 gives an insight into international and European rules of the ODS market. Section 4 provides a comment and further questions for research. Section 5 concludes.

For this inventory study we shall examine in particular the existing regulations of the European Community concerning ODS trading. The goal is to determine to what extent it makes sense to research – as a next step – the practical legal experiences with the ODS market, in order to identify learning points that are useful for the institution of new international, European or domestic schemes for emissions trading, especially where greenhouse gases are concerned.

2 LEGAL CONSIDERATIONS OF EMISSIONS TRADING

Concerning the legal acceptability of the introduction and execution of emissions trading, many questions have to be answered. This section gives a short, but not exclusive, overview, as it is meant to be an illustration of legal aspects that arise when considering instituting an emission-trading scheme.[10] Of course these aspects have to be discussed ultimately in the specific national, international or – when relevant – European law context.

Legal Qualification of the Tradable Right

The tradable right must be seen as permission given by the government, based on a power granted by statute, that some exactly defined pollution may be caused. As it is the government that uses its legal power to set a cap on overall emissions and imposes the legal rule on polluters that a specific pollution can occur only with a government-distributed permit or right, it is clear that the rights must be seen in the context of administrative law. Therefore, decisions according to the institution and realization of the permit market must fulfil principles of administrative law.

The specific modelling of the tradable right and the conditions for the tradability determine the freedom of choice for the participants in the market.[11] For example, not only a geographic tradability but also a trade involving another period can be considered. The definition of the tradable right and the trade conditions should be given by statute, in combination with necessary compliance provisions.

Private law is also an important area, especially when the effectiveness and feasibility of the permit market are considered. When citizens or organizations sell and buy their emission rights, this relationship must be seen in the context of private law. Limitations for the tradability following from private law rules must be considered when instituting a permit market.[12] Many questions about private law have to be answered when the tradability of permits are instituted by law, such as whether it is allowable to lease the rights. Perhaps it is a wise measure when the statute prescribes elements of the contract or imposes conditions which the contract has to fulfil. Generally, when introducing a new emission-trading scheme the legislature must think about the provisions in the (administrative) act in relation to the relevant private law aspects.

Start of the Market

The starting-up of the market is a very crucial phase, because it has a great influence on the position of the legal identities.[13] The government must establish rules and provisions about the way the tradable permits will be distributed among the legal identities. These rules and provisions must fulfil the principles of administrative law. One of the main principles is that it must be clear to the legal identities under what conditions a tradable permit can be obtained. The start-up of the market also has to fulfil other general administrative law principles, such as a careful consideration of all the relevant interests, and open and prudent procedures. Of course private organizations or market mechanisms (like an auction) can be used for the factual start-up, but this all has to be done according to administrative criteria for the start-up and distribution, set by the government.

The permit market can be started up in different ways, and looking at the development of the emission-trading concept many creative ways can be distinguished.[14] The following main distinctions can be drawn, whereby combinations are possible:

- *An auction* The administration sells by public auction a number of permits to the legal identities. For example, in the US the auction possibility has been afforded (in a limited concept) by the Clean Air Act concerning acid rain measures.[15]
- *A free distribution of individual permits* Subject to administrative criteria the permits will be distributed among the legal identities. The 'grandfathering' method means that the existing polluters get rights according to their historical or actual emission levels.[16] We shall see that the ODS market has been based on this grandfathering principle. In the case of the international greenhouse gas emissions trading, according to the Kyoto Protocol the emission rights will be distributed for free among the participating parties (Annex I - states).
- *A free distribution according to general technical standards (performance rate system)* Another way of creating a permit market is setting general (technical) standards, but giving the legal identities the choice of (i) creating the actual emission reductions on their own, or (ii) buying realized extra emission reductions achieved by other firms. With

this method no cap will be imposed, so there is no theoretical guarantee that an overall emissions reduction will be reached. Especially in the case of economic growth, when more firms start up or when production increases, overall emissions can grow. The Dutch government is considering setting up such an emission-trading scheme for NO_x (cost-even mechanism).[17]

Withdrawal of Existing Rights and the Duty to Compensate

Another important aspect is that the earlier (not tradable) rights to pollute must be withdrawn (or transformed) in a proper way when the permit market is set up. This will not be necessary when the permit market is simply introduced alongside the grandfathering concept by making the actual permits tradable.[18] But this concept may be in conflict with the meaning of the polluter pays principle.

Article 1 of the first protocol of the European Convention of Human Rights (ECHR), for example, regulates the right of every person to the peaceful enjoyment of his or her possessions. A right allocated by the government can – in certain circumstances – be seen as a possession in the meaning of this article. It is not strictly forbidden to withdraw such a right, but there can be a duty for a state that is party to the Convention to financially compensate the owner of the right.[19] When the existing right cannot be qualified as a property right according to Article 1, Protocol 1 ECHR, compensation can also be made when the withdrawal of the permit is seen as a regulatory restriction on the use of property rights.

The duty to compensate also has to be considered when the government wants to withdraw (fully or partly) allocated tradable rights in an emission-trading scheme.

Administrative Criteria for Intervention in the Transfers

In general, from the point of view of efficiency it must be a goal to keep administrative costs of regulatory instruments as low as possible. Therefore, it is important for a smooth working of the emission-trading instrument that the participants are not prohibited from trading their permits by administrative rules that are not necessary. In the traditional command-and-control approach we are accustomed to a strong governmental influence looking after the way polluting activities will be carried out. The concept of the emission-trading approach uses the market mechanism and entails little governmental influence. A central question is whether administrative intervention is necessarily needed for influencing or controlling the trading activities. Each governmental influence or intervention must be motivated on the basis of the goals, that the government has to protect. Examples of these goals are:

- the achievement of an environmental policy goal – especially when there are also local effects of the polluting activity for which administrative decisions will be needed according to the acceptance of these activities;
- the purpose of informing third parties about the working of the emission-trading system and to let them participate in decision procedures;
- the control and enforcement of the rules according to emissions trading.

The government has to clarify the substantive and procedural conditions under which the transfers of the rights may take place. Thereby, consideration must be given to the governmental actions that will be needed to effect the transfer, for example:

- Is permission needed for a transfer?
- Must a transfer be announced to the government?
- Is registration needed, and how will it be modelled?

With respect to emissions trading, the guiding principle must be that as few administrative interventions as possible must be made. The attractiveness of the system is the tradability, and this should not be disturbed if it is not necessary. In this respect, attention must be paid to smooth and fast but legally acceptable administrative procedures. Also the possibility of using electronic systems for decision-making procedures or the registry of transfers must be examined.[20]

Position of the So-called Third Parties (Non-governmental Organizations)

In the field of environmental law, attention should be paid to the position of third parties such as environmental organizations. Also by drafting and implementing the emission-trading system consideration should be given to what provisions must be made to these third parties, which represent the voiceless environment.[21] The Rio Declaration and the Treaty of Aarhus in particular have laid down principles concerning access to environmental information and public participation in decision-making procedures.

The following points should be considered:

- What environmental information is accessible to third parties?
- When have third parties the right to participate in the administrative

decision-making procedures?
- When and against which administrative decisions can third parties seek administrative or judicial review?
- What role can third parties play concerning compliance with the rules for the permit market?

Clearly, each obstruction to a smooth-working trading system is undesirable from the point of view of an efficient working permit market. But a balance has to be found between the position of the third parties and the functioning of the system.

Lowering the Cap

No one can predict future environmental circumstances and quality, not even the government. However, the government has to make laws and instruments through which it can influence future environmental behaviour. Looking at the concept of the permit market, a situation can arise when the government must lower the cap earlier than has been foreseen by law. Therefore, consideration must be given to legal provisions for intervening in the market to lower the cap. This is especially necessary, because in certain circumstances the government has a legal duty to financially compensate the persons or firms who are damaged by the withdrawal of tradable rights.

Consideration must be given to:

- the period for which the (future) rights will be given. If the period has ended, and there has been no need to withdraw all rights earlier, then there is no problem. On the other hand, if the duration of the rights is short, the uncertainty for the market participants about their future must not be ignored. Also extra costs can be incurred by the government and participants because of the need for frequent new allocations of the rights;
- the reasons for which it could be necessary to withdraw rights prematurely, and the conditions under which the administration can execute its power to withdraw rights;
- the possibility of 'banking', which means that permits that are not used can be used in a later period.

Compliance and Enforcement Provisions

Compliance provisions can be distinguished in:

- the monitoring and reporting provisions;

- the enforcement tools.[22]

The effectiveness of any emission-trading scheme will largely depend upon its compliance and enforcement provisions. Because (under most circumstances) payment must be made for a right to emit, there is an inclination to disobey the rules. The question is how high will the extra costs of the monitoring and compliance regime be in the case of emissions trading, compared to other regulatory mechanisms?

Another issue is that when an emission-trading scheme is introduced, the usefulness of the common compliance provisions must be considered. It is very likely that specific enforcement tools will have to be instituted, such as the power to deduct rights in the case of non-compliance.[23]

Compatibility with Existing Environmental Statutes

Overall, the accordance with existing law, especially the environmental statutes or regulations, has to be studied. For example, the Dutch government asked the Council of State (Raad van State) to advise on the compatibility of the NO_X cost-even mechanism – which includes emissions trading – with the Dutch Environmental Management Act (*Wet milieubeheer*).[24] According to this advice, the cost-even mechanism as it was proposed cannot be based on the existing national (and European) environmental regulations.

The question is how emissions trading can be built upon and/or operate alongside existing policies and regulations, and if not, what new provisions are necessary and possible. In particular, the compatibility of the permit market with technological criteria such as the ALARA (as low as reasonably achievable) principle is debateable as it is desirable to reach maximum efficiency or effectiveness (for the purpose of effectiveness the cap is an interesting element of emissions trading; under the performance rate system, which conforms better to technological criteria, no direct cap would be instituted).[25] Since the procedure for instituting acts is often lengthy, compatibility with the existing environmental statutes is an important aspect for the possibility and/or attractiveness of introducing a permit market.

Compatibility with Other Existing Laws, Such as Trade Rules

To avoid unwanted and unexpected limitations to the feasibility of the permit market, consideration must be made to other legal aspects that can influence the working of the instrument. For instance, it is important to examine the legal provisions concerning international trade.[26] Of course, states within the European Community have to consider compatibility with EC law when introducing a national emission-trading scheme.[27]

3 OVERVIEW OF THE EUROPEAN PERMIT MARKET FOR OZONE-DEPLETING SUBSTANCES

Some Legal Topics of a Complex Regulatory Framework

In order to take measures against the depletion of the ozone layer, regulations have been made by different authorities on different levels. International, European and national regulations have been developed. These regulations have often been changed because of the growing concerns about the condition of the ozone layer. Looking at these regulatory measures, it becomes clear that a complex regulatory framework has been instituted.[28] Also a study of the acid rain emission-trading scheme as it is instituted in the US pointed out that the regulatory provisions were detailed and complex.[29] It is too early to conclude that this is a characteristic of the permit-market concept, but it is a point to be noted.

In this inventory research of the legal experiences with ODS trading, we shall examine the main points of the emission-trading regulations at the level of the European Union. In addition, selected legal items will be discussed. A supplementary survey will be made of the international regulations and the national legal provisions in the different countries (within and outside the European Union).

This section will give an overview of the main aspects of the international and especially the European regulatory provisions regarding the permit market. In addition, a comment will made about some selected aspects. These are:

- the start of the market;
- provisions for lowering the cap;
- administrative criteria for and/or intervention in the transfers; and
- compliance provisions.

International Tradable Environmental Permits

The international legal bases for the measures against the depletion of the ozone layer are the Vienna Convention for the Protection of the Ozone Layer (22 March 1985) and *the Montreal Protocol on Substances that Deplete the Ozone Layer* (16 September 1987).[30] The Vienna Convention established a framework for the adoption of measures against the expected damage caused by the modification of the ozone layer. In the Montreal Protocol the parties agreed to take measures concerning the emissions of ozone-depleting substances (for example, chlorofluorocarbons (CFCs)). Each treaty party had to freeze the production and consumption of several CFCs at the level of

1986. Furthermore, a phased reduction had to be reached in respect of the 1986 level (20 per cent in 1993 and 50 per cent in 1998). The Protocol entered into force on 1 January 1989. From 1 January 1990 it was forbidden for the participating parties to import the substances from countries that had not agreed to the Protocol.

The Montreal Protocol allows the transfer of production rights under certain conditions. In its definitions (Article 1) 'industrial rationalisation' means the transfer of all or a portion of the calculated level of production of one party to another, for the purpose of achieving economic efficiencies or responding to anticipated shortfalls in supply as a result of plant closures. So, a (limited) permit market was installed in the field of international environmental law!

The adoption of legal ozone measures on the international level can be seen as an important application of the precautionary principle.[31] Because of growing knowledge about the poor condition of the ozone layer, the Protocol has been revised several times in a short period. In 1990 new provisions were made. Among other things, a stronger phase-out schedule was agreed. The latest new provisions to the Montreal Protocol have entered into force on 28 July 2000.[32]

An Implementation Committee was established, which has both a dispute resolution element, as well as an implementation element.[33] Also – for the first time in international environmental law – provisions were made to assist developing countries in meeting the costs of compliance with the Convention and Protocols.[34]

The international agreements and efforts to protect the ozone layer have had a substantial impact.[35] For example, by 1995 the global production of (regulated) CFCs was down 76 per cent from the base year 1988. But the implementation of the measures was not without problems, such as the illegal trade in ozone-depleting substances.[36]

ODS Trading: The First European Market for Environmental Emission Rights

In 1978 a resolution on fluorocarbons was adopted, urging that all appropriate measures should be taken to ensure that industry situated in the Community did not increase its production capacity.[37] In 1980 legally binding measures were taken.[38] Member states were required to take all appropriate measures to ensure that industry did not increase its production capacity of certain ODSs.

With Decision 88/540/EEG the European Community became Party to the Vienna Convention and the Montreal Protocol (the member states are also parties). There are several types of community legislation. In the case of the measures for the ozone layer 'regulations' that are directly applicable by law in the member states have been decided. On 1 January 1989 the first European regulation for the implementation of the Montreal Protocol entered into force, and since then many new regulations have appeared.[39] With the European regulation on ozone-depleting substances a system of transferable emission rights was used for the first time in the history of European environmental law.[40] The European regulations have more stringent rules (for example, phase-down schedules) than the international provisions in the Montreal Protocol.[41]

The European ODS regulations are not easy to understand because of the detailed and complex character and the frequent changes that have been made. These changes were necessary because of the fast-changing knowledge about the consequences of the gases for the ozone layer. The latest regulation was adopted on 29 June 2000 (Regulation 2037/2000).[42] This regulation applies among other things to the production, the use, the placing on the market, the import and export of ozone-depleting substances. A remarkable point is that the new regulation – as a revision of the former regulation – was seen to be necessary 'in the interest of legal clarity and transparency'. Only some major points of the tradability of ODS rights will be mentioned below, as far as it is necessary for the discussion of selected legal aspects.

Allocation of Rights: Start of the Market and Phase-down/out

The start of the market has to be effected by an allocation of rights. For example, according to Regulations 594/91 and 3093/94, rights have been distributed by the grandfathering method. For example: Article 10 of Regulation 594/91 allocates production rights with reference to the production level in the year 1986. In addition, the regulations stipulate how much of these production quota may be used in subsequent periods, starting in 1991 and ending in 1997. From 30 June 1997 no gases – subject to regulations – may be produced. So the regulation includes a more rigorous phase-down.

This scheme of allocation of production quota and phase-down or phase-out has been applied in the different European regulations concerning several ozone-depleting substances, with different time periods. Also in the latest Regulation 2037/2000 the grandfathering method was used: for methyl bromide the base year is 1991,[43] and for the newly regulated hydrochlorofluorocarbons (HCFCs) the base year is 1997.[44]

Regulation 2037/2000 has been amended by Regulation 2039/2000 as a result of a discussion about the base year for allocating rights for placing substances (HCFCs) on the market. Basically, this concerned the concept of grandfathering. Originally, Regulation 2037/2000 took 1996 as the base year for allocating quotas of HCFCs. The following consideration was given as reason for the amendment:

Since 1996 the HCFC market has changed considerably with respect to importers and the maintenance of 1996 would result in a large number of importers being deprived of their import quota. As a general rule, quotas should be based on the most recent and representative figures available, in this case those for 1999, and so the maintenance of 1996 could be considered arbitrary and might even result in a breach of principles of non-discrimination and legitimate expectations.

Regulation 594/91 has been amended by Regulation 3952/92[45] to achieve a faster elimination of ozone-depleting substances. The phase-out that was originally provided for 30 June 1997 was advanced to 31 December 1994.

On 15 December 1994 a new regulation was adopted which replaced the former regulations. Regulation 2037/2000 introduces new rules about the production and use of substances (especially methyl bromide) and their phase-out. Thereby, for certain cases a more rigorous phase-down than adopted in former regulations has been regulated because of the earlier than anticipated availability of technologies for replacing ozone-depleting substances. Regulation 2037/2000 limits the production of methyl bromide (Article 3(2)) with a phase-out at 31 December 2004, and HCFCs (Article 3(3)) with a phase-out at 31 December 2025.

Before 31 December 2002, the Commission must review the level of production of HCFCs with a view to determining:

- whether a production cut ahead of the year 2008 should be proposed, and/or
- whether a change to the levels of production should be proposed.

The preamble of Regulation 2037/2000 states that even after the phase-out of controlled substances the Commission may under certain conditions grant exemptions for essential uses. For example, Article 3(1) provides the necessary authority to the Commission.

Some concern has been expressed about the problems that small and medium-sized enterprises in particular may have with the switch to new technologies or alternative products, necessitated by the phase-out of the production and use of controlled substances. The preamble states that the member states should therefore consider providing appropriate forms of assistance specifically to enable these enterprises to make the necessary changes.

Criteria for Transfer of Emission Rights: 'Industrial Rationalization'

The CFC regulations use the definition of industrial rationalization as the criterion for the possibility of a transfer of emission rights. Regulations 594/91, 3093/94 and 2037/2000 allow the transfer of production rights for industrial rationalization purposes. Regulation 2037/2000 defines industrial rationalization as follows:

> [T]he transfer either between Parties or within a Member State of all or a portion of the calculated level of production of one producer to another, for the purpose of optimising economic efficiency or responding to anticipated shortfalls in supply as a result of plant closures.

This definition makes it clear that transfers can only be made between producers. Environmental organizations or other organizations may not participate in transfers of production rights.

A distinction can be made between trades within a member state, trades between member states and trades between a member state and a third party.[46]

Trades within member states

Article 3(8), Regulation 2037/2000 regulates a transfer within a member state:

> To the extent permitted by the Protocol, the competent authority of the Member State in which a producer's relevant production is situated may authorise that producer to exceed the calculated levels of production ... for the purpose of industrial rationalisation within the Member State concerned, provided that the calculated levels of production of that Member State do not exceed the sum of the calculated levels of production of its domestic producers ... for the period in question. The competent authority of the Member State concerned shall notify the Commission in advance of its intention to issue any such authorisation.

Important elements of the regulation are:

- *A reference is made to the Protocol* For clarity concerning the possibility of trading, producers who want to make a transfer have to refer to the conditions stipulated in the Montreal Protocol. It is likely that this reference also includes new provisions to the Protocol. This exacerbates uncertainty on future trade conditions.
- *There must be an authorization* There is one formal condition to authorization: the competent national authority must notify the Commission in advance of its intention to issue any authorization. Furthermore, there are no formal requirements, although many questions about the execution of the decision-making process can be asked: should the decision be written and published, what conditions must the decision procedure fulfil, are third parties to be allowed to participate, when can the transfer occur, and should there be the possibility of judicial review? Also, which period has to be taken into account between notifying the Commission and deciding about the authorization?

- *There must be a 'competent authority in the member state'* This authority must decide about authorization. The member state has to make a legal provision concerning the institution of this authority. No conditions about this competent authority are stipulated in the regulation.
- *Industrial rationalization* This is the main and only substantive criterion. Clearly, there can be no other reason for refusing the authorization.

Transfers between member states

Article 3(9), Regulation 2037/2000 provides for a transfer between Member States:

> To the extent permitted by the Protocol, the Commission may, in agreement with the competent authority of the Member State in which a producer's relevant production is situated, authorise that producer to exceed the calculated levels of production ... for the purpose of industrial rationalisation between Member States, provided that the combined calculated levels of production of the Member States concerned do not exceed the sum of the calculated levels of production of their domestic producers ... for the periods in question. The agreement of the competent authority of the Member State in which it is intended to reduce production shall also be required.

An important difference here, with compared transfers within a member state, is that the European Commission has the power to decide about authorizations. The European Commission can only give authorization together with the agreement of the competent authorities of the relevant member states.

Transfers with third parties

Article 3(10), Regulation 2037/2000 provides for the transfer between a member state and a third party:

> To the extent permitted by the Protocol, the Commission may, in agreement with both the competent authority of the Member State in which a producer's relevant production is situated and the government of the third Party concerned, authorise a producer to combine the calculated levels of production laid down in paragraphs 1 to 9 with the calculated levels of production allowed to a producer in a third Party under the Protocol and that producer's national legislation for the purpose of industrial rationalisation with a third Party, provided that the combined calculated levels of production by the two producers do not exceed the sum of the calculated levels of production allowed to the Community producer under paragraphs 1 to 9 and the calculated levels of production allowed to the third Party producer under the Protocol and any relevant national legislation.

An important new element in reference to these provisions is that the specific relevant national legislation of the third party is also part of the conditions for obtaining approval for the transfer.

Compliance and Enforcement Provisions

Regulation 2037/2000 makes several provisions for reporting, inspection, monitoring and enforcement.

Reporting

The producer has the duty to report every year before 31 March, data about each controlled substance in respect of the period 1 January to 31 December of the preceding year.[47] A copy has to be sent to the competent authority of the member state concerned. Information has to be given about any increase in production authorized under Article 3(8), (9) and (10) in connection with industrial rationalization. According to Article 19(5) the Commission should take appropriate steps to protect the confidentiality of the information submitted. No specific rules have been given about the right of citizens or environmental organizations to have access to the reports.

Inspection: a strong position for the Commission

For the purpose of inspection some important powers have been given to the Commission. In carrying out the tasks assigned to it by the regulation, the Commission may obtain all the information from the governments and the competent authorities of the member states *and* from producers. When requesting information from a producer the Commission should at the same time forward a copy of the request to the competent authority of the member state within the territory where the production is situated, together with a statement of the reasons why that information is required.[48]

Furthermore, the competent authorities of the member states should carry out any investigations that the Commission considers necessary under the regulation.[49] This provision constitutes a strong position for the Commission, as the member states have little or no discretion in the matter.[50] In addition to this, subject to the agreement of the Commission and of the competent authority of the member state where the investigations are to be made, the officials of the Commission will assist the officials of that authority in the performance of their duties.[51]

The Commission must take appropriate action to promote adequate exchange of information and cooperation between national authorities and between national authorities and the Commission. The Commission must also take appropriate steps to protect the confidentiality of information obtained with the inspection.

Penalties

Article 21 stipulates that member states will determine the necessary penalties applicable to breaches of the regulation. They have to be effective, proportionate and dissuasive.

4 COMMENT AND QUESTIONS

In this section the selected legal aspects will be discussed, in order to identify learning points for designing and executing an emission-trading scheme. In addition, it will show which questions still have to be answered to obtain more insight into the (legal) feasibility of the ODS market. This knowledge will be useful for designing and implementing emission-trading schemes for greenhouse gases.

Grandfathering: Fairness and the Polluter-Pays Principle

For a domestic greenhouse gas emission-trading programme, how should the allocation of the tradable rights to citizens and firms be made? Such an allocation is also needed for a European emission-trading scheme, although it is an option to let the member states make the allocation and, in addition to this, to link these national schemes to get a Europe-wide system.[52] The European Climate Change Programme concludes that a mixture of allocation methods may be the most practical way forward.[53] In this respect, many deliberations were made in the US about the way the national ODS market was to be started. Finally, they opted for an allocated quota system (grandfathering) in combination with a regulatory fee.[54]

Concerning the permit market for ODS, we have seen that both the international agreements and – in line thereby – the European ODS regulations, use the grandfathering mechanism for allocating rights. This means that the tradable rights are distributed for free according to administrative criteria based on historical polluting behaviour. The regulations for the ozone-depleting substances use a simple (and in this respect – an attractive) method: a base year is designated, and the emission levels in that base year are the criterion for the allocation of the rights, without exception.

An important negative effect of this way of grandfathering could be that the allocation of rights is unfair, especially when judged according to the polluter-pays principle and according to an equitable distribution of the efforts needed for reaching the environmental policy goal. As with grandfathering the existing polluters get rights according to their historical emission levels, the biggest polluters being given most of the rights. This is contrary to the policy of favouring those who have or have demonstrated good environmental behaviour. On the other hand, it can be difficult to find administrative criteria for a fair distribution of the emission rights that can be used feasibly. Furthermore, the (part) withdrawal of existing rights can lead to an obligation of the administration to pay a financial compensation.

On the basis of this inventory overview it is too soon to conclude that the distribution of rights for the ODS market has been unfair and/or contrary to the polluter-pays principle, although there are strong indications. A more thorough study could give more insight into the legal and political experiences with the grandfathering method, whereby the arguments for this method should be studied more carefully. An important aspect to which attention could be given is that existing firms were limited by the European ODS regulations in their activities, and therefore were confronted with a legal restriction on their existing rights to produce CFCs. What are the legal experiences with this, and have there been any legal actions taken by firms in order to obtain financial compensation? Furthermore, it would also be interesting to get a better idea of the compatibility of the existing grandfathering method with trade law.

In sum, more knowledge about the legal possibilities of using the grandfathering method, such as its compatibility with the polluter-pays principle, the legal acceptability and feasibility of withdrawal of existing rights, and the compatibility of the allocation with trade rules is useful for developing new emission-trading schemes such as domestic greenhouse gas trading schemes.

The Legal Acceptability of a Rigorous Phase-out

Firms that produce ozone-depleting substances were confronted by strong measures to reduce this activity. In this way the ODS market is an example of a strong regulatory mechanism for reducing air pollution. In addition, the - short - history of the ODS market shows that phase-out schemes were strengthened by new regulations. For example, the phase-out date for certain substances was in the first instance 30 June 1997 but was changed to 31 December 1994. Also the latest regulation has a provision for the Commission to review the production level with a view to determining whether a change or cut in production levels is necessary.

The main legal questions that arise when such measures are imposed are whether the government (i) has the legal right to impose these strong or stronger measures, and (ii) has to compensate regulated firms. It would be interesting to research these aspects further in the light of the ODS market, for example:

- What consideration has been given and/or provisions made on the

European (and international) level about imposing strong measures (especially when they are more rigorous than the international agreements) and the possibility of making compensation to firms?
- What legal actions have been taken (or are considered) against these measures, and - more hypothetically - what successful legal actions could have been taken by the regulated firms?
- How have member states (or other Treaty parties) provided 'appropriate forms of assistance' specifically to enable firms to make the necessary changes? In the latest regulation it has been stated that member states should make this undertaking.

In sum, a survey of the complications of imposing and enforcing the phase-down scheme can give us more information about the legal possibilities of this method. In particular, the deliberations about and experiences with (legally required) financial compensation should be examined. This knowledge can also be useful when, in the context of emissions trading for greenhouse gases, a phase-down has to be made.

Clear and Simple Administrative Interventions with Trades

In the international agreements about the protection of the ozone layer a criterion for transfers has been set down. According to the European regulations, this criterion is a condition for each transfer, and an appointed governmental body has to decide whether each transfer respects the criterion. The criterion leaves room for some discretion.

So, a transfer of production rights may only occur if it is approved by a governmental authority. In the case of an internal state transfer, a national authority has the power to decide about the approval. In the case of external state transfers, the Commission has the power to decide, but the approval of (the authority of) both involved states is also necessary.

Two important questions arise. For a smooth and effective working of a permit market, every unnecessary restriction for trades should be avoided. Yet, the international provisions have instituted the criterion of industrial rationalization. The question is, what problems have occurred by using this criterion? For example, to what extent have uncertainties caused by this criterion prevented firms from emissions trading? What interpretations have been given by the authorities, how much information must have been given by the participants to the administrative body before a decision could be made, for instance about 'optimizing economic efficiency', and what uncertainties or difficulties have occurred? The main question is: what was the reason for using this criterion and is there any reason for using a similar or other substantive criterion for trades in greenhouse gas emission-trading schemes?

Legal feasibility of emissions trading 165

How were the procedures executed? How much time was needed to give a decision about the approvals? Was the decision procedure open to third parties (because the criterion of industrial rationalization allows some discretion, involvement of third parties could be required)? What were the possibilities for judicial review? To what extent were/are the decision procedures of the member states different? In sum, the procedural elements for obtaining approval for a transfer – such as the decision-making process and (the possibility of) judicial review – should be examined further: it would be interesting to examine how they were instituted and how that influenced emission-trading activities.

For the climate change emission-trading schemes, a principal question is whether a substantive criterion is necessary. In order to keep the emission-trading scheme simple, substantive criteria for the acceptability of transfers must be avoided.[55] Yet, for activities under joint implementation and the clean development mechanism, approvals will be necessary. For the joint implementation activities the Kyoto Protocol prescribes that a project needs the approval of the parties involved.[56] Also, activities under the clean development mechanism will be subject to approval and certification.[57] It is not clear yet what approvals will be necessary for greenhouse gas emissions trading. At this moment it cannot be fully excluded that some substantive criteria will be necessary in the case of emissions trading for greenhouse gases. For example, it might be interesting to know whether a substantive criterion as is used in the case of the ODS market really had (or has) a negative effect on the actual willingness or potential to trade. Perhaps some learning points for the flexible mechanisms can be gathered by studying the specific legal provisions (and when available, the experiences with them) in the case of the international transfers of emission rights with respect to ODS trading. Also experiences with the procedural elements of ODS trading can serve as a learning point for (European) greenhouse gas emission-trading schemes, in cases where approval of individual transfers is necessary.

Compliance and Enforcement Provisions

The level of compliance with the regulatory scheme constitutes the effect of the measures. Especially in the case of emissions trading, attention to compliance is important because of the financial benefits firms can get if they do not obey the law. Therefore we should study the compliance and enforcement provisions in the case of the ODS regulations, and especially their effectiveness.

There are indications that illegal activities have occurred.[58] Research should focus, for example, on the following questions:

- What are the experiences with the yearly report mechanism? The production quota concern a whole year, and there is an obligation for firms to report after that year. Suppose that a report or an inspection points out that a firm has produced too much in the previous year in respect of the relevant rules: what legal enforcement actions have or could have been taken? When enforcement actions were actually taken, what can be learned from that?
- What are the experiences with the strong position taken by the Commission for obtaining information about compliance behaviour? What else can be said about these inspection powers, especially in relation to the position of the member states?
- What provisions have been made in national law for executing the inspection powers, and what are the experiences with it?
- Regulation 2037/2000 refers only to the criteria under which penalties are incurred. What penalties have been provided in national law, to what extent do they differ among the member states, and what are the experiences with it?
- The Green Paper states that an effective functioning of emissions trading requires a certain degree of harmonization of the rules of monitoring, reporting and verification.[59] From this point of view, are the provisions made by states in the case of the ODS market sufficiently harmonized?[60]

As with greenhouse gas emissions trading, the effectiveness of the compliance and enforcement provisions will also be of great importance, so any information about this topic would be welcome.

5 ODS TRADING AS A FIELD OF LEARNING

When in the context of the climate change policy emission-trading schemes are to be designed, many questions will arise about the legal compatibility and feasibility. Because emissions trading at the European or international levels is a new regulatory tool, a lack of knowledge exists about the way this instrument will work in practice. On the international and European levels, the particular permit market for ozone-depleting substances is the only one that already exists. By studying the factual legal experiences with ODS trading, knowledge will increase about the way an emission-trading system functions in practice, and which legal aspects are in this context important to consider. Otherwise, we should bear in mind that the ODS market is a specific and limited one, and that every environmental problem may have its own characteristics for which it is necessary to make specific provisions.

This inventory research has focused on some legal themes from the *European* ODS market. This has shown us that by studying more extensively the legal provisions for and experiences with this permit market we could obtain more insight into the feasibility of essential elements of an emission-trading scheme, such as the primary allocation of rights, the phase-down of emissions, the administrative intervention with trade, and the compliance and enforcement provisions. We should also study other topics and experiences with ODS trading, for example the question of how private law facilitates or impedes the smooth trade of the permits. Another topic to be looked at would be the incompatibility of the ODS policy with the concept of integral management of the environment. What do the ODS experiences teach us about this, and how serious is it when a specific policy does not fit in with this concept of integral management?

In particular, the compliance and enforcement provisions and experiences in the case of ODS trading should be studied more extensively. Because the phase-down scheme was rigorous, it is likely that it was difficult for firms to obey the rules. One of the interesting research questions is whether the enforcement provisions – made by the states according to the ODS regulations – were sufficient, and what improvements can be recommended for new emission-trading schemes such as those for greenhouse gases.

When new emission-trading schemes to address environmental problems are designed, new insights into alternative methods for regulating or influencing the polluting behaviour – such as electronic decision making or new enforcement powers – should also be used together with knowledge about the experiences with already existing schemes. To obtain a better idea about the legal feasibility and attractiveness of new permit markets for greenhouse gases, knowledge about the practical functioning of ODS trading can help to some extent. Although the climate change problem is a specific policy field that differs from the ozone-depletion problem, ODS trading is the only example of emissions trading we have in practice in Europe and on the international level from which we can learn.

NOTES

1. The author thanks Geert van Calster, Catholic University of Leuven, for his comments.
2. European Commission (2001).
3. Hobley (2001).
4. The Dutch government has researched the possibilities of implementing a CO_2 emissions trading scheme but concluded that it is better aiming at a European emission trading system.
5. European Commission (2000).
6. See ECCP (2001).
7. Ibid., p. 12.
8. Dales, (1968).
9. An overview of four permit markets instituted in the US can be found in Peeters (1992).

10. This section is derived from Peeters (1992). See also Backes (2000).
11. Illustrated in Peeters (1992, pp. 81-3).
12. For a survey of the compatibility of the trade of (environmental) permits with Dutch private law, see Kamminga, (2001, pp. 339-45).
13. For considerations about the relation between the different ways of starting up the permit market and competitive distortion and state aid issues, see Woerdman (2001).
14. For more details, see Peeters (1992).
15. Ibid., p. 299.
16. Consider the real grandfather who favours his firstborn grandchild. The later-born grandchildren automatically get less than the older one, because the oldest one has had a longer period to relate to their grandfather.
17. TK Kamerstukken 26 578, 'Verhandelbare emissies als instrument in het milieubeleid', TK 1999-2000.
18. Further information about the transaction of existing regulations in a permit market can be found in Peeters (1992, pp. 110-19).
19. For more information about Article 1, Protocol 1 ECHR, see Loof, et al. (2000).
20. See, for example, Cason and Gangadharan (1998). More generally, the Dutch government is considering a new statute on electronic decision making: Commissie wetgeving algemene regels van bestuursrecht, 'Aanvulling van de Algemene wet bestuursrecht met regels over verkeer langs elektronische weg tussen burgers en bestuursorganen (Wet elektronisch bestuurlijk verkeer)', The Haag, 18 April 2001.
21. For international law principles concerning environmental information; see Sands (1995, pp. 616-20).
22. The Green Paper (European Commission 2000) and the ECCP report (2001, p. 11) distinguish: monitoring, reporting, verification and enforcement.
23. This provision in an international context is part of the compliance system on which agreement has been reached in Bonn (European Commission, 2001).
24. Raad van State (2000).
25. Peeters (1992, pp. 110-18). For emissions trading combined with technological standards, also see Backes (2000, pp. 22-4), and Backes and Teuben (2001).
26. This is illustrated by Grimeaud (2001)
27. Woerdman (2001) addresses the issue of permit allocation in relation to state aid.
28. Sands (1995, p. 259).
29. Peeters (1992, pp. 288-313).
30. For a description, see Sands (1995, pp. 259-71) and more recently Kiss and Shelton (2000, pp. 507-12).
31. Kiss and Shelton (2000, p. 508).
32. There is an overview of the amendments at www.unep.ch/ozone.
33. This dual function is an important innovation in international and environmental law according to Carpenter et al. (2000, p. 18).
34. Kiss and Shelton (2000, p. 509).
35. Ibid., p. 511.
36. Ibid.
37. *Official Journal* 1978 C 133/1. For a short overview of the European Community actions on the protection of the ozone layer, see Jans, (2000, pp. 366-67).
38. For a short overview of the community policy according to the protection of the ozone layer, see Jans (2000, pp. 366-7), and in Dutch Jans et al. (2000, pp. 525-9).
39. The lawfulness of Regulation 3093/94 was disputed (C-314/95, Gianni Betatti). The European Court of Justice was prepared to examine the compatibility of this measure with the environmental objectives and principles of the EC Treaty. EC institutions have had wide discretionary powers in shaping the Community's environmental policy (Jans, 2000, p. 21).
40. Jans, et al. (2000, p. 526).
41. Kiss and Shelton (2000, p. 511).
42. For some changes, see Regulations 2038/2000 and 2039/2000.
43. Article 3(2).
44. Article 3(3).

45. Adopted on 30 December 1992, entered into force on 1 January 1993, withdrawn by Regulation 2037/2000.
46. Regulation 2037/2000 also provides for other ways of transfer of production rights (for example Article 3(6); Article 4(5). Because of their inventory character these are not discussed in this research.
47. Article 19.
48. Article 20(1)(2).
49. Article 20(3).
50. Jans (2000, p. 367).
51. Article 20(4).
52. ECCP (2001, p. 10).
53. Ibid., p. 11.
54. Peeters (1992, pp. 265-8).
55. See also the statement of the ECCP: 'An EU emissions trading system must be environmentally effective and economically efficient, simple and transparent' (ECCP, 2001, p. 11).
56. Article 6(1)(a) Kyoto Protocol.
57. Article 12(4), (5), (7).
58. In the Netherlands at least one firm was suspected of offending the ODS rules, *Handhaving* 2000/6, p. 20, en Tweede Kamer der Staten-Generaal, vergaderjaar 1995-2000, Aanhangsel 925 p. 2011. See also the European Commission, 'Illegal trade in ozone depleting substances', posted at www.europa.eu.int/comm/environment/ozone/illegal_trade.htm. Among other things, a working group made recommendations about national legislation to punish illegal production.
59. European Commission (2000).
60. See also ECCP (2000, p. 13), 'The working group recommends coordination to guarantee a minimum level of sanctions in an EC-scheme in order to achieve an environmental outcome'.

REFERENCES

Backes, Ch.W. (2000), *Duurzame groei?*, Deventer: Tjeenk Willink.
Backes, Ch.W. and R. Teuben (2001), 'Legal aspects of the Dutch approach on CO_2 reduction', Paper presented at the Conference on Institutions and Instruments to Control Global Environmental Change, Maastricht University, 21-22 June.
Carpenter, Chad, Charles Di Leva and Kilaparti Ramakrishna (2000), 'A legal review of the key provisions and background to the Kyoto Protocol and the Buenes Aires Action Plan', in Prodipto Gosh (ed.), *Implementation of the Kyoto Protocol*, Asian Development Bank.
Cason, Timothy N. and Lata Gangadharan (1998), 'An experimental study of electronic bulletin board trading for emission permits', *Journal of Regulatory Economics*, **14**, 55-73.
Dales, J.H. (1968), *Pollution, Property and Prices*, Toronto: University of Toronto Press.
European Climate Change Programme (ECCP) (2001), 'Long report June 2001', posted at www.europa.eu.int/comm/environment/climat/eccp.htm.
European Commission (2000), *Green Paper on Greenhouse Gas Emission Trading within the European Union*, COM (2000)87, 8 March.
European Commission (2001), 'Summary of the key elements of the Bonn Agreement on climate change', posted at www.Europa.eu.int/comm/environment/climat/pressbckgnd.htm.

Grimeaud, David J.-E. (2001), 'To adopt climate change measures: overview on the balance between environmental protection and international trade', Paper presented at the Conference on Institutions and Instruments to Control Global Environmental Change, Maastricht University, 21-22 June.

Hobley, Anthony (2001), 'Emissions trading in the United Kingdom: an overview', *Environmental Liability*, **9**(1), 3-10.

Jans, Jan H. (2000), *European Environmental Law*, Groningen: Europa Law Publishing, August.

Jans, J.H., H.G. Sevenster and H.H.B. Vedder (2001), *Europees milieurecht in Nederland*, Boom juridische uitgevers, The Hague.

Kamminga, T.A. (2001), 'De handel en wandel binnen het privaatrecht. Verhandelbaarheid van publiekrechtelijke rechten', *Nederlands JuristenBlad*, 339-45.

Kiss, Alexandre and Dinah Shelton (2001), *International Environmental Law*, 2nd edn, New York: Transnational Publishers, Inc.

Loof, J.P., H. Ploeger, and A. van der Steur (eds) (2000), *The Right to Property. The Influence of Article 1 Protocol no. 1 ECHR on Several Fields of Domestic Law*, Maastricht: Shaker Publishing.

Peeters, Marjan (1992), *Marktconform milieurecht? Een rechtsvergelijkende studie naar de verhandelbaarheid van vervuilingsrechten*, Zwolle: Kluwer.

Raad van State (Council of State) (2000), Voorlichting van de Raad van State van 26 oktober 2000, supplementary document to Tweede Kamer, *Verhandelbare emissies als instrument in het milieubeleid*, Brief van de Minister van VROM, 1999-2000.

Tweede Kamer (TK), Kamerstukken 26 578 nr. 3, *Verhandelbare emissies als instrument in het milieubeleid*, Brief van de Minister van VROM, 1999-2000.

Sands, Philippe (1995), *Principles of International Environmental Law*, Vol. 1, Manchester: Manchester University Press.

Sevenster, H.G. (1998), 'Milieubeginselen uit EG-Verdrag toetsteen voor CFK-verordening', in *Nederlands Tydschrift voor Europees Recht*, no. 10, 226-8.

Woerdman, Edwin (2001) 'Developing a European carbon trading market: will permit allocation distort competition and lead to state aid?', Paper presented at the Conference on Institutions and Instruments to Control Global Environmental Change, Maastricht University, 21-22 June.

8. CDM in climate policies in the Netherlands: a promising tool?[1]

Rianne de Leeuw and Ekko C. van Ierland

1 INTRODUCTION

At the Kyoto Conference in 1997, the government of the Netherlands agreed to a quantified emission limitation and reduction commitment (QELRC), which requires it to reduce greenhouse gas (GHG) emissions by 6 per cent by 2010 as compared to 1990 levels (van Ierland et al., 1999). Under a business-as-usual scenario, GHG emissions will increase from 218 Mtonnes (Mt) CO_2-eq. in 1990 to 256 Mt CO_2-eq. in 2010. Thus, the 6 per cent reduction target requires a reduction of 50 Mt CO_2-eq. per year compared to unchanged policy (CPB, 1999).

The Kyoto Protocol describes three 'flexible mechanisms' by which parties can obtain emission reductions abroad, as a supplement to domestic actions. The flexible instruments are international emissions trading (IET), joint implementation (JI) and the clean development mechanism (CDM). These mechanisms have the potential to lower the costs of compliance, and so increase the likelihood that the Protocol will be ratified and the targets met (OECD, 1999). Where the contribution of Backes and Teuben in this volume (Chapter 6) focuses on the legal aspects of the domestic measures of voluntary agreements and national emissions trading, this chapter focuses on the opportunities of the clean development mechanism for Dutch climate policy. To date, the details of the CDM have not been specified in international policy making. Despite the intention to take decisions on all mechanisms at COP6 in November 2000 (UNFCCC, 2001), political debates did not lead to an agreement on the implementation of the Kyoto Protocol (Senter, 2001b). The COP6+ conference in Bonn in July 2001, however, did lead to an agreement. The Netherlands had already started preparations for a national CDM programme before this conference (Jepma, 2000; Senter, 2001b).

This chapter addresses the question whether the CDM is a promising tool for reducing GHG emissions, and what the main difficulties are. First, Section 2 focuses on the position of CDM in Dutch climate policy and the conditions that the Dutch government imposes on the instrument. Section 3 deals with the characteristics of CDM that are relevant with respect to these conditions. Section 4 contains several case studies illustrating the points made in the previous sections. Finally, in Section 5 the results are discussed and some conclusions are drawn.

2 THE CLEAN DEVELOPMENT MECHANISM IN DUTCH CLIMATE POLICY

Introduction: the Clean Development Mechanism

Article 12 of the Kyoto Protocol defines a clean development mechanism. Under CDM, countries with QELRCs (Annex I countries) may fund emission reduction projects in countries without targets (known as non-Annex I countries). The donor country can thus gain certified emission reductions (CERs) if it adheres to certain criteria for certification. Below are listed the main criteria for certification of CDM emission reductions, as imposed by the CDM authorities (based on Stewart, 1999). The CERs can be used to comply with the Annex I countries' QELRCs in the first commitment period (2008-12). CERs gained in the pre-commitment period can be 'banked' (Parkinson et al., 1999). The CDM will be supervised by an executive board, and a 'share of the proceeds' from project activities will be used to assist particularly vulnerable developing countries in meeting the costs of adaptation (UNFCCC, 2001).

1. Designing a project
   - prove additionality of the emission reductions;
   - conduct a baseline study;
   - describe the sustainable development content of the project;
   - follow approval and registration procedure.
2. During the project:
   - monitoring and reporting requirements;
   - follow verification and certification procedure of credits;
   - report transfers and holdings.
3. Other:
   - payment of levy (covering administrative expenses, adaptation expenses non-Annex I countries);
   - adhere to operational elements, for example, value and risk sharing with host country.

As reflected in the first set of conditions in the list, the CDM has two objectives: (i) to contribute to sustainable development in the host country and assist in contributing to the ultimate objective of the UNFCCC, and (ii) to help industrialized countries stay within their emission budget (UN, 1997).

Many elements of CDM are still under debate. Key issues include the question of supplementarity (whether a limit should be placed on the use of the flexible mechanisms); how to establish a baseline; which projects should be eligible for CDM (particularly whether sink projects such as reforestation and nuclear energy projects should be permitted); who would be liable if a party that has transferred a part of its assigned amount under emissions trading is found to be in non-compliance; and who should be on the CDM executive board (UNFCCC, 2001; JIN, 2000).

Climate Policies in the Netherlands

The government of the Netherlands plans to follow two action paths in order to reduce its GHG emissions. First, 50 per cent of the required reductions will be achieved nationally[2], through the implementation of domestic policies (for example, renewable energy, improving energy efficiency in coal-fired power plants and reducing energy demand in important sectors) (VROM, 1999). Since the costs of national emission reductions may come to more than €100 per ton CO_2-eq., the Dutch government intends to achieve the other 50 per cent of the emission reductions through the flexible mechanisms mentioned in the Kyoto Protocol, so as to improve cost efficiency (Senter, 2001a).

The government has two *main conditions* for the flexible instruments: first, it will only use the flexible mechanisms if reductions abroad will be *cheaper* than reductions at home, and, second, that the emission reductions it buys should be *qualitatively good* (VROM, 2000). This is reflected in the Terms of Reference of ERUPT (Senter, 2000a), the tender procedure the Dutch government commissioned Senter to design in order to integrate joint implementation into its climate policy. This procedure consists of three phases: an exclusion phase, a selection phase and a contract-awarding phase. Figure 8.1 shows the criteria in these phases (for a more detailed description, see Senter, 2000a). Note that detailed information is needed to check all the criteria.

As shown in Figure 8.1, the criteria in the first place focus on the price of the ERUs. Further, they focus on the feasibility of the project and the capacity of the supplier to successfully complete it on one hand, and the emission reduction potential on the other. These are a direct reflection of the main concerns of Dutch policy on flexible instruments.

The tender procedure for CDM is still in the design phase, and is expected in the fall of 2001.

Originally, Dutch CDM policy resided under the Ministry of Foreign Affairs, Minister of Development Cooperation (VROM, 2000). Because the CDM had to fit the existing policies of this minister, specific requirements

I. Exclusion criteria:
- integrity of the supplier (financial standing, no illegal activities, payment of taxes, etc.).

II. Selection criteria:
a. financial and economic standing (supplier must be financially able to deliver claims);
b. registration in professional or trade register;
c. technical capacity (supplier should have the technical capacity to realize the project by which ERUs are generated).

III. Contract-award criteria:
a. price per ERU;
b. project feasibility (to reach its primary purpose, not so much the potential to generate GHG reductions);
c. commitment of supplier (should be able to carry out responsibilities of the contract);
d. emission reduction potential (Senter will assess the baseline study, validation report, letter of approval, suitability of the technology proposed).

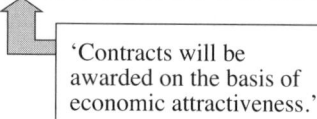

'Contracts will be awarded on the basis of economic attractiveness.'

Source: Based on Senter (2000a).

Figure 8.1 Exclusion, selection and contract-award criteria in the ERUPT tender procedure

were expected for CDM projects. Already in the Dutch AIJ programme (PPP-JI; Pilot Projects Programme on Joint Implementation), emission reduction was in no instance the only project goal for projects in non-Annex I countries (VROM, 2000). The goals of the development policy were applied to all projects (combat poverty, improve the position of women and children, and sustainable development). This is illustrated in figure 8.2.

In 2000, the funds for CDM were transferred to the Ministry of Housing, Spatial Planning and the Environment (VROM). This was done to prevent the impression that CDM substitutes for 'regular' development assistance funds (VROM, 2001b). For 2001, 200 million guilders (NLG) is available, for 2002

CDM in climate policies in the Netherlands 175

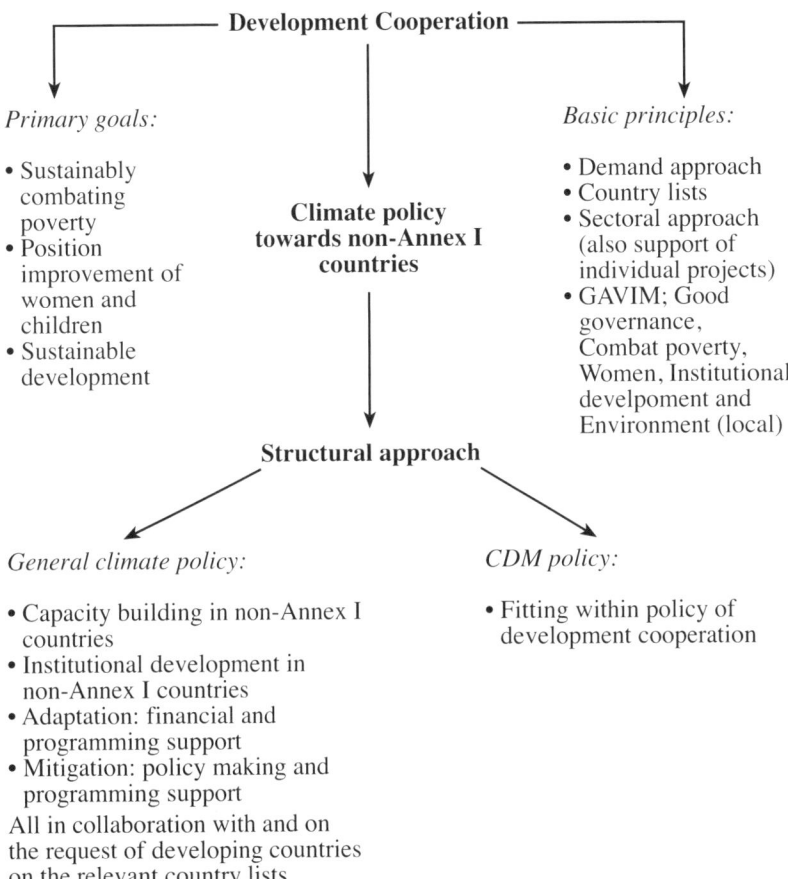

Figure 8.2 The former position of CDM under the Minister of Development Cooperation

this amount is NLG 300 million, and from 2003 to 2010, each year NLG 125 million is available to the Ministry of VROM for CDM projects. These funds are for project-related activities only, that is, to buy CERs. They are not for general climate activities like (non-project-related) capacity building and adaptation support. For these types of activities, the Ministry of Foreign Affairs/Development Cooperation is responsible.

Referring to figure 8.2, the 'general climate policy' activities still reside under the Minister of Development Cooperation, whereas the 'CDM policy' does not. The starting point of the Ministry of VROM in implementing CDM is obtaining CERs cost-effectively, through projects that stimulate sustainable development in the host country. The criteria that the Ministry of VROM will pose to CDM projects are under design, and will mainly reflect the Kyoto principles, international regulations ('good governance') and other boundary conditions of the Ministry. The latter are expected to relate to, for example, risk management, the lifetime of projects and selection of host countries. The primary aim is to obtain emission reductions through CDM in a cost-effective manner. Although the Minister of Development Cooperation might have put additional demands on CDM projects to reflect the general policy of the Ministry, that will not be the case now that the Ministry of VROM coordinates the instrument. VROM will leave the actual execution of the procedures to intermediary organizations (VROM, 2001a,b).[3]

Activities Undertaken in Developing JI and CDM

The Netherlands has been actively participating in the development of the Kyoto mechanisms. Within the framework of the Activities Implemented Jointly pilot phase (AIJ), the Netherlands established a pilot programme for JI and CDM, as did many other countries. The AIJ pilot phase ended in 2000, and in the same year the Netherlands launched a fully-fledged initiative for joint implementation, called the Emission Reduction Procurement Tender (ERUPT). In the autumn of 2001 a tender procedure for CDM was launched, called CERUPT.

A first call for ERUPT was issued in July 2000. Out of 26 expressions of interest from all over the world, nine were selected and interested parties were invited to submit a proposal. In April 2001, the Minister of Economic Affairs signed contracts for five projects, thereby buying approximately 4 Mt of CO_2 emission reductions spread over five years (0.8 Mt per year). The purchases involve NLG 79 million (Senter, 2000b; 2001c). The average price per tonne of CO_2 reductions is thus approximately NLG 20 (approx. US$8). The CERUPT tender procedure specifies maximum prices the government wants to pay for emission reductions achieved through CDM. These range from €3.30 for various projects such as fuel switch and methane recovery to €5.50 for renewable energy projects.

3 CHARACTERISTICS OF CDM IN RELATION TO CDM POLICY IN THE NETHERLANDS

Quality of Emission Reductions: Additionality and Sustainable Development

The Dutch government's quality concern means that it will only approve and implement those projects that can reasonably be expected to pass the test of validation and certification (VROM, 2000). Thus, a project at least needs to fulfil all criteria as proposed by the CDM authorities to be accepted in the Dutch CDM tender procedure. As shown in the list in Section 2, above, there are two aspects inherent to the project design and set-up, that is, *additionality* and *sustainable development*. Further, the specific Dutch criteria that will be attached to the tender procedure are relevant. This section focuses on the additionality and sustainable development content of CDM projects.

The issue of additionality could well be the most controversial issue of the clean development mechanism. The UNFCCC (2000) states that:

> [R]eductions in anthropogenic emissions by sources (and anthropogenic enhancement of removals by sinks) should be additional to any that would occur in the absence of the project activity, keeping in view that business-as-usual projects shall not be eligible as CDM projects, while ensuring that overseas development assistance and the other existing financial commitments of Parties included in Annex I are not used for the acquisition of certified emission reductions ... .

The Ministry of VROM distinguishes three types of additionality (VROM, 2001a): (a) investment additionality (projects are only financially profitable with CDM funds), (b) financial additionality (CDM funds may not substitute for Overseas Development Assistance funds), (c) technical or environmental additionality (projects have to achieve a better environmental performance than the regular performance in the relevant sector or country). The Ministry has decided on minimum criteria for project design and baseline documents to be able to assess additionality.

Additionality is assessed through a baseline study. The baseline for a JI or CDM project represents the estimated future emissions at the project location in the absence of the investment under the project (Woerdman and Van der Gaast, 2001). It is an important concept, since it is used to calculate the emission reductions achieved by a project, by comparing the emissions in the baseline with the emissions in the project. In this way the baseline determines the allocation of emission reduction certificates. Many analysts regard the development of methodologies for setting baselines as the most difficult task in CDM rule making (Stewart, 1999). This is mainly because of the hypothetical character of a baseline situation. During the past few years, numerous baseline approaches have been developed, ranging from project-specific to highly generalized top-down approaches (see, for example, Michaelowa, 1998; OECD, 1999; Ellis and Bosi, 1999; Meyers, 1999; van Ierland et al., 2000; Bandsma et al., 2000).

During the AIJ pilot phase of the Netherlands, a baseline study was also demanded. Most Dutch project developers chose to apply a project-specific baseline approach (JIRC, 2000). Overall, AIJ projects were characterized by project-based baselines (JIN, 2001). JIRC (2000) re-estimated/verified the baseline of ten Dutch AIJ projects to address problems involved in baselines and verification and to make recommendations about these. One of its recommendations is to use project-specific baselines and to develop more precise guidelines for baseline setting.

The second quality concern discussed here is the sustainable development content of a CDM project. Article 12 of the Kyoto Protocol states that the CDM shall assist non-Annex I parties 'in achieving sustainable development'. In setting rules and guidelines regarding sustainable development, the main issues are (Stewart, 1999):

- whether each host country should determine what its sustainable development objectives are or whether international sustainable development criteria should be developed and applied; and
- in the case of criteria: should the host country apply these to determine whether the project contributes to its sustainable development, or the CDM operational entities.

The Dutch pilot phase required investors to have a Letter of Approval for each project from the relevant host-country government, stating (among other things) that the project meets the sustainable development goals of the host country (JIRC, 2000). The ERUPT tender procedure also demands a Letter of Approval. For CDM, although the Ministry of VROM will in the first instance leave the decision to the host country, there will be a check on the sustainability aspects of the projects (VROM, 2001b).

Costs of Emission Reductions

Flexible instruments in general, and CDM specifically, are only attractive if the emission reductions they generate are more cost-effective than domestic action. For the Netherlands, CDM has to provide cost advantages for emission reductions in the range of 25–50 Mton/yr. In each tender, the government will contract those projects that meet the imposed conditions and offer CO_2 reductions at the lowest price. The exact 'cut-off price' (the price above which projects are not attractive for the Dutch government) depends on the marginal costs curves of domestic and other options, as well as the expected spin-off effects of domestic measures, like technology development and employment. According to van der Linden et al. (2000), at 25 Mton/yr reduction, the theoretical marginal national economic costs (costs of the last tonne of reductions) are approximately US$15 per tonne CO_2. The policy mix of domestic reduction options actually chosen by the Dutch government, however, also includes some measures that are more expensive than US$15 per tonne. This illustrates that other factors besides costs play a role in policy making. The average costs of the policy mix to achieve 25 Mton/yr of reductions domestically are US$20 per tonne CO_2-eq. If also the second 25 Mton/yr were realized domestically, the average costs would rise to US$30, and the marginal costs for the last reduction unit would be as high as US$45 per tonne CO_2. These figures do not include a possible spin-off. If these effects are taken into account, with a generous estimation of US$15 per tonne, the author concludes that reductions abroad should cost a maximum of US$15 to be clearly cheaper than domestic measures.

The price the Dutch government would have to pay to obtain reduction units through CDM may differ from the costs involved in realizing these units, and depends on the economic principle of supply and demand. Supply and demand in turn are based on the economic reduction costs curves per country, and determine where and by how much emissions are reduced. Sijm et al. (2000) show that there is an enormous potential supply of emission reduction options in Eastern Europe and non-Annex I countries compared to Annex I reduction requirements. A considerable fraction of this potential could be harnessed at a low or even negative cost. Given a set of assumptions and constraints, an annual abatement potential in the non-Annex I countries in the first budget period (2008–12) at costs of up to US$50 per tonne CO_2-eq. has been projected of approximately 2.3 Gigatonnes (Gt) of CO_2 (1990 US$). The study further suggests that approximately 1.7 Gt CO_2-eq. per year would be available during the 2008-12 budget period at net marginal costs below US$10 per tonne CO_2. The potential of 'no-regret' options is estimated at some 0.8 Gt. If emission credits are banked from projects implemented during the 2000–2008 period, the annual reduction potential will increase accordingly. Estimates of the CDM potential of studies vary widely from less than 30 Mt CO_2 to more than 1600 Mt (Sijm et al., 2000). The main abatement potentials are, in decreasing order: energy efficiency measures in the power sector and demand-side energy efficiency measures (together 66 per cent), fuel switch (from oil or coal to natural gas; 17 per cent) and renewable energy (14 per cent).

Various studies have been undertaken to estimate the equilibrium price of emission credits. For example, ECN (1999) estimates a price of US$4 (including no-regret) to US$15 (excluding no-regret) per CER (energy options only).[4] Several top-down models (in van der Linden et al., 2000), each with its own assumptions and restrictions, suggest that the market price on a world-scale market for reduction units is somewhere between US$2.6 and US$12 per tonne CO_2-eq.[5] These studies are based on numerous assumptions and limitations, which are not discussed here in detail. The price estimates given above serve to indicate the possible market price for emission reduction units the Dutch government may be confronted with. The World Bank uses a price of US$10 in its Prototype Carbon Fund (World Bank, 2000).

The price estimates are based on total world demand and supply of emission reduction units, and thus do not represent the price for reduction through CDM. The main assumptions in the studies are a full use of the technical potential of CDM, world trade and a perfect market (no limits to trade, no strategic behaviour, no transaction costs and so on). However, if the technical potential of CDM is not used fully, and/or the trading scale becomes smaller, price estimates rise rather sharply.[6] Further, according to ECN (1999), CDM can reduce the global trading price from US$18–29 to US$4–15 per tonne of CO_2. This implies that emission reductions in non-Annex I countries will be relatively cheap in comparison with others.

Discussion

The previous sections lead to the conclusion that CDM in principle is a promising option for climate policies in the Netherlands. The governing structure of the instrument in Dutch climate and foreign policy is clear and specified, and experiences with the JI tender procedure show that such a procedure can indeed generate various promising projects. The supply potential of 2.3 Gt of CO_2 (below US$50 per tonne CO_2-eq.) is more than enough to meet the government's aim to achieve 25 Mton CO_2 reductions (-3 per cent) a year through CDM and JI. Further, studies are indicating that CDM could generate emission reductions at a cost lower than US$15 per tonne CO_2, the price under which emission reductions achieved through flexible instruments are considered to be economically efficient.

Several important notions remain. First, although simulation studies may underestimate the technical potential of CDM because of a lack of data, there are doubts whether this potential can be used (either fully or partially) in practice, and whether a world-scale trading scheme is realistic. Simulation studies generally assume perfect markets, the ability to fully use the reduction opportunities in non-Annex I countries, and neglect transaction costs. It has to be seen in practice whether all the low-cost reduction options can be implemented. Other factors that influence the equilibrium price and/or attractiveness of the instrument in practice, and that usually are not accounted for in simulation studies, are the investment climate in the host countries, the local capacity and expertise, whether or not no-regret options qualify for CDM and – more in general – the rules and modalities that will be specified for CDM in international policy making (sinks, banking, baselines and so on) (van der Linden et al., 2000). Second, the actual implementation of CDM may lead to many practical complications at the project level. The case studies in the next section highlight this.

4 CASE STUDIES

This section discusses four CDM case studies in which baselines were established and emission reductions calculated. The main questions underlying these case studies are how baselines and emission reductions can be calculated, what the price of the emission reductions is for these projects and which complications they identify for the practical implementation of CDM in Dutch policy.

Case Study 1: Renewable Energy in the Philippines

This case study involves the electrification of a village in the Philippines that is not connected to the grid. The electrification will be realized using renewable energy sources. The project involves a pilot, possibly paving the way for more extensive electrification. The project includes two phases. In the first phase, a PV/LPG hybrid (photovoltaics combined with a liquefied petroleum gas generator for back-up) will be installed in combination with a mini-grid. In the second phase, the hybrid will be replaced by a biomass generator, which will be plugged into the grid. Total project lifetime will be 20 years; one year for the first phase and 19 for phase two.

In short, the baseline comprises two parts: (i) the emissions of the fossil fuels that are replaced by the project (mainly kerosene for lighting), and (ii) the emissions of the biomass, if it had not been burned in the biomass generator. Without the project, this biomass would be left to decompose, producing CO_2 and CH_4 (methane). Thus, the baseline is defined as consisting of fossil fuel emissions and the emissions of rotting biomass. The project's emissions consist of emissions from the burning of biomass and the remaining use of conventional fossil fuels. It is expected that the project will replace 90 per cent of the energy use; the other 10 per cent also has to be calculated as project emissions.

Four alternative baseline calculations have been conducted: top-down historic trend; bottom-up historic trend; top-down projected situation; and bottom-up projected situation. They are based on a model (de Leeuw, 2000; van Ierland et al., 2000) that identifies two aspects in establishing the baseline emissions of a project, that is, the data level and the reference used. First, the data level can (at the extremes) vary from project specific (bottom up) to top down. A top-down approach uses national data as a basis to calculate emissions at the project site. A bottom-up approach uses project-specific information. Second, the reference for the baseline can (at the extremes) be a historic trend or a future projected situation. One calculates the baseline using historic trends (extended into the future) for the variables. It is then assumed that these trends are a good indication of what would happen in the future if the project were not implemented. When a future situation is used as a reference, one defines a situation that is expected and/or striven for, and calculates the baseline emissions accordingly. This allows for a baseline that anticipates a change in trends (very useful in, for example, rapidly developing countries). The four resulting baseline approaches more or less span the field of possibilities found in literature (for example, Michaelowa, 1998; OECD, 1999; Ellis and Bosi, 1999; Meyers, 1999; Bandsma et al., 2000). In practice, mostly project-specific baselines have been used until now (JIN, 2001). Investors as well as verifying companies can use the model to assess or review the emission reductions associated with a project.

Figure 8.3 and Table 8.1 summarise the results, and provide a brief description of the method by which the various baseline calculations were derived. Figure 8.3 shows the baseline and project emissions over the years, and Table 8.1 gives the totals for the total project lifetime.

In case (1), the baseline approach is top down and based on historic trends. It is thus based on national data for the Philippines.[7] The baseline was calculated by extrapolating both the national trend in emissions per GDP per capita and the trend in GDP per capita. From these trends, the extrapolated historic trend of emissions per capita during the years 2000 to 2019 was derived. Finally, this latter trend was multiplied by the number of people in the village (assumed to be constant at 200) in order to arrive at the estimated emissions of the village over the project lifetime. To this number, the emissions from rotting biomass were added.[8] Case (3a) uses the same approach, except that the emissions per GDP per capita and the GDP per capita are not assumed to develop according to a linear trend, but are projected to converge to the current values in the Netherlands (as a benchmark). This calculation therefore uses a strong growth *baseline* scenario. The *project* emissions, however, are based on a moderate growth scenario. It is not surprising that the resulting emission reductions are large. Assuming equal growth in the 'with project' situation gives smaller emission reductions, as represented by case (3b).

Baseline cases (2) and (4) are almost the same. Both include the emissions of the current use of fossil fuels. Because of a lack of data, it was not possible to calculate a trend on the project level, so fossil fuel use is assumed to be constant over the 20 years of the project. Also, both cases (2) and (4) incorporate the emissions from rotting biomass. The difference is that case (4), which reflects a projected future situation, includes emissions from the use of inefficient diesel generators in the baseline. The project company expects that these would have been used had the electrification project not been carried out.

Emission reductions and costs

Table 8.1 shows that the estimates of the emission reductions of the project

Table 8.1 Results of calculations of baselines and emission reductions for the Philippines case study

Scenario	Calculation method for baseline	Baseline emissions	Project emissions (bottom up)[1]	Emission reduction scenario
(1) Baseline: top-down historic trend Project emissions: bottom up	Linear extrapolation historic national trend (CO_2-eq./(GDP/cap)) and GDP/cap + rotting biomass emissions	8880	3970	4910
(2) Baseline: bottom-up historic trend Project emissions: bottom up	Emissions of current use kerosene in village + rotting biomass emissions	7700	3970	3730
(3) Baseline: top-down projected future situation Project emissions: bottom up	National (GDP/cap) and (CO_2-eq./(GDP/cap)) converging to current Dutch values over the 20 years of the project + rotting biomass emissions	11210	3970 (3a) 6070 (3b)[3]	7240 (3a) 5140 (3b)
(4) Baseline: bottom-up projected future situation Project emissions: bottom up	Emissions of current use kerosene and expected use of diesel generators village and rotting biomass emissions[2]	9260	3970	5300

Notes:
All figures in tonnes of CO_2-eq.; figures represent totals over entire lifetime.
1. Project-level data was provided by the company (based on its experiences with other projects and its expertise).
2. Baseline estimate provided by the project company, based on measurements, experience from other projects and assumptions.
3. Under 3(b) it is assumed that project emissions grow according to the growth rates assumed in the baseline calculation. Reduction 3(a) represents the calculation as specified in scenario (3).

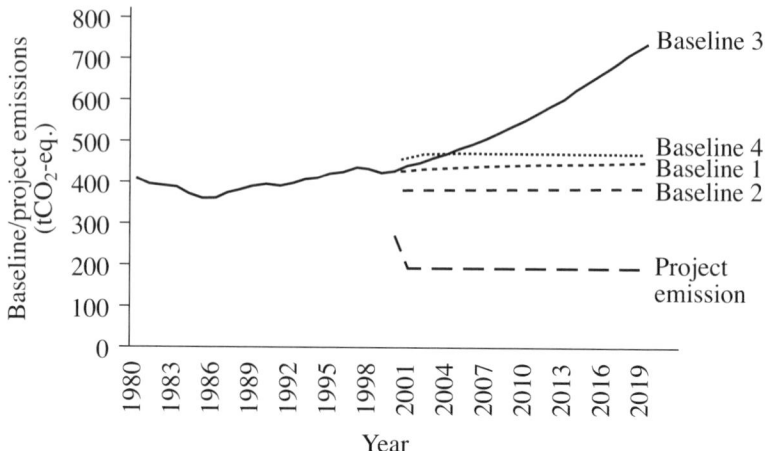

Figure 8.3   Project emissions and four baselines (tonnes of CO_2-eq.)

vary from 3730 to 5300 tonnes CO_2-eq. The estimated reduction of 7240 (3a) is not considered here, since it represents an unlikely situation (that is, high growth in the baseline, small growth with the project, which means the project is an obstruction to development).

Unfortunately, it was not possible to calculate the emission reduction costs per tonne under various scenarios, since no financial data for this single pilot project were made available. The larger project of which the pilot is part is expected to have a positive net present value (NPV) at an interest rate of 7 per cent and a moderate rate of return, without CDM revenues. The project is thus expected to have a negative reduction cost for greenhouse gas emission reductions. However, the expected payback period of approximately 10 years is very long.

Discussion

The baseline values that were calculated in the case study range from 7700 to 11 210 tonnes CO_2-eq., thus showing a large difference in the results of various baseline approaches. One should note, however, that the baseline and emission reduction calculations are only rough indications of the 'real' emission reductions that can be associated with the project. Many assumptions had to be made in the calculations, including a 20-year lifetime and no leakage.

Despite the inaccuracies, the model approach shows that the 'real' emissions reductions of the project will probably be in the range of 5000 tonnes CO_2-eq. Strictly speaking, the project has a negative reduction cost, because the NPV is positive. However, the financial performance of the project does not seem to be very attractive, prompting the project company to choose another, more profitable project. It could thus be argued that the emission reductions are additional, that is, that the project would not be implemented if the company did not have the prospect of CDM revenues.

Case Study 2: Fuel Switch in Egypt

This case study involves a fuel switch project for a number of cities and one industrial area in Egypt. Natural gas will replace three fuel sources, namely LPG (for domestic and commercial use), heavy fuel oil (for industrial use) and light fuel oil (for commercial and industrial use).

The project includes two phases. In the first phase, a natural gas pipeline system will be constructed from a high-pressure grid to the region under consideration. The project will connect three industrial users in an industrial area and approximately 25 000 domestic and commercial users to the natural gas grid. In the second phase, the industrial use of gas will be expanded, and an additional 58 000 domestic and commercial users will be connected to the grid. Both construction phases should take three years. The lifetime of the entire project will be, according to the company, 20 years (including construction).

As in the Philippines case study, four baseline cases were calculated. Table 8.2 summarizes the results of the baseline and the emission reduction project. For the calculations we were largely dependent on information provided by the project company, for example, for data on expected gas sales. The definition and calculation of baseline cases (1) and (3) is the same as in the Philippines project. Cases (2) and (4) are equal in this project. According to the model, the difference in approach between these bottom-up cases is that (2) uses the historic trend as a reference, while (4) uses a projected future situation. In the data provided by the project company, however, the projected future fossil fuel use was based on historic use, causing the two cases to coincide.

Table 8.2 shows some remarkable results: cases (1) and (3) show an emission increase due to the project. This could have two causes:

- The project boosts energy use so much that the emission reductions per unit of energy consumed are more than compensated for by a large increase in energy consumption (large rebound effect). If this is the case, the calculated emissions increase is realistic.
- In calculating the baseline scenario, one or more unrealistic assumptions are used. If this is so, the baseline scenario and calculated emissions increase are unrealistic.

The first cause is unrealistic because the expected gas sales used in the 'with

Table 8.2 *Results of calculations of baselines and emission reductions for the Egypt case study*

Scenario	Calculation method for baseline	Baseline emissions	Project emissions (bottom up)[1]	Emission reduction
(1) Baseline: top-down historic trend Project emissions: bottom up (emissions of expected gas sales)	Linear extrapolation historic national trend (CO_2-eq. per GDP per cap) and GDP per cap	16.3	21.1	−4.8
(2)/(4) Baseline: bottom-up, projected situation is based on historic trend Project emissions: bottom up	Emissions of fossil fuel equivalent of anticipated gas sales; based on current use and ratio of LPG, heavy fuel oil and solar	28.5	21.1	7.4
(3) Baseline: top-down projected future situation Project emissions: bottom up	National (GDP per cap) and CO_2-eq. per GDP per cap, converging to current values in the Netherlands over the 20 years of the project	18.3	21.1	−2.8

Notes:
All figures in Mt of CO_2-eq.; figures represent totals over entire lifetime.
1. Project emissions were calculated as the emissions caused by the expected gas sales.

project' scenario are based on current fuel use and do not incorporate a large growth in energy use. The project could only produce more emissions than the baseline situation if in the baseline emission *reductions* were to take place compared to the current situation. However, looking at the Egyptian national trends in emission intensity of energy consumption and total emissions per capita, this seems unlikely.

Then, only cause two remains, indicating that very likely one or more of the assumptions in the calculations are unrealistic. Indeed, one of the assumptions seems incorrect: it is implicitly assumed that the ratio of households to commercial and industrial users involved in the project is comparable to the national ratio. According to data provided, the whole region has almost two million inhabitants, with five major cities and *one* existing industrial zone. The project, however, covers three towns (82 000 households) *and* the industrial area. Even assuming an average household size of ten people, not even half the population in the region is involved in the project, whereas all the regional industry is. This implies that the ratio of households to industry in the project is much smaller than the national average. This leads to a serious underestimation of the baseline emissions for the project and therefore the calculation shows an emissions increase due to the fuel switch project.

It is concluded that the top-down baseline cases (1) and (3) calculated for this project are highly unrealistic. To correct this, it would have been necessary to have information on regional GDP. However, this information was not available. The model could only generate one baseline that may be considered realistic: case (2).

Emission reductions and costs
Since only case (2) gives a result that can be considered realistic, the emission reductions under this scenario were used in the cost-effectiveness calculations. The cost-effectiveness calculations are based on the first phase of the project only, since no cash flows for the second phase were available. If only phase one is executed (that is, connecting three industrial and 25 000 domestic and commercial users to the natural gas grid) the resulting emission reductions are 4.70 Mt CO_2-eq. over 20 years (calculated in the same way as the reductions for phases one and two in Table 8.2). For matters of confidentiality, only some general financial information is given.

Without revenues from CERs (certified emission reductions, to be gained in a CDM project), the NPV of the project is around US$15 million at a 10 per cent interest rate, with an internal rate of return of over 15 per cent. The (undiscounted) payback period is slightly less than five years. If one assumes a CER[9] price of US$10, NPV would increase to over US$30 million (at 10 per cent), with the internal rate of return exceeding 20 per cent. The (undiscounted) payback period would drop to four years.[10]

Discussion

The case study shows extremely diverging outcomes of emission reductions for top-down and bottom-up baseline scenarios. It confirms the statement of Ellis and Bosi (1999) that the local and regional variability of the different baseline components will determine the extent to which standardization of the baseline is feasible. In the Egyptian case, the regional variability in industry concentration and GDP caused a standardized, top-down approach to fail. When constructing a baseline, one should be very careful to make sure that project or regional characteristics resemble national characteristics fairly closely.

If the project is granted CERs, the financial performance will improve considerably, moving to a higher level of financial attractiveness. However, as the project seems to be profitable without credits, it can be questioned whether the emission reductions are additional.

The emission reductions calculated in a top-down approach might be overstated for this project, since the top-down approach implicitly assumes that the gas sales replace all current fossil fuel use. In practice, this is unrealistic. It would be more accurate to use an estimated 'replacement percentage', if available.

Case Study 3: Waste Management in Ecuador

The case study for Ecuador deals with the reduction of GHG emissions of waste management in Quito, the capital of Ecuador. The study (Benitez, 1999) focuses on the various options for reducing emissions as compared to uncontrolled landfilling. Cases considered are sanitary landfilling with energy recovery and a system of integrated solid waste management. The study is based on an analysis of various options that are considered with regard to cost effectiveness and environmental impact, including emissions of GHGs, liquid effluents, land degradation and recovery of resources. Valuable materials that can be obtained from the waste stream are dry recyclables, compost and energy.

Under scenario 1 (that is, uncontrolled landfilling) the private costs are low and there is no treatment of the biogas (methane) and liquid effluents. Under scenario 2 (that is, sanitary landfilling with energy recovery) the gas is collected and used for electricity generation by means of piston internal combustion and gas or steam turbines. The efficiency in conversion is about 30 per cent. An alternative use of the landfill gas is to first remove the toxic contaminants from the gas and later to distribute the gas for domestic use (Benitez, 1999). Scenario 3 (that is, integrated solid waste treatment) includes the magnetic separation of metals and manual sorting of waste on conveyor belts for paper, plastic, glass and non-ferrous metals.

Comparison of the scenarios enables us to establish the reduction of GHG emissions as compared to the baseline, which is considered to be scenario 1, and the additional net costs of reducing greenhouse gas emissions under scenarios 2 and 3. Table 8.3 shows the additional costs and the total emission reductions for scenarios 2 and 3 as compared to scenario 1.

Table 8.3 Total costs and emission reductions for scenarios 2 and 3 compared to baseline scenario 1

Scenario	Total cost of emission reductions (million US$)	Total emission reductions (Mt CO_2-eq.)	Cost of CO_2 emission reductions (US$/tonne CO_2-eq.)
Sanitary landfill	206	8.0	25.8
Integrated solid waste treatment	316	12.9	24.5

Source: Benitez (1999).

Results show that both a sanitary landfill and an integrated solid waste treatment system can reduce GHG emissions at a cost of about US$25 per tonne. At the same time, both systems result in additional positive externalities, such as reducing liquid effluents and the use of space. Trading of GHG by means of CDM offers scope for increased cost effectiveness, if emission reduction options in industrialized countries were to cost more than US$25 per tonne CO_2-eq.

The question to what extent the baseline in this project is acceptable depends on the judgement whether it is likely that in the coming 20 years a system of sanitary landfill or integrated solid waste treatment will be introduced in Quito, even if climate policies do not play a role. This judgement is necessary in order to define a proper baseline on a project basis. It is, however, extremely difficult – and the case study clearly emphasizes this aspect – to identify a scientific approach to establish whether or not these systems would be introduced without climate policies.

Case Study 4: Agroforestry in Mexico

The case study on forestry and agroforestry measures in the Central Highlands of Chiapas in Mexico is based on de Jong et al. (2000). The research analyses the costs of reducing greenhouse gas emissions by forestation, fallow management and agricultural and pasture management in Mexico. The study takes into account the private costs of these management options and the opportunity costs of agricultural income forgone.

It also pays extensive attention to the risk of carbon leakage, that is, the option that additional deforestation will take place at another location, outside the project boundaries. If this type of carbon leakage occurs, a project-based approach should pay due attention to it, in order to avoid a severe overestimation of the emission reductions of the project.

The project shows that reforestation can result in relatively large reductions of CO_2 emissions at relatively low costs of US$5-20 per tonne. Within this cost range, forestry and agroforestry measures in the relevant study area could mitigate from 1 to 42 tonnes C, with a maximum economic supply of carbon sequestration of around 55 tonnes C at US$40 per tonne C. If sufficient warranty could be granted that indeed the forestation project is additional and that no carbon leakage would take place, the project would be useful for CDM. Also in this case, however, it is not very clear whether or not reforestation would occur even if no climate policies were implemented. Again, it is emphasized that it is difficult – if not impossible – to identify an unbiased scientific method to make a judgement about the additionality of the project.

CONCLUSION AND DISCUSSION

This chapter illustrates that CDM in principle is a promising option for climate policies in the Netherlands. It has the potential to deliver a reduction of 25 Mt CO_2-eq. a year at a cost below US$15 per tonne. However, the practical application of the instrument faces many complications. It is hard to ensure that the instrument only certifies those projects that truly contribute to the reduction of greenhouse gases, that is, that go beyond business-as-usual emission reductions. The case studies reveal that serious problems are related to the criterion of additionality. Another important topic is the monitoring of the projects and the danger of carbon leakage, if the monitoring takes place only at project level. Although developing countries are obliged to prepare communications and inventories, this obligation is made dependent on the financial assistance they receive from the developed countries. If CDM is restricted to the Netherlands' development assistance focus countries, this may provide facilities for collecting and providing the necessary information. The incentive to cheat in CDM can be dealt with by liability provisions, which may even lead to a situation where the companies prefer to be cautious about results, rather than to overestimate them. Even insurance provisions could be included. In all cases, however, a well-defined system of monitoring and sanctions will be required and the problem of the hypothetical (and thus immeasurable) nature of baselines is not solved.

Thus, many difficulties will have to be overcome before the government of the Netherlands can implement CDM while adhering to its own criteria. The combination of efficiency and high-quality emission reductions might prove to be conflicting. It will require either a political decision to favour one criteria over the other or a very specific and restrictive set of conditions. However, measures to enhance the quality of emission reductions generated through CDM will almost inevitably increase transaction costs. This will increase the price of emission reductions generated through CDM, which might lead to only a minor contribution of CDM to climate policies in the Netherlands.

Consideration might be given to applying emission reductions abroad only through joint implementation. Here, the countries involved are required to report national emissions of greenhouse gases on the basis of a complete inventory. Deviations from the reduction target, caused by an overestimation of the reduction potential of individual projects, can be identified and corrected on a national level. For CDM countries, such a reference target does not exist, implying that there is no mechanism to correct for overestimation/ certification of the emission reductions of individual projects. Even for JI, however, it remains extremely complicated to monitor emissions and address additionality in a consistent an unbiased manner across various projects. Experience under the Montreal Protocol reveals that the former Eastern Bloc countries have major problems in complying with their obligations – which could be a precedent for the climate regime. This really makes both joint implementation and the clean development mechanism extremely difficult to apply in practice. A well-designed international agreement that may gain the full support of both developing and industrialized countries might require more transparency than CDM and joint implementation can offer.

NOTES

1. The authors would like to thank anonymous referees for their useful comments.
2. Although this might change, after the decision at COp6+ that there will be no obligation to achieve at least 50 per cent of the assigned reduction requirement through domestic action.
3. Note that this is a draft document, stating main ideas and viewpoints. The contents will change over time.
4. Main limitations of the study: transaction costs are excluded, costs for the adaptation fund have not been taken into account, the possibility of 'banking' is not taken into account, the entire life of the project has been taken into account for the cost calculations, and only CO_2 is considered. The main goal of the report was to support the decision making on the Executive Report on Climate Policy II; Co-operation with other countries.
5. Asian Development Bank (Zhang, 1999): US$2.6 (also other GHGs); MIT (Ellerman et al., 1999): US$6.5 (only CO_2); University Pierre Mendes (Cirqui et al., 1999): US$6 (only CO_2); Stanford Energy Modeling Forum US$6–12. Both Zhang and Ellerman have doubts about

the potential of the instruments in practice. The parties in the Stanford Energy Modeling Forum doubt whether a trade on a world scale is realistic.
6. Annex I trading only: MIT (Ellerman et al., 1999) US$35; University Pierre Mendes (Cirqui et al., 1999): US$17; Stanford Energy Modeling Forum: US$15-35 per tonne CO_2.
7. Derived from the Internet sites of the US Energy Information Agency at www.eia.doe.gov.
8. This figure was provided by the project company and is derived from emission factors found in literature.
9. One CER is 1 tonne of CO_2-eq.
10. Other assumptions: the CERs are paid for at the end of the year in which they are realized; the exchange rate used is: 0.34 US$ per £E (Egyptian pound: exchange rate at 16 May 2000.)

REFERENCES

Bandsma, J., W. van der Gaast and C.J. Jepma (2000), *Baseline: Criteria and Issues: Overview of discussion and results of previous workshops on the criteria of baselines*, Prepared for the Expert Workshop to Develop Initial Guidelines on Baseline Determination, Amsterdam, 17-19 January, Paterswolde, The Netherlands, Joint Implementation Network.

Benitez, P. (1999), *Economic and Environmental Aspects of Municipal Solid Waste Management in Quito, Ecuador*, Wageningen University, Environmental Economics Group, Wageningen.

Centraal Planbureau (CPB) (1999), 'Effecten van de Uitvoeringsnota Klimaatbeleid', (Effects of the executive report on climate policy), Working Paper, Centraal Planbureau, The Hague.

Cirqui, P., M. Silva and V. Laurent (1999), 'Marginal abatement costs of CO_2 emission reductions, geographical flexibility and concrete ceilings: an assessment using the Poles model', Institut d'Économie et de l'Energie, Université Pierre Mendez, Energy Policy 27585-601.

de Jong, B.H.J., R. Tipper and G. Montoya-Gómez (2000), 'An economic analysis of the potential for carbon sequestration by forests; evidence from Southern Mexico', *Ecological Economics*, **33** (2), 313-27.

de Leeuw, G.J. (2000), 'The clean development mechanism for Dutch investors: a preliminary study of options, conditions and issues', Thesis, University of Twente/ Cartesius Institute/Wageningen University, Environmental Economics Group.

ECN (Netherlands Energy Research Foundation (1999), 'Potential and cost of clean development mechanism options in the energy sector; inventory of options in non-Annex I countries to reduce GHG emissions', in co-operation with AED (Alternative Energy Development Inc, Silver Spring, USA) and SEI (Stockholm Environmental Institute, Boston, USA), on behalf of the Netherlands Development Cooperation (DGIS), Petten, ECN, December.

Ellerman, A.D., H.D. Jacoby and A. Decaux (1999), *The Effects on Developing Countries of the Kyoto Protocol and CO_2 emissions trading*, Cambridge, MA: Massachusetts Institute of Technology.

Ellis, J. and M. Bosi (1999), *Options for Project Emission Baselines*, OECD/IEA information paper, Paris, Organization for Economic Cooperation and Development, and International Energy Agency.

Jepma, C.J. (2000), 'Editor's note: a break in The Hague?', *Joint Implementation Quarterly*, **6** (4) (PP 1), Paterswolde, JIN.

JIN (Joint Implementation Network) (2000), 'COP6 Special; COP6 Discussions', *Joint Implementation Quarterly*, **6** (4), 8-9, Paterswolde, JIN.

JIN (Joint Implementation Network) (2001), E-mail contact, 1 August.
JIRC (Joint Implementation Registration Centre) (2000), *Setting a standard for JI and CDM; Recommendations on baselines and certification, based on AIJ experience*, The Hague: JIRC.
Meyers, S. (1999), 'Additionality of emission reductions from clean development mechanism projects: issues and options for project-level assessment', Berkeley (CA), Ernest Orlando Lawrence Berkeley National Laboratory.
Michaelowa, A. (1998), 'Joint implementation – the baseline issue: economic and political aspects', *Global Environmental Change*, **8** (1), 81–92.
OECD (Organisation for Economic Co-operation and Development) (1999), *Status of Research on Project Baselines under the UNFCCC and the Kyoto Protocol*, OECD working papers Vol. VII, no. 25, Paris: OECD.
Parkinson, E.A., K. Beggs, P. Bailey and T. Jackson (1999), 'JI/CDM crediting under the Kyoto Protocol: does "interim period banking" help or hinder GHG emission reduction?', *Energy Policy*, **27**, 129–36.
Senter Internationaal (2000a), *ERUPT (Emission Reduction Unit Purchasing Tender): Terms of Reference*, The Hague: Senter Internationaal.
Senter Internationaal (2000b), 'Nine companies put in offer for ERUPT', Press release 15 November, Website: http://www.senter.nl/erupt/news151100.htm.
Senter Internationaal (2001a), 'The Netherlands approach to boost CO_2-abatement: joint implementation as part of the Netherlands' obligations under the Kyoto Protocol', Website: http://www.senter.nl/cop6/ji.htm, January.
Senter Internationaal (2001b), 'ERUPT Survey: Coming up next year: ERUPT-CDM', Website: http://www.senter.nl/erupt/enquete/enquetevragen.asp, January.
Senter Internationaal (2001c), 'Jorritsma buys Kyoto reductions in Central and Eastern Europe', Press release, www.senter.nl/erupt/news170401.htm, April.
Sijm, J.P.M., F. Ormel, J.W. Martens, S.N.W. van Rooijen, M.H. Voogt, M.T. van Wees and C. de Zoeten-Partenset (2000), 'The role of joint implementation, the clean development mechanism and emissions trading in reducing greenhouse gas emissions', Petten, The Netherlands: ECN (ECN-C-00-026).
Stewart, R. (lead author) (1999), 'The clean development mechanism: building international public–private partnership: a preliminary examination of technical, financial and institutional issues', Based on the deliberations of the Ad Hoc Working Group on the CDM, United Nations.
United Nations (1997), *Kyoto Protocol to the United Nations Framework Convention on Climate Change*, United Nations.
UNFCCC (United Nations Framework Convention on Climate Change) (2000), 'Mechanisms pursuant to Articles 6, 12 and 17 of the Kyoto Protocol; modalities and procedures for a clean development mechanism; Submissions from Parties; Note by the Secretariat', Prepared for the sixth conference of the Parties, agenda item 7(c). FCCC/CP/2000?MISC.2.
UNFCCC (United Nations Framework Convention on Climate Change) (2001), 'The sixth session of the UNFCCC Conference of the Parties; Mechanisms: Issues in the negotiating process', Website: http://cop6.unfccc.int.modules/none.asp?pageid=113, United Nations.
van der Linden, N.H., J.R. Ybema, M. Beeldman, and S.N.M. van Rooijen, (2000), 'Een samenvattende analyse van potentiëlen en kosten van broeikasgasreductie-opties in binnen- en buitenland' (Summarizing analysis of the potentials and costs of greenhouse gas emission reduction options domestically and abroad), Petten, The Netherlands: ECN (ECN-C-00-015). Executive summary in English.

van Ierland, E.C., P. Mensink, C.J. Brink, P. Kabat, G. Nabaures, F. Brouwer, C. Kroeze, J. Pluimers and L. Slangen (1999), 'Socio-economische aspecten van klimaatverandering en landbouw; Quickscan LNV Agenda Klimaat' (Socioeconomic aspects of climate change and agriculture; quickscan of the climate agenda of the ministry of agriculture, nature and fishery), Wageningen University.

van Ierland, E.C., G.J. de Leeuw and J. Krozer (2000), 'Joint implementation, clean development mechanism and the baseline: an economic analysis', Paper presented at the 2000 Taipei Conference on Policies for Greenhouse Gas Reduction and Pollution Control in Asian Pacific, 30 November-1 December, Taiwan.

VROM (Ministry of Housing, Spatial Planning and the Environment) (1999), *Uitvoeringsnota Klimaatbeleid deel I: Binnenlandse maatregelen* (Executive report on climate policy. Part I: domestic actions), The Hague: Ministry of VROM; Central Directory of Communications.

VROM (Ministry of Housing, Spatial Planning and the Environment) (2000), *Uitvoeringsnota Klimaatbeleid deel II: Samenewerking met het buitenland* (*Executive report on climate policy, Part II: International Cooperation*), The Hague: Ministry of VROM; Central Directory of Communications.

VROM (Ministry of Housing, Spatial Planning and the Environment) (2001a), 'Implementatie van het Clean Development Mechanism; de werkwijze en achterliggende principes' (Implementation of the clean development mechanism; manner of execution and underlying principles), Concept/Working document, The Hague, June.

VROM (Ministry of Housing, Spatial Planning and the Environment (2001b), Telephone contact with Mrs Hernaus, Department Clean Development Mechanism.

Woerdman, E. and W.P. van der Gaast (2001), 'Project-based emissions trading: the impact of institutional arrangements on cost-effectiveness', in *Mitigation and Adaptation Strategies for Global Change* (forthcoming / accepted for publication).

World Bank (2000), Website Prototype Carbon Fund, www.prototypecarbonfund.org/html/vcc.htm.

Zhang Zhongxiang (1999), 'Estimating the size of the potential market for all three flexibility mechanisms under the Kyoto Protocol', University of Groningen.

9. Optimal institutional arrangements and instruments for the promotion of energy from renewable sources

Jan C. Bongaerts and George Dogbe

1 INTRODUCTION

This chapter deals with the instrumentation of the promotion of electricity from renewable energy sources (RES) in the member states of the European Union (EU). Although the EU can be seen as an important 'player' in the area of international policy making on global changes, one has to take into account that – with respect to important issues such as the promotion of RES in general and electricity from RES in particular – the EU has little or no powers and, as a result, any real contribution will have to come from the action of the member states. That is why studying their policies is of interest.

The chapter is structured as follows. We start with a statistical section highlighting the energy situation in general and with respect to electricity from RES in particular. Section 3 is devoted to some general background issues influencing the promotion of electricity from RES (such as, for example, the liberalization of electricity markets.) Section 4 deals with demand management instruments. Section 5 deals with supply promotion packages. In the final section, some conclusions are drawn.

2 STATISTICAL OVERVIEW OF ENERGY FROM RES IN THE EU

Renewable energy is being promoted as an alternative energy resource in all EU member states as they strive to be energy sufficient, as well as reduce their greenhouse gas emissions from fossil fuel consumption. There have therefore been campaigns to reduce energy consumption within member states. Although the EU has constantly consumed more than 16 per cent of the world's total energy output since 1980 (Table 9.1), its output has always been between 8 and 10 per cent, making it the world's largest energy importer and

Table 9.1 EU15 primary energy consumption and production, 1980-1999

Year	Consumption (BTU × 10^{15})	% of world output	Production (BTU × 10^{15})	% of world output
1980	53.53	18.5	25.7	8.9
1981	51.78	18.2	26.7	9.4
1982	50.87	18.0	27.3	9.7
1983	50.82	17.8	28.5	10.0
1984	52.16	17.3	28.5	9.5
1985	53.76	17.4	31.0	10.1
1986	54.86	17.2	31.5	9.9
1987	55.79	17.1	31.4	9.6
1988	56.29	16.6	31.2	9.2
1989	56.92	16.5	30.0	8.7
1990	56.96	16.2	29.7	8.5
1991	58.08	16.7	29.5	8.5
1992	57.58	16.5	29.2	8.4
1993	57.94	16.5	29.6	8.4
1994	57.74	16.2	30.3	8.5
1995	59.77	16.4	30.7	8.4
1996	61.05	16.4	32.1	8.6
1997	61.42	16.1	31.7	8.3
1998	62.58	16.3	31.7	8.3
1999	62.73	16.5	31.9	8.4

Source: Energy Information Administration (EIA).

the second largest energy consumer.[1] It is obvious that the EU is overdependent on energy imports. It has constantly imported between 42 and 50 per cent of its total energy demand (Figure 9.1). Hence, the vision of energy self-sufficiency within the EU seems impossible since in terms of resources and proven technologies for competitive energy production, the EU has few natural endowments. In the absence of any immediate alternate energy source, the EU's energy dependence may reach more and more perturbing levels unless energy consumption rates begin to show a downward trend.

RES have been identified as the possible sources of energy that could redeem the EU from this calamity. But the big question is: what is the role of RES in the present energy balance and how does it fit into the picture in the near and long terms? The RES share of the total energy balance has been quite

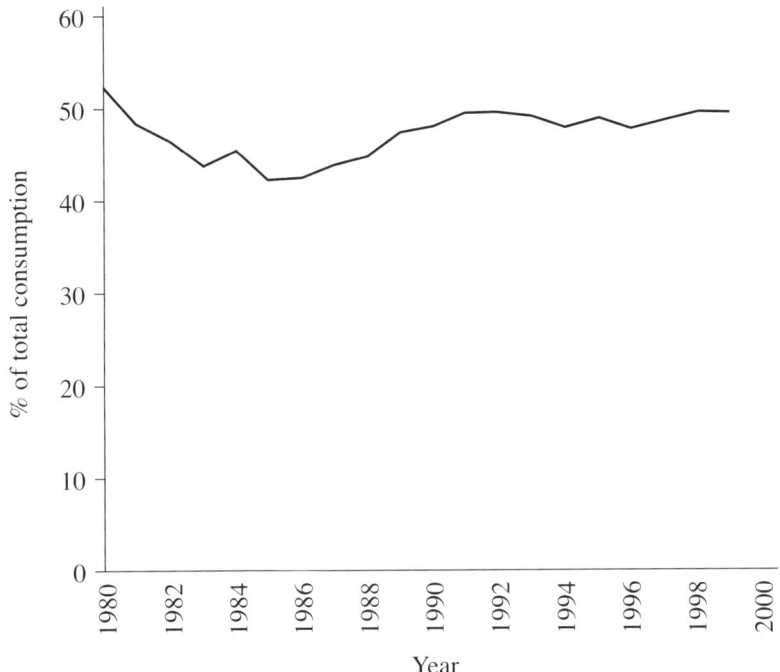

Source: Energy Information Administration (EIA), Vienna.

Figure 9.1 EU15 energy imports (% of total energy demand), 1980-1999

stable between 5 and 6 per cent for the last 20 years (Figure 9.2). The EU has set a target to double this share of renewable energy from 6 to 12 per cent in the year 2010. A look at member states shows that France, Sweden and Italy contribute the highest margin to total renewable energy. A snapshot in 1999 shows that France and Sweden together contributed about 42 per cent of total EU-wide renewable energy with Italy contributing 15 per cent (Figure 9.3) but then in relative terms (share of inland primary energy consumption), Sweden (33.9 per cent), Austria (31.3 per cent), Finland (17.6 per cent) and Portugal (8.6 per cent), are the leaders within the EU. Denmark seems to have had quite a dramatic improvement in the use of energy from renewable sources over the last two decades, from near zero in 1980 to more than 5 per cent in 1999 (Table 9.2).

In terms of technology of renewable energy generation, biomass (58.2 per cent) contributes the highest source (Figure 9.4) followed by large hydropower (34.7 per cent) and small hydropower (3.6 per cent). Wind sources still contribute less than 2 per cent despite many achievements in research in this

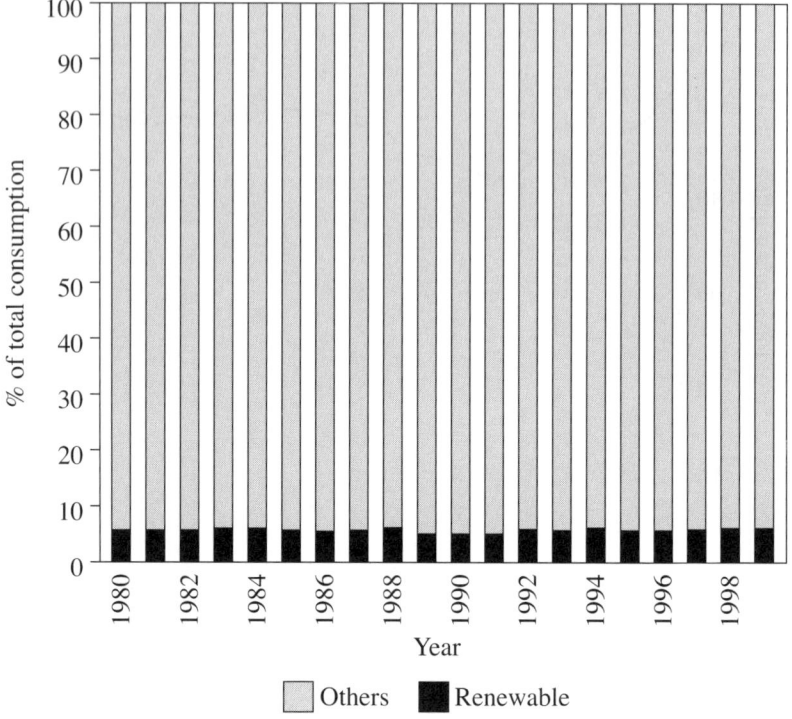

Source: Energy Information Administration (EIA), Vienna.

Figure 9.2 Share of renewables in the total energy balance EU15, 1980-1999

area. The present share of solar sources is a little over 1 per cent. Photovoltaic sources have not been able to make any imprint in the renewable energy balance: total installed capacity within the EU was only a little below 27 MW in 1997. Table 9.3 shows installed capacity from the various technologies in the member states. Germany is the most diversified in terms of technologies of generation of renewable energy. Biomass, however, has considerable importance for almost all member states with France disposing of the largest capacity. Germany and Denmark show supremacy in wind technology accounting for about 68 per cent of installed wind capacities.

About 32 per cent of energy from renewable sources is used in the production of electricity. Electricity's share in the total energy balance is fairly flat in the short term. In 1999 it constituted 12.3 per cent of the total energy balance and, as a result of this slow growth rate will constitute about 13.7 per cent in 2010 (Figure 9.5). However, the only institutional arrangement to

Table 9.2   Share (%) of renewable energy in inland energy consumption

Country	1980	1990	1999
Austria	28.1	28.7	31.3
Belgium	0.3	0.4	0.5
Denmark	0.0	0.9	5.1
Finland	10.7	9.9	17.6
France	8.5	6.5	7.5
Germany	1.7	1.7	2.5
Greece	4.7	1.7	3.9
Ireland	2.7	1.9	2.0
Italy	8.7	5.6	7.3
Luxembourg	0.7	0.6	2.3
Netherlands	0.3	0.3	1.3
Portugal	19.2	13.6	8.6
Spain	9.5	6.8	5.5
Sweden	29.5	35.3	33.9
United Kingdom	0.5	0.7	1.4

Source: Energy Information Administration (EIA), Vienna.

promote renewable energy sources at EU level (Directive 96/92/EC on the internal market for electricity)[2] is centred on electricity from renewable sources. In view of this more or less stable fraction, renewable energy policies should also be developed in other sectors that have a rather rapidly growing energy demand; for instance an EU directive targeting the share of biofuels in the total energy demand in the transportation sector could be one of such that would enhance the achievement of its 2010 objective for renewable energy use.

Currently, renewable sources constitute about 15 per cent of the total electricity production of which hydro sources (including small hydro, <10 MW) constitutes the highest share. It is interesting to note that, of the renewable electricity produced, hydro sources continue to decrease (Figure 9.6) but then the 1999 share of about 83 per cent is still substantial. On the one hand, a falling share of hydro sources in renewable electricity production is, however, an indication that other technologies in renewable electricity production are gaining entry into the market and given the right support in terms of investment in research and development (R&D) and policies their full potential could be realized. On the other hand, an 83 per cent contribution from hydro sources in renewable electricity generation still suggests that other technologies have not caught up and still have a long way to go. The situation

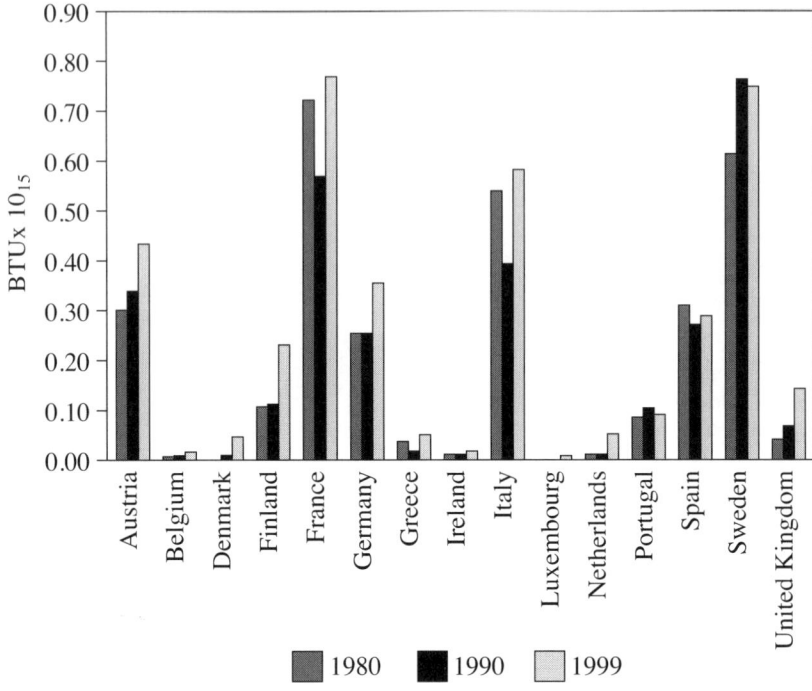

Source: Energy Information Administration (EIA), Vienna.

Figure 9.3 EU renewable energy, production by member states

at the level of the member states is shown in Table 9.4. Here, while Germany, the UK, Finland and Denmark have made remarkable improvements in other technologies (showing declining dependence on hydro sources) in their domestic renewable electricity production, Austria, Sweden and France still heavily depend on hydro sources (more than 96 per cent). This trend is also observed in absolute terms (Table 9.5).

One major issue that brings about the need to promote energy from RES is the generally higher cost of production in comparison with the generation of electricity from fossil fuel sources. RES are generally at a competitive disadvantage because most of the external costs associated with the generation from fossil fuels is not reflected in prices. In addition, many conventional sources still benefit from substantial government subsidies and these together keep prices of energy from conventional sources artificially lower than they would have been. Thus individuals do not pay for the full social cost of fossil fuel production. For less economically developed technologies, such as photovoltaics, another reason for the considerably higher cost of generation is

The promotion of energy from renewable sources 201

Table 9.3   Capacities of renewable energy installed in gross MW, 1997

Country	Wind	Small Hydro	Large Hydro	Photo-voltaics	Biomass*	Solar thermal*	Geothermal*
Austria	20	924	10,350	1	5,741	411	8
Belgium	7	47	65	0	1,124	15	8
Denmark	1,060	10	0	0	2,937	41	8
France	3	1,848	19,025	0	66,076	160	814
Germany	1,900	1,296	2,979	5	11,992	354	33
Greece	27	42	2,484	0	3,776	1,302	29
Ireland	46	68	162	0	642	1	0
Italy	100	2,076	17,797	12	6,459	116	744
Luxembourg	2	27	0	0	10	0	0
Netherlands	336	37	0	2	959	51	0
Portugal	20	212	3,748	0	11,718	74	8
Spain	406	1,180	11,189	5	16,807	121	43
Sweden	108	1,058	15,390	0	9,282	0	0
UK	330	29	1,411	0	2,229	111	2

Notes:
*Heat/electricity cogeneration.
Biomass: Energy crops (wood), energy crops (ethanol/million litres), energy crops (biodiesel/ million litres), forest residues, solid agricultural waste, liquid agricultural waste, municipal solid waste, municipal digestible waste, solid industrial waste, liquid industrial waste, landfill gas.

Source: Compiled from the European Renewable Energy Exchange (EuroRex) website.

the structural discrimination of new technologies.[3] Their very low market share does not allow for economies of scale to be realized thus leading to higher per unit cost and reducing competitiveness. Table 9.6 shows the cost of production per kWh of electricity in certain International Energy Association (IEA) countries from different sources. Generally, electricity from RES costs two to three times more than from conventional sources (and in particular electricity from photovoltaic sources costs a lot more) hence the need to promote them by either fiscal or incentive means if they are to play a meaningful role in the total energy balance.

Conclusions on the Statistical Overview

A critical look at the technologies of energy production makes the EU's target of 12 per cent an ambitious one. In fact, due to environmental problems associated with large hydro (which accounts for more than a third of RES as

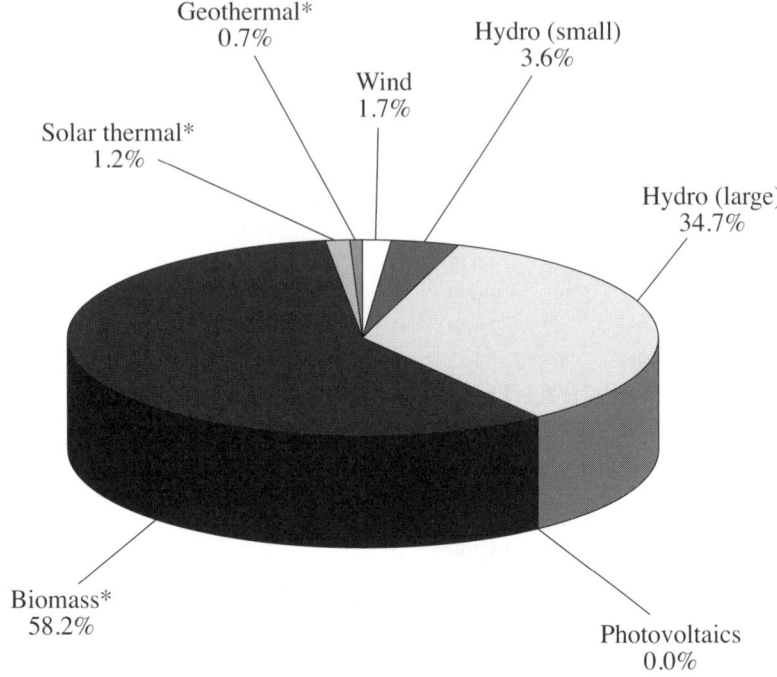

Notes:
*Heat/electricity cogeneration.
Biomass: Energy crops (wood), energy crops (ethanol/million litres), energy crops (biodiesel/million litres), forest residues, solid agricultural waste, liquid agricultural waste, municipal solid waste, municipal digestible waste, solid industrial waste, liquid industrial waste, landfill gas.

Source: Compiled from the European Renewable Energy Exchange (EuroRex) website.

Figure 9.4 Share of renewable energy production by technology, EU15, 1997

shown in Figure 9.4), this target does not consider large hydro expansion as potential source. In the White Paper for a Community plan for renewable energies[4] hydropower is estimated to increase by 14.1 per cent (including small hydros) from 1995 figures. This will mean that to reach the 12 per cent target, generation from other renewable sources of energy should increase to (4.91 x 10^{15} Btu) about 612 per cent of 1999 production if energy consumption increases at the lower bound of current rates of 1 per cent to 2 per cent per annum. However, in terms of both costs and output efficiencies, certain renewable technologies, such as photovoltaics, have very little role to play in the near and medium terms. This will mean that most of the expected growth

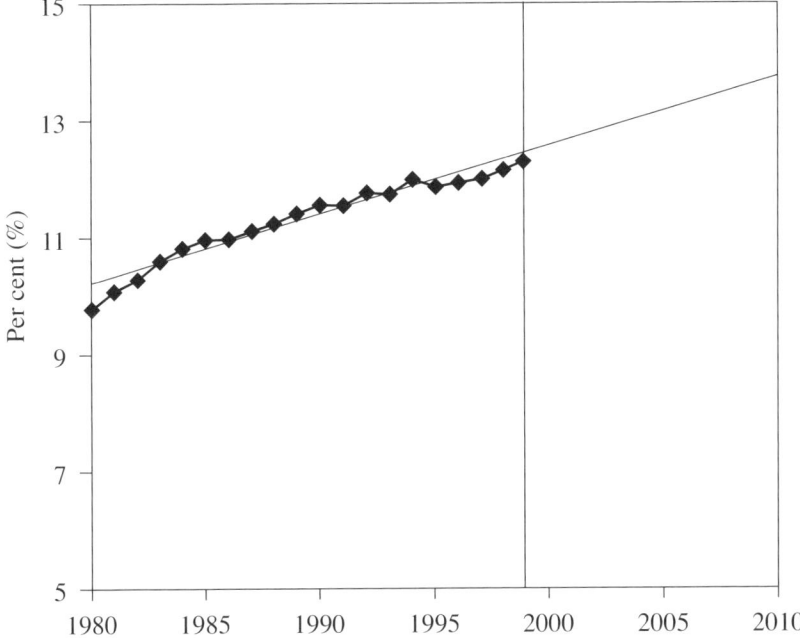

Source: Energy Information Administration (EIA), Vienna.

Figure 9.5   Electricity consumption as a share of total energy consumption, EU15

would have to come from wind and biomass. The big question then is: are policy instruments already in place within the EU and also within member states to drive this ambitious target to reality? The sections that follow address these different policies.

3 BACKGROUND FRAMEWORK FOR POLICIES FOR RENEWABLE ENERGY SOURCES

A Complex Issue

The promotion of RES is, at first sight, a pretty complex multitude of issues. Although all RES are directly or indirectly related to the sun, they are present in various shapes. Moreover, a wide variety of technologies are required to make the energy from RES available for consumption. Finally, an increase of the supply of this energy requires an increase of demand. Hence, one can make

Table 9.4 Share (%) of hydro in inland renewable electricity generation

Country	1980	1990	1999
Austria	98.9	96.7	96.0
Belgium	49.0	29.5	25.7
Denmark	100.0	4.0	0.6
Finland	100.0	100.0	57.2
France	98.4	96.2	96.7
Germany	78.4	78.2	56.0
Greece	100.0	99.9	93.5
Ireland	100.0	100.0	75.8
Italy	92.2	90.6	86.5
Luxembourg	74.8	68.6	84.1
Netherlands	0.0	11.5	2.0
Portugal	96.3	93.2	86.1
Spain	98.8	97.4	86.3
Sweden	98.7	97.6	96.3
United Kingdom	100.0	79.5	39.3
EU15	95.4	93.5	82.9

Source: Energy Information Administration (EIA), Vienna.

a distinction between instruments for the promotion of supply and of demand.

There is a great variety of instruments for the promotion of supply. In as much as the natural resources themselves are concerned, instruments can be embedded in agricultural policies (for example, for the promotion of biomass as a fuel). For the 'supply' of water for hydropower, instruments relating to infrastructure policy can be considered, such as the building of dams or storage power stations. For wind and solar energy, support can be given in the shape of physical planning (for example, production of 'aeolic' and solar light atlases and regulations on building permits for plants). Depending on whether municipal wastes are considered as a renewable energy source (as in the case of Austria, the Netherlands, Denmark and the UK), instruments relating to environmental policy can be applied (for example, setting priorities for certain types of waste treatment, such as incineration and the use of energy from incineration). For virtually all RES, some instruments of physical planning are required.

At the level of the promotion of energy-related technologies, grants for research and technological development (RTD) or for investment projects or even for running costs may be taken into consideration.[5] Fiscal instruments affecting the rate of depreciation can be a part of these grants. Lastly, in order

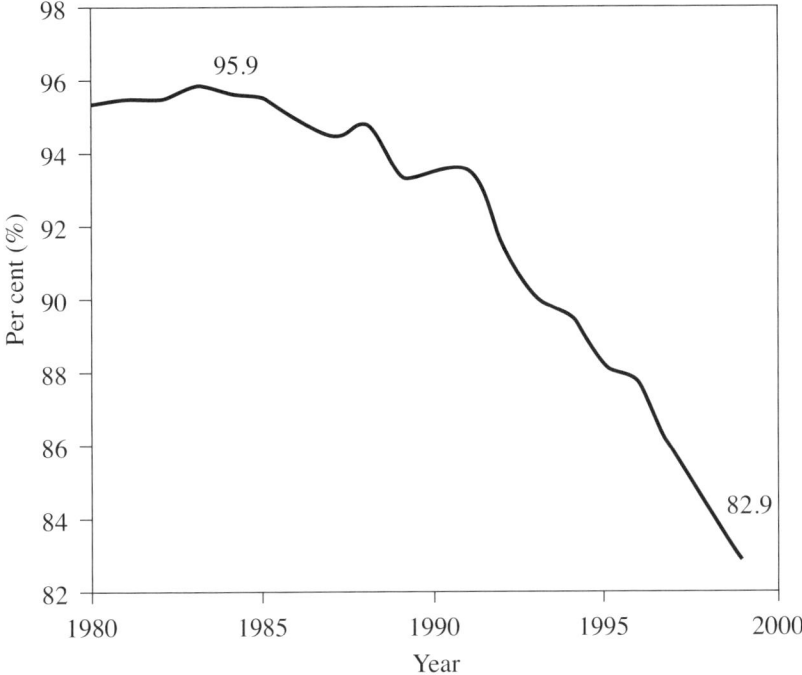

Source: Energy Information Administration (EIA), Vienna.

Figure 9.6 Share of hydro in renewable electricity production, EU15

to promote consumption, fiscal instruments, such as differential tax treatment regimes and legislatively determined feed-in rates for electricity may be set. Other market-based instruments such as tradable greenhouse gas (GHG) certificates and tradable 'green' power certificates may also be considered.

This summary statement is far from complete and bears only a slight resemblance to the complexity of the real world of RES promotion policies that are in place in the member states of the European Union. In sections 4 and 5, we try to present an overview of these policies. Given the complexity of the issue, we shall mainly be concerned with electricity from RES.

Liberalization of the Electricity Markets

Even though the promotion of electricity from RES is different from the liberalization of the electricity markets and, in several member states, started well before the adoption of EU Directive 96/92/EC on the internal electricity

Table 9.5 Total renewable inland electricity generation (bn kWh)

Country	1980 Total	1980 Non-Hydro Sources	1990 Total	1990 Non-Hydro sources	1999 Total	1999 Non-Hydro sources
Austria	28.8	0.3	32.2	1.1	41.8	1.7
Belgium	0.6	0.3	0.9	0.6	1.3	1.0
Denmark	0.03	0.03	0.7	0.7	4.4	4.4
Finland	10.1	0.0	10.8	0.0	22.2	9.5
France	69.4	1.1	54.9	2.1	74.0	2.4
Germany	23.9	5.2	21.9	4.8	34.0	15.0
Greece	3.4	0.0	1.8	0.0	4.8	0.3
Ireland	0.8	0.0	0.7	0.0	1.1	0.3
Italy	48.8	3.8	34.6	3.2	51.8	7.0
Luxembourg	0.1	0.0	0.1	0.0	0.4	0.1
Netherlands	1.0	1.0	1.0	0.9	4.7	4.6
Portugal	8.2	0.3	9.7	0.7	8.3	1.2
Spain	29.5	0.3	25.8	0.7	27.7	3.8
Sweden	58.9	0.7	73.6	1.8	71.9	2.7
United Kingdom	3.9	0.0	6.4	1.3	13.5	8.2
EU15	287.4	13.1	275.1	17.9	362.0	61.9

Source: Energy Information Administration (EIA),Vienna.

market (IEM Directive)[6], there is little doubt that the opening up of those markets constitutes an advantage for electricity from RES. Ending monopolies and giving access to the grids to all electricity generators creates a potential for electricity from RES. Without access to the grids, there is no opportunity for new operators. The question remains whether the technologies put in place for electricity from RES are competitive with existing technologies and, as a consequence, whether additional (policy) measures specifically dedicated to the promotion of electricity from RES are required. Hence, the liberalization of electricity markets is a necessary, but it may not be a sufficient condition.[7]

Moreover, one has to take into account that the liberalization of the market as set out by the EU IEM Directive in itself does not immediately open the market completely in the sense that any consumer is free to select his or her supplier and vice versa. With the exception of Germany, Finland, the UK and Sweden, the member states limit this flexibility of consumers to choose their suppliers to consumers of a minimum threshold.[8] Even though in principle the EU IEM Directive sets out a timetable within which member states must have a reached a minimum market opening of 33 per cent and an expected

Table 9.6 Comparative cost of renewable and other electricity, selected IEA countries (cents/kWh)

Country	Projected of new baseload plants[1]			Renewable electricity costs		
	Coal	Gas	Nuclear	Wind	Biomass	Other renewable
Canada	3.6–5.2	3.2	3.9–4.6	NA		
Denmark	3.6	4.3–5.5	–	5.5 (offshore)	11.9 (straw)	
France	5.7	3.2	4.8	6.5[3]		33.2[2] (non-grid PV)
Germany	NA	3.3	NA	5.8–11		
Italy	5.1	3.5	–	7.5		
Japan	7.4	7.8	7.6	13.3[2]		11–14.4 (Hydro < 0.5 MW)
New Zealand[4]	5	2.8	–	4.5–6.3		3.6–5.7 (geothermal)
Netherlands	5.4–5.9	2.6–2.8	–	9.0[2] (1995)		
UK (England and Wales)	3.9				5.1[3] (MSW)	
US	3.3–3.5	2.3–2.6	4.5		4.1 (biomass)	

Notes:
1. 10 per cent discount rate (1996 US$). Source: *Projected Costs of Generating Electricity*, IEA/NEA, 1998.
2. 1997 cost estimates from national administrations.
3. Average bid prices for successful projects, 1997.
4. Source: *New Zealand Energy Outlook*, Ministry of Commerce, February 1997.

Source: *Renewable Energy Policy in IEA Countries*, Vol. 2.

maximum of 100 per cent, certain articles in the directive nevertheless give some leeway for member states to protect the generation of electricity from sources that have been 'faithful' in the supply of power. Clearly, this could limit competition. The scope of this chapter does not give us the opportunity to go into detail on this issue.[9] One point, however, needs to be noted.

Article 10(1) of the EU IEM Directive allows room for public service obligation (PSO) to be imposed in the general economic interest on electricity undertakings. Although member states have to define this within a framework compatible with the EC Treaty defining PSO based on *security* and *safeguard* of supply, which of course is allowed by the EU IEM Directive, could be a big source of protection for the generation of electricity from certain sources, especially in member states in which that generation is principally dominated by particular sources. For France (which has actually made use of the PSO clause) this means protection for nuclear sources (75 per cent), for Austria this means hydro (68 per cent) and for Germany, coal (55 per cent).

The French concept of public service is particularly broad and encompasses the organization of the electricity system in its entirety. It has PSO defined to give it the right to select the *primary energy sources and production technique*[10] to safeguard supply. Although Germany has a 100 per cent market opening, in principle it may have less than 45 per cent of the market opened to competition as it pays billions of German marks every year to the coal industry (whose production cost is 3-4 times world market price) to maintain jobs and ensure security of supply. It thus becomes difficult to reconcile competition and security of supply and in this context the market could be described as being liberalized but 'protected' to a certain degree.

One appropriate instrument for the promotion of electricity from RES, which is contained in the EU IEM Directive, sets the obligation for the grid operator to give priority to electricity from RES. If put into practice, this instrument calls for specific procedures for the remuneration of electricity from RES by the grid operator. But, as we shall see, there are many more measures and, typically, member states have adopted more or less complex 'packages' of policy instruments for the promotion of electricity from RES.

Instruments for the Promotion of Electricity from RES

Why promote electricity from RES?

In this chapter we do not explicitly address the question of necessity or desirability of electricity from RES. At this stage it may suffice to quote a paragraph taken from the explanatory memorandum of the European Commission's proposal for an EU Directive of the European Parliament and of the Council on the promotion of electricity from renewable energy sources in the IEM:[11]

The promotion of renewable sources of energy is a high Community priority, for reasons of security and diversification of energy supply, for reasons of environmental protection and for reasons of social and economic cohesion. This was outlined in particular in the Commission's White Paper on Renewable Energy Sources which was endorsed by the Council and the European Parliament.[12]

From these few lines we can see that RES in general and electricity from RES in particular may contribute to a decline of the EU's dependency on imports of (fossil) energy sources.[13] It is interesting to note that, in the view of the European Commission, this reason comes first, before environmental protection. Third, since most technologies for RES and electricity from RES imply a decentralization of the energy industry (that is, technologies have to 'move' to the geographic locations which are best suited for them), they offer the potential for economic and social development, in particular in rural areas.

Other arguments in favour of the promotion of electricity from RES relate to the creation of employment, the development of rural areas and the generation of new export industries. These are, among others, explicitly contained in the UK government policy for the promotion of electricity from RES. A UK policy paper on the agores web site,[14] states: 'In the context of industrial policy there could also be additional gains, including assisting the renewables sector of the UK industry to become more competitive, exporting goods and services and providing new employment'. Clearly, this extremely brief mention of the rationale for the promotion of electricity from RES does not pay adequate tribute to the continuing debate on this issue. Lack of space and the specific focus of this chapter prevent us from dwelling upon the matter. At this stage, it may be sufficient to conclude that there is overall agreement on setting a high priority on the promotion of electricity from RES in all the EU member states. In this context, one may also note that the protection of the environment (in particular, but not exclusively, the management of climate change) is only one of several objectives.

Promotion of electricity from RES within a market-based approach[15]
Browsing through the various documents on the policies for the promotion of RES and its putting into practice in the EU member states available at the agores web site, one faces the issue of investigating the relationship between electricity from RES and the liberalization of electricity markets. In several member states, the (ultimate) objective of the promotion of electricity from RES consists in the creation of an industry generating this electricity which, in the context of a market economy, is competitive and, hence, economically sustainable in the long run.

This is the case for many member states, such as Austria, Germany, Greece, the Netherlands, Spain and the UK.[16] It is most prominent in Denmark where the creation of a 'fully functional green electricity market' is the ultimate policy objective.[17]

In a policy paper[18] produced for the agores web site by EVA, the Austrian energy agency, the man principles of the government's policy are summarized as follows:

> The Federal Energy Concept states the following principles:
>
> - The energy policy is basically market oriented. Therefore, government interference is mainly limited to cases where market forces do not work.
> - Long-term sustainable development, including environmental compatibility, is a predominant goal in energy policy strategies.

In the Netherlands, a paper produced by NOVEM[19] describes the government's policy objectives as follows:

> The steps in the action programme for 1998-2000 [in the framework of an overall objective of meeting 5 per cent of the nation's energy demand from renewables in 2010 and 10 per cent in 2020 ... [jcb-gd] are classified into three themes:
>
> - improve the price performance ratio;
> - promoting market penetration;
> - addressing administrative bottlenecks.

In a document prepared for the agores web site on the situation in the UK, the aims of government policy are described as follows:[20]

> - Assisting the UK to meet national and international targets for the reduction of emissions including greenhouse gases;
> - Helping to provide secure, diverse, sustainable and competitive energy supplies;
> - Stimulating the development of new technologies necessary to provide the basis for continuing growth of the contribution from renewables into the longer term;
> - Assisting a UK renewables industry to become competitive for home and export markets and in doing so provide employment in a rapidly expanding sector;
> - Contributing to rural development.

While it almost goes without saying that similar policy views are held in the northern part of the EU (where member states began to liberalize their electricity markets during the mid-1990s[21]), the EU IEM Directive has also led to policy reviews in the South of the EU, noticeably in Greece, Italy and Spain. In the case of Greece, the opening up of the grid to independent producers and to auto-producers greatly improved the opportunities for electricity from RES.[22]

Given this market economy reference framework, most instruments actually used in the EU member states either intend to improve the competitiveness of supply through a decrease of the costs (of various items, but typically investment costs) or else tend to create a market for 'green' electricity. In the latter case a distinction can be made among so-called renewable obligations, 'green electricity' certificates and feed-in tariff schemes. All three instruments can be applied independently of one another but combinations are also possible. In the following section, we investigate the policy situations in the member states.

4 INSTRUMENTS FOR DEMAND MANAGEMENT

Renewable Obligations

Demand management instruments tend to 'create' a market for electricity from RES through a (regulatory) obligation for (certain groups of) customers to purchase a certain amount of 'green' electricity. In other words, through a regulatory statute a 'renewable obligation' is being created which has to be fulfilled by generators and traders – sometimes even by consumers – by a given deadline. In most member states that have introduced (or will introduce) a renewable obligation, the policy approach is dynamic in the sense that the obligation is to increase over time. Obligations can be formulated in terms of electricity consumption (in MWe, hence, in absolute amounts) or as relative shares of total electricity consumption or in terms of installed 'green electricity' capacity (in MWe, which is the case in Ireland).

In *Italy*, together with the new legislation on the reform of the electricity sector (in line with the EU IEM Directive), substantial provision was made for the promotion of 'green electricity'. Besides introducing new budget lines for financial aid for research and technological development, for pilot plants and real investment projects, the new law (legislative decree 79 of 16 February 1999) determines a legally binding share of 2 per cent of electricity from RES as of 2002, to be produced by new or retrofitted RES power plants. The share is to be increased in the years after 2002.

It is calculated on the basis of the amount of electricity produced by fossil fuelled plants minus electricity from co-generation (or combined heat and power – CHP), exports and electricity consumption by power plants themselves. In a direct relationship with this obligation, a market for tradable green certificates (in unit sizes of 100 MWh of electricity or multiples) will be created for all industrial agents affected by the new law, that is, generators, traders and importers of electricity.

In of the *Netherlands*, there is no statutory renewable obligation, but the government has initiated a policy concept through which – hopefully – 'green' electricity is considered to become a premium product, which will capture its own market. As of 1 July 2001, any electricity consumer will have the freedom to purchase such green electricity from her or his preferred supplier. In other words, every generator of electricity from RES will be able to identify customers and sign contracts with them. The electricity is subject to a 'green' label[23] which makes it a standardized product and, as such, recognizable for producers and consumers.[24] An important incentive, however, is given by the fact the so-called regulating energy tax (or 'ecotax') will be reimbursed to purchasers of 'green' electricity.

In the *United Kingdom*, there will be an obligation for a certain amount (or share) of the total production of electricity to be generated by RES technologies.[25] The UK government expects the contribution from all renewables to be around 5 per cent by 2003 and at 10 per cent by 2010. The additional costs will be borne by the suppliers and passed on to the consumers. Producers can meet their obligation in three distinct ways: (i) actual delivery of 'green electricity'; (ii) participating in a tradable 'green electricity' certificates market; and (iii) buying out of their obligation through a 'penalty' payment to OFGEM, the Office of Gas and Electricity Markets.[26]

Similarly, in *Denmark*, since the reform of the market for 'green electricity' in 2000, annual quotas for 'minimal' shares of electricity from RES technologies are being set. The main objective of this policy consists in creating a 'fully functional green market' in the sense that the liberalization of the electricity markets as required by the EU EIM Directive is taken as the starting point for nothing less than the creation of a market for electricity from RES (and not just, say, a supply-side stimulation programme). Denmark's policy is probably the most 'outspoken' in this context in the sense that the objective is to create a competitive 'green electricity' industry. This will have to become economically viable without subsidies, earning an income from sales of electricity and of 'green electricity' certificates.[27]

In *Austria*, after the entry into force of the EU EIM Directive, the Austrian Parliament adopted the *Elektrizitätswirtschafts- und Organisationsgesetz - ElOWG* in July 1998, which transposes the directive into Austrian law. For electricity from RES, the law states that, by 2005, 'new' RES (that is, excluding (large) hydropower and waste incinerators) should make up to 3 per cent of all electricity sold to end-users. For this purpose, the generators may sign contracts for delivery with all customers (not only the so-called eligible customers). Hence, for electricity from RES, the market is to be fully liberalized.

In *Ireland*, there is no renewable obligation in terms of consumption but there is the so-called 'Alternative Energy Requirement' (AER), operating since 1994. The AER is an obligation for expanding installed capacity for electricity from RES. The obligation is met through a series of public tenderings[28] of additional capacity for electricity from RES on the basis of the lowest bids. Winners of these competitions receive the right to sell electricity from RES to the grid (operated by the Irish Electricity Supply Board – ESB) for a period of 15 years. The tender terms for the fifth round can be inspected on the web site of the Irish Department of Public Enterprise.[29]

One part of the conditions for tendering concerns the setting of maximum prices for electricity from RES with bids exceeding those prices to be refused any consideration. In this way, the competition sets an upper limit to prices for electricity from RES. It could be argued that a tendering scheme will favour technologies operating at lowest costs only and, hence, in the long run, do not contribute to the creation of a level playing field for all technologies for electricity from RES. For Ireland – given its geographic location – this means that wind power has an advantage. In order to take such considerations into account the fifth round (of 2001), which comprises an additional 255 MWe (quite a sizeable amount for a small country[30]), 210 MWe is dedicated to (large-scale) wind power (that is, > 3 MWe [*sic*]), 40 MWe is reserved for small-scale (wind power) and 10 MWe and 5 MWe are earmarked for biomass and hydro, respectively.

'Green Electricity' Certificates

Often seen as a corollary instrument to the creation of renewable obligations, tradable 'green electricity' certificates act as entitlements (or even access rights) to the grids in the sense that producers and traders of electricity may either opt for actual physical delivery of 'green electricity' to the grid or else opt out (in total or in part) by purchasing such certificates.[31] The main policy reasons for the introduction of this instrument are seen in their accuracy (with respect to goal setting and achieving) and their flexibility, in the sense that operators with the lowest costs for the generation of electricity from RES will enjoy incentives to try to sell 'green electricity' certificates in excess of the amount they need themselves. Several member states are considering, or are close to setting up, such tradable 'green electricity' certificates schemes.

In *Italy*, the 'green electricity' certificates market will be introduced together with the 2 per cent renewable obligation in 2002. Operators with 'green electricity' capacity installed after 1 April 1999 will be endowed with the right to 'create' such tradable certificates (in units of 100 MWh of electricity or multiples) for a period of eight years. The intention is that, in a fully liberalized market, generators and traders of electricity have the options to derive income from two markets: one for electricity and the other for 'green electricity' certificates.

As a corollary to the developments around the creation of a liberalized market for 'green electricity' in the *Netherlands*, TenneT, the Dutch grid operator, has set up a fully electronically operating trade exchange for 'green certificates'. These are issued to producers of (or 'holders of contracts for') green electricity in sizes of 1, 10, 100 and 1000 MWh of electricity. According to EnergieNed, first moves by companies to find customers for 'green' electricity seem to have met with some success.[32]

In the case of *Denmark*, a market for 'green electricity' certificates will also be created in order to promote the renewable obligation and, as in the case of the other member states, individual producers or traders will be subject to a fee in the case of non-compliance. The difference with the other member states consists in the fact that, in Denmark, consumers not fulfilling their obligation will also be subject to a fee. This is set at DKK 0.27 per kWh and, as a result, it functions as a de facto price ceiling for 'green certificates'. As in the case of Italy, the 'green certificates' market is supposed to offer an additional source of income to generators and traders of electricity from RES.

In the *United Kingdom*, the tradable 'green electricity' certificates will be seen as one of three options for generators and traders of electricity to fulfil the renewable obligation. (The other two options are actual delivery of electricity from RES and payment of a 'penalty' fee to the Office of Gas and Electricity Markets.) The price of the certificates will depend upon the difference between (a) the costs of generation of 'conventional' electricity + the Climate Change Levy (effective as of April 2001) and (b) the costs of generation of electricity from RES (which is exempt from the Climate Change Levy).[33]

In this context, the right to 'green electricity' certificates is not necessarily only linked to the ownership of capacity for the generation of electricity from RES. In reality, it is linked to the delivery of electricity from RES. Hence, suppliers will be able to obtain renewable electricity and be entitled to stock or sell the accompanying 'green electricity' certificates in various ways:

- on the basis of capacity which they own;
- on the basis of capacity owned by others with whom they have individual contracts;
- on the basis of capacity owned by others with whom they have collective contracts;
- on the basis of capacity owned by others with whom they have contracts through intermediaries.[34]

Feed-in Tariffs

In the case of instruments related to feed-in tariffs, the objective is to increase the share of electricity from RES through an increase of installed capacity. Grid operators will typically be obliged to purchase any electricity from RES at regulated prices irrespective of 'identified' customers. In other words, on the basis of a specific regulatory regime, operators of grids are obliged to purchase electricity from RES at given prices contained within the regulation or set by regulators on the basis of such a regulation. In practical policy shaping, many details have to be arranged, such as the amount of electricity from RES (in kWh) which benefits from the regime for a given period (that is, *capping* or *no capping*), the sharing of the costs of connection to the grid between the grid operator and the supplier of electricity from RES, the fees themselves (using criteria such as the costs of generation of electricity from RES, the maturity of technologies, geographical locations of plants, the actual availability of capacity of plants, time and period price differentiations (for example, time-of-day or weekday and weekend differentiations in the feed-in tariff), decline of tariff over time (to account for depreciation and for progress of technologies), the time period of the feed-in regime itself, and so on.

As a result, independent operators generating electricity from RES do not necessarily have to worry about identifying and contracting with final customers since their entire production is fed into the grid according to these regulated conditions. This is an essential feature of most feed-in tariff schemes and it results in the fact that competition among operators of plants for the generation of electricity from RES is restricted or even non-existent. This is clearly different from a situation in which an electricity market is fully liberalized for 'green' electricity and in which operators of plants for the generation of electricity from RES (or traders) are supposed to be in the business of having to identify and persuade their proper customers.

Feed-in regimes are in place in many EU member states: Belgium, Austria, Denmark, Germany, Greece, Italy, Luxembourg, Portugal and Spain. The case of *Germany* with its act on renewable energy (*Erneuerbare Energiegesetz*) has reached international significance due to its having been challenged in the European Court of Justice as presumably violating the Treaty's regime on state aid. Meanwhile, the case has been decided 'in favour' of the German law.[35] The German *Erneuerbare Energiegesetz* sets compulsory tariffs payable by the grid operators to the generators of electricity from RES. The individual remunerations depend upon the type of fuel and or technology (for example, biomass for electricity or wind power), the 'productivity' of the plants (lower rates for wind than from photovoltaic electricity), certain geographical conditions (for example, near shore and inland wind power) and age (or, more specifically, duration of depreciation) of plants.[36]

Austria's situation with respect to RES in general and electricity from RES in particular, is unique, since this member state has a long tradition with RES and a high share of RES in total energy consumption. Already in 1990, this share stood at 25.7 per cent and it is rising continuously.[37] In particular, 70 per cent of all electricity generated comes from hydropower plants, with probably

10 per cent to 15 per cent being supplied by small hydropower plants (which are legally defined as < 5 MW, but, in practice, they are typically in the range of 100 to 300 kWe). The other important RES is biomass for heating (now enjoying a revival through new technologies) and for electricity generation (often as combined heat and power) and also, more recently, for electricity generation after wood gasification. Austria promotes these developments with feed-in tariff schemes with seasonal (winter/summer) and time-of-day differentiations (night/day and weekends).[38] In combination with the requirements for 'new' electricity from RES resulting from the 3 per cent obligation (see above), the governors of the states (*Bundesländer*) are obliged to set appropriate minimum feed-in tariffs.

Denmark's feed-in tariff system must be understood from the point of view of the primary objective, which is the creation of a functioning 'green electricity' market. Hence, feed-in tariffs are in place when plants are first opened and are gradually reduced and phased out. As of 2002, the tariff structure is such that, typically, after 10 years of operation, the feed-in tariff for a plant will be phased out (that is, reduced to zero). The remuneration for the electricity from RES will then consist in the market price for electricity + revenues from trading 'green electricity' certificates. Briefly,[39] the tariff structure will be as follows:

- plants constructed before end of 2002: fixed feed-in tariff for ten years + certificate;
- after ten years: market price of electricity + certificate;
- plants constructed from 2003: market price of electricity + certificate.

In *Spain*, the feed-in tariff system is pretty complex, as it is different for large and small generators (the cut-off capacity between both being set at 50 MWe). Large producers will be paid the hourly price of the market + an 'electricity from RES' premium. Small operators will receive a fixed price (instead of the market price) + an 'electricity from RES' premium + a premium on availability of capacity. (The latter part sets incentives to reliability in terms of generation and grid supply and, hence, helps to avoid investments in stand-by capacity.)

In *Portugal*, again the obligation to transpose EU IEM Directive 96/92/EC triggered a change in electricity policy. For electricity from RES a special decree law was adopted which installs – in addition to other arrangements, such as a regime of operating licences – a feed-in tariff system which is based upon the following elements:

- The avoided costs for the Public Power System due to the starting-up and operation of the power plant, including (*i*) the avoided investment cost on new

power plants and (*ii*) the transport, operation and maintenance, including fuel cost.
- The environmental benefits from the use of endogenous energy resources.[40]

It is of some interest to note that the environmental costs are based on the valuation (in terms of a unit price per tonne) of the amount of CO_2 emissions that are avoided.

5 INSTRUMENTS FOR SUPPLY MANAGEMENT

Introduction

It remains an open question whether feed-in tariffs and renewable obligations are 'only' demand-side management instruments. Clearly, their purpose is to stimulate new demand in the first place. But in doing so, they may make life easier for (would-be) generators of electricity from RES. In particular, in those cases in which such instruments improve the reliability of investment projects cost and revenue calculations, they contribute to an improvement of the supply conditions, through lowering (financial and other) transaction costs. We prefer to leave these instruments within the category of market demand management instruments because we consider only those instruments as typically favouring supply which have a direct impact on the costs of investments projects and on the running costs of operations. In that respect these instruments mainly comprise fiscal instruments in the shape of lower taxes, tax rebates and subsidies.

Fiscal Instruments

When investigating fiscal measures, typically (or traditionally), a distinction is made between taxes and subsidies. We feel that this chapter should not be dedicated to a classical economic analysis of such measures, as this is fairly well established. Hence, we do not address issues such as (the efficiency of) earmarked or non-earmarked taxes, subsidies for RTD, pilot and demonstration plants, for investment projects and even for operating (running) costs. In the real-world situation of the EU member states, all varieties occur.[41]

In this context, note that all member states have programmes for cost relief for electricity from RES in place. It would go much beyond the scope of this chapter to give a detailed overview of all of them. This is simply beyond our capacities and – given the changes and modifications which are likely to occur – the outcomes of such an investigation would not be very long lasting. However, a very rough and superficial outline of the structures of the fiscal packages in all 15 member states allows for some interesting comparisons. (See Table 9.7 below.)

Moreover, it may also be helpful to identify the motivations of the member states for adopting fiscal measures (or better 'packages of measures') for the promotion of electricity from RES. In other words: the way in which these packages are put together seems to reflect various 'philosophies' with respect to the promotion of electricity from RES. An inspection of the various policies of the member states reveals that this is indeed the case. Due to unclear indications, however, parts of our results are more speculative than certain.

Energy efficiency

A first motivation consists in setting a negative incentive for energy consumption as such, for example, in order to promote a rational use of energy or an increase of the productivity of energy. Presumably, this motivation is present (*expressis verbis* or *passim*) in the energy tax schemes of all the member states. Clearly, this motivation may also be carried forward by other instruments, such as energy labelling of products and (accompanying) dissemination of relevant information, but it seems that taxation of energy is the prime instrument. Although energy conservation and energy efficiency do not feature as issues in this chapter, we should not leave them out of the picture since it is important to keep in mind that RES in general and electricity from RES in particular should not be seen as excuses for inefficient uses of energy.

Raising revenues for the promotion of RES

A second motivation consists in raising revenues for subsidies and grants for the promotion of RES. This means that the energy taxes are earmarked. Such a system is in place in *Austria*, where the revenues of these taxes are returned as grants for investments projects for environmental projects including electricity from RES (see below). In the *United Kingdom*, the revenues of the 'penalty' payments for non-compliance with the renewable obligation will be 'recycled' with the purpose-improved compliance.

Keeping a level playing field for the promotion of RES

A third motivation for setting up an energy tax system consists in keeping a level playing field for the various energy technologies, in particular for RES-based technologies (and CHP technology – which is not necessarily an RES-based technology but can be very energy effective). In this case, either a (differential) tax system or a (package of) subsidies or a combination of both instruments serves to keep the costs of generation of electricity for all technologies within a defined (small) range. This motivation has a long-standing tradition in *Denmark*, and, to a lesser extent, in *Germany* (in particular in respect of the protection of CHP technologies which currently 'suffer' from high natural gas prices and - liberalization-induced - low electricity prices[42]).

A report on the Danish policy for RES explains this motivation as follows:[43]

> A host of energy taxes has been applied during the last 25 years. An important one is the 'energy tax' from 1986, which serves the purpose of maintaining a stable price relation between fossil fuels and renewables. In the latter half of the 80s the international price level of oil decreased considerably. In order to protect the domestic market for natural gas and renewable energy, a fossil energy tax was introduced to maintain the fairly high price level of oil.

As a consequence of this motivation, the Danish energy tax scheme is not earmarked. (See, again, the Danish energy agency report which states that: 'A *minor* part of the taxes and fees is used for promotion of renewable energy in a number of ways' - our emphasis.)

Expansion of the capacity for the generation of electricity from RES
Clearly, this motivation does not come as a surprise. Presumably, it is present in *all* EU member states. The typical mechanism consists in putting together fiscal instrument packages which lower the costs of investment in technologies and installations for electricity from RES in order to increase their competitiveness.

As a corollary, differential tax systems can be set up with (partial) reimbursements of tax payments by operators of technologies for electricity from RES to such operators. In other words: while there is a 'general' tax on electricity, producers of electricity from RES are (partially) reimbursed. This system is in place in Austria, Finland, Germany and the Netherlands.

Some Examples of Grants for the Promotion of Electricity from RES

The following paragraphs contain descriptions of grant systems currently in place in some members of the European Union. The description is not complete and the reporting is only for illustrative purpose.

Finland has a policy in place for the promotion of RES in general and electricity from RES in particular which is a combination of energy taxes and subsidies for investments and running costs of plants generating electricity from RES. Originally, the energy tax was related to the carbon content (and levied as a tax per tonne of CO_2) - with a refund for small-scale operations fuelling biomass (for example, peat). The subsidies covered (and still cover) aid to investment projects for new plants for electricity from RES but also support of production (fixed as a maximum amount per kWh of electricity.) The latter (that is, subsidies of running costs) is rather unique for the European

Union. This may be explained by the fact that the Finnish context is somewhat peculiar in the sense that electricity prices are generally low (largely because of the abundance of hydropower) and – after the liberalization of the electricity markets – they declined further, with even conventional thermal condensation power stations fuelling coal losing ground.

As a result, the existing energy tax scheme, which was based on the carbon content of fuels, has been replaced. A new electricity tax law was enacted with electricity as its tax base. Producers of electricity from RES with (wood) biomass, wind power and small-scale hydropower (< 1 MW) are, however reimbursed. In addition, biomass plant and small-scale and hydro plant operators are subject to a lower (than the normal 22 per cent value added tax) (VAT) rate in the sense that they are being returned part of their VAT payments.

These policies are put in place to 'correct' for the fact that, given the low electricity prices, only hydropower and – to some extent – biomass-fuelled CHP technologies are competitive. (Given climatic conditions, there is a steady demand for electricity *and* heat.)

Sweden has a grants scheme for financial support of investments in technologies for electricity from RES, noticeably for wind power, bio-fuelled CHP and small-scale hydropower. The grants are given as a share per installed kWh of electricity but with an overall upper limit not exceeding 25 per cent of the total investment costs of the project.[44]

Spain has a budget line in place for demonstration projects for solar energy (only with experienced companies) and also runs regional programmes, which involve a transfer of grants to the *Communidades autónomas*. These grants are combined with grants from the EU structural funds and – in the case of RES – are spent for the promotion of RES in small and medium-sized enterprises.

Belgium, as a federal state, has programmes at national and regional levels. In addition to accelerated depreciation rates (up to 13.5 per cent) for capital expenditures, companies may benefit from grants for investments in RES technology coming under broad framework programmes (for example, for economic expansion in general) or from dedicated programmes. Examples are programmes run by the regions which allow for grants to investment projects up to 20 per cent of the investment costs, the exact level depending on the programmes and on the location (in terms of the EU's regional and structural funds regulations).[45]

An interesting interim measure in *Austria* consisted in a voluntary agreement between the minister of economic affairs (as minister of energy) and the electricity utilities to pay a bonus to independent operators of wind turbines and photovoltaic systems newly installed between 1996 and the end of 1998. This measure was referred to as the 'three-year agreement'.

Summary Table

After examining some examples of fiscal packages for the promotion of electricity from RES in some member states, we have designed an overall picture for all EU member states. Table 9.7 contains a summary overview of the use of demand management instruments and the structure of the fiscal measures packages for the promotion of electricity from RES in the EU member states.

6 CONCLUSIONS

Table 9.7 summarizes merely the bare and basic characteristics of the policies put in place for the promotion of electricity from RES in the EU member states. Moreover, since the table does not address 'indirect' instruments, such, as energy taxation schemes, it reflects only part of the picture, that is, it does not take into account the potential effects of such taxes on investment decisions with respect to electricity from RES.

Looking at the issue of the optimality of these packages is problematic but the empirical overview of Table 9.7 suggests a few tentative conclusions, which should be examined further.

- The first critical point deals with the question: how much electricity from RES do we need? From a strict resources policy point of view, one would have to inspect the level and the degree of the depletion of fossil fuels and – eventually – arrive at a 100 per cent replacement. If markets for these resources and for RES are functioning properly, price signals will indicate the growing scarcity of fossil fuels and – as a consequence – trigger off switches to electricity from renewables. According to this point of view, no policy for the promotion of electricity from RES is needed. In other words: any attempt to change the 'course of events' would be suboptimal.
- The need for a policy for the promotion of electricity from RES is, however, inspired by other motivations, such as a decrease of the EU's dependency on imports from fossil energy sources from foreign producers, climate protection, economic and social cohesion, the creation of new industries, new employment and new exports. This means that the optimality of the instrumentation would have to be assessed on the basis of these objectives. This is fairly difficult, if not impossible. At best, one can attempt to design the policy packages cost-effectively in terms of their effect on the objectives.
- In this context, we see that demand management instruments may

Table 9.7 Schemes for electricity from RES in the EU member states

Country	RTD	Investments							Demand management			Comments on investments
		Capital cost instruments					Operating cost instruments					
		F	TR	ACCD	FEX	IS	FIT	DT	RO	FIT	GEC	
Austria	Yes	✓										
Belgium	NA	✓	✓				✓		✓	✓		
Denmark	Yes	✓	✓		✓		✓			✓		
Finland	Yes	✓	✓				✓		✓	✓	*✓	
France	Yes	✓		✓			NA	NA				
Germany	Yes	Fixed to a percentage of investment cost					✓	✓		✓		ACCD to a limited extent Each state has its own programme
Greece	NA	✓	✓				✓		✓	✓		
Ireland	No**		✓			✓			*✓			
Italy	No**	✓	✓	✓			✓		✓	✓	*✓	No subsidies for planning cost
Luxembourg	NA	✓		✓			✓	✓				
Netherlands	Yes		✓	✓			✓		✓		✓	'Covenant' on market penetration

Portugal	No**	✓ □			✓ □			
Spain	No**	✓ □			✓ □			
Sweden	Yes	✓ □				✓ □	□	□
UK	Yes	At present no obvious commitment	✓ □		*✓ □	□	□	*✓ □ No climate change levy

Notes:
**No specific RTD project in place but, probably, some RTD funds available through other general RTD programmes; clearly EU programmes available.
*Not yet in place.

ACCD = Accelerated depreciation rates
DT   = Differential (excise or energy) tax rates
F   = Financial aid for investment
FEX = Financial aid for exports outside of the EU
FIT = Feed-in tariff scheme
GEC = Green electricity certificates
IS   = Subsidies on interest payments
NA   = No available data
RO   = Renewable obligation
RTD = Research and technological development
TR   = Tax rebate

Source: Own compilation from the agores webside and selected sources.

probably be more effective than supply-demand instruments. This is not a question of public resources (with demand management instruments typically not using any public funds, whereas supply-side instruments require fiscal measures) but of their structure. Examples such as 'green electricity' certificates – sometimes in combination with renewable obligations – or Ireland's AER illustrate this point. While supply-side instruments are rather unspecific as to the total effect on supply, with demand management instruments one can meet the objectives almost by definition. Moreover, supply-side instruments typically depend on public budgets, which are always limited.

- As an illustration of the aspect of effectiveness, note that the liberalization of electricity markets is an important precondition for the promotion of electricity from renewables. Clearly, liberalization is not identical with the promotion of electricity from RES – it is a necessary but not sufficient precondition. What we see is that member states with a deliberate policy preference for liberalization are also able to put in place a policy for the promotion of electricity from RES with the intention of creating a 'green electricity' industry which is economically viable and competitive both in respect of conventional industries and in respect of the 'green electricity' technologies themselves.[46]

Even so, there is a risk of 'unequal' treatment or, in other words, of preventing all 'green electricity' technologies from enjoying equal conditions of development. This may happen (and it does happen) in the case of 'sequential' maturity of various technologies, where some of them (noticeably wind power in near shore areas) can be used at lower costs (due to higher productivity) than others (for example, photovoltaics).[47] In this case, the promotion policy needs to be fine-tuned in order to correct for this lack of a level playing field. This is typically the case with feed-in tariff schemes, which are mostly pretty complex.

- In contrast, however, one can turn this issue upside down and state that such a level playing field is not necessary if the (politically determined) targets can be met without it. To take a practical example, it is fairly obvious that for a member state such as Denmark, the best options are in wind power and – in second place – in biomass, and certainly not in photovoltaics. A similar argument would be true for the north of Germany, whereas for Saxonia and Austria, the significance of biomass might precede that of wind power. As a result, an effective policy for the promotion of electricity from RES would focus on these technologies and leave out others such as photovoltaics. Equity

arguments probably set incentives to policy makers to nevertheless include them in their packages.
- One important political development, which has been dealt with *passim* in this chapter,[48] concerns the Kyoto Protocol and its implementation within the European Union, in particular after the negotiations of the Bonn Agreement to the Kyoto Protocol, which was adopted by the Ministerial Segment of the Conference of the Parties ('COP6') on 23 July 2001. A first assessment of the outcome of this agreement[49] by the prestigious Wuppertal Institute points out that, among others, 'the use of the mechanisms [that is, the so-called Kyoto mechanisms of emissions trading, joint implementation and clean development mechanism] shall be supplemental to domestic action and domestic action shall thus constitute a significant element'.[50] Moreover, the assessment states 'the European Union successfully ruled out the use of nuclear power both in JI and CDM activities'. Taking together these two outcomes of the Bonn Agreement, one may be tempted to conclude that the European Union, as the major global player in the climate change debate, is bound to continue and reinforce its political moves for the promotion of RES in general and electricity from RES in particular.
- Finally, it is difficult to assess the 'side-effects' of policies for the promotion of electricity from RES. One important side-effect relates to the fact that most technologies for electricity from RES operate decentrally and have to move to the areas where they are best suited. This may have an impact on the economic and social development of rural areas (and, in the context of the EU, especially of remote areas), which is difficult to quantify. And what about 'citizens' "energy parks" where people from a given geographical location put up joint capital funds to operate their own electricity from RES plants? This might not only be considered as a financial operation but also create a new type of entrepreneurship which cannot be fully assessed on the basis of a cost-benefit analysis alone.

NOTES

1. European Commission (2000, p. 19).
2. Directive 96/92/EC of the European Parliament and of the Council of 1996 concerning common rules for the internal market in electricity, *Official Journal* L 027, 30 January 1997, pp. 20 - 29. See also: http://europa.eu.int/comm/energ/en/elec_single_market/index_en.html.
3. Explanatory Memorandum, German Act on Granting Priority to Renewable Energy Sources (Renewable Energy Sources Act).
4. Communication from the Commission: Energy for the Future: Renewable Sources of Energy - White Paper for a Community Strategy and Action Plan, COM (97) 599 final, of 26 November 1997. See: http://europa.eu.int/comm/energy/library/599fi_en.pdf.

5. While the impact of RTD grants may be limited in scope, its significance can have a strategic value. This is illustrated in a paper by Taishi Sugiyama, who argues that, as an alternative to energy efficiency, energy conservation and the promotion of RES, for reasons of a broader context of climate policy making, attention should be given to CO_2 recovery and storage technology (see Sugiyama: 2000). Visitors to the Japanese pavilion at Expo 2000 in Hanover will recall that attention was paid to this technology.
6. Denmark's policy for the promotion of RES resulted from the oil crisis in 1974 and was already started by 1976.
7. This is also the view of J.P. Painuly of UNEP (United Nations Environment Programme), who considers a liberalized market 'to provide a better environment for a healthy growth of renewable energy technologies in the long term'. But as the title of his paper suggests, he identifies several barriers to entry for RES, (see Painuly, 2001 www.elsevier.nl/locate/renene).
8. For a comparative overview of progress made by member states, see Eurelectric (2000).
9. For more information on this issue, see: Communication from the Commission to the Council and the European Parliament: Completing the Internal Energy Market, COM (2001) 125 final, of 13 March 2001, available for download at http://europa.eu.int/comm/energy/en/internal-market/int-market.html and the accompanying package of proposed measures.
10. NN: Analysis of the Electricity Sector Market Liberalization in the European Union Member States pursuant to the Directive 96/92/EC on the Internal Market in Electricity, Study realized for the EU-Japan Centre for Industrial Cooperation: Posted at //www.eujapan.com/europe/marklibr.htm.
11. The text of this proposal can be found on the agores web site managed by the Directorate General for Energy and Transport of the European Commission. See: www.agores.org.
12. See: Communication from the Commission: Energy for the Future: Renewable Energy Sources - White Paper for a Community Strategy and Action Plan (COM (97) 599 final; Council Resolution of 8 June 1998 on renewable sources of energy (*Official Journal* no. C 198, 24.6, 1998) and Resolution of the European Parliament on the Communication from the Commission (A4-0207/98).
13. See Section 1, above.
14. See: NN: UK renewable energy policy information for the agores web site at www.agores.org.
15. In this chapter we restrict ourselves to policy measures for the promotion of electricity from RES taken at the level of the member states of the European Union. This leaves out other public actors, such as regional governments and local governments. Readers with an interest in activities by local governments in the framework of RES and climate change policy may consult the web site of ICLEI, the International Council for Local Environment Initiatives, which has launched a global Cities for Climate Protection campaign covering 8 per cent of all greenhouse gas emissions. See: www.iclei.org/co2/index.htm.
16. In this context, one might argue that this dependency and the 'carbon intensity' of the energy sector can also be reduced through energy efficiency and energy conservation. While this chapter does not explore this potential, readers are referred to a paper by Gill Owen who argues that increasing energy demand in the countries he studied (the UK, Denmark and the Netherlands) counterbalances energy and CO_2 savings (see Owen, 2000). A similar conclusion is reached by Godfrey Boyle who, on the basis of a global model of carbon emissions, concludes that energy efficiency improvements can only 'buy time' [*sic*] in the carbon abatement process and that, ultimately, a shift to carbon-free energy sources is required (see Boyle, 2000).
17. See, in particular, section 4 of this chapter on instruments for demand management.
18. EVA (Austrian energy agency): 'Austria' - Statement for the EnR Renewable Energy Working Group, available on the agores web site at www.agoret.org.
19. Kwant and Wink: (2000). NOVEM is the Dutch organization for energy and environment (Nederlandse Organisatie voor Energie en Milieu).
20. See EC Altener programme: UK renewable energy policy information for the agores web site, at www.agores.org.

The promotion of energy from renewable sources 227

21. In Finland, the first legislative step was taken in 1995, the opening up of electricity markets began in 1997 and liberalization was completed in 1998. In Sweden, the reform process started in 1990 and the electricity markets were fully liberalized on 1 January 1996.
22. Greek energy agency: 'Greece' – statement for the agores web site at www.agores.org. An independent producer is a generator working for the market (that is operating with the intention of selling all electricity output to customers through the grid). An auto-producer mainly generates electricity for own consumption with the option of selling excess power to the grid.
23. Finland also has a green label, but this was introduced by a private organization (Finnish association for nature conservation) for voluntary adoption by electricity generators. (See the agores web site paper on Finland quoted above.) In Germany, at least three different labelling schemes, also developed by private organizations, exist. For more information, see www.uni-muenster.de/energie/ and www.eurosolar.de (links).
24. For an overview on the creation of green electricity labels in the US from a 'business' point of view, see: Wiser (1999).
25. See the UK report prepared for the agores web site, quoted above.
26. The UK report states that the revenues from this payment for 'non-compliance' will be 'recycled' to suppliers with the objective of setting further incentives to actually meet the renewable obligation.
27. See the next subsection.
28. Referred to as 'Competitions for renewable energy based electricity generating capacity'. See: www.irlgov.ie/tec/energy/alternative/competition.htm.
29. See the web site quoted in note 28 above.
30. See Table 9.3 on installed capacity in section 2, above.
31. There is a vast literature of markets for tradable emission certificates. From an economic point of view, authors typically argue that the flexible characteristic of such markets allows for a cost minimization approach towards the solution of the emission problem under investigation. They also argue, however, that the 'rules of the game' or the institutional framework must not block this flexibility. For a selection of recent views and outcomes on this issue, see: Yates and Cronshaw (1999); www.idealibrary.com; Toman et al. (1999). Morthorst (2000); Albrecht and François (2001); Bahn (2001); Voogt et al. (2000). For information about the well-known tests on trade with certificates carried out by Shell and BP, see: www.shell.com and continue with <issues and dilemmas>, <climate change> and the *<Shell Tradable Permit System (STEPS)>* and www.bp.com/key_issues/index.asp and continue with <climate change>, <emissions trading> and <BP's ET system>. For a view on a somewhat different but related issue, see Glasbergen and Groenenberg (2001).
32. See the NOVEM paper quoted above and *Energie Nederland*, Vol. 4, No. 4 (20 March, 2001), No. 5 (10 April 2001) and No. 6 (1 May 2001). *Energie Nederland* is the magazine of EnergieNed, the federation of the Dutch energy industry.
33. For an assessment of the Climate Change Levy, see Pocklington (2001).
34. See the UK policy paper prepared for the agores web site. at www.agores.org.
35. The European Court of Justice decided on this case on 13 March 2001 (C - 379/98).
36. There are various sources of information about the promotion of electricity from RES in Germany, but the documents on the agores web site are definitely outdated. See, however, www.eurosolar.de and www.uni-muenster.de/Energie/.
37. See section 2, above, for further details.
38. See the Austria policy paper prepared for the agores web site quoted above.
39. See: Odgaard (nd), for detailed information, in particular in Annex 3.
40. Quotation from: Portugal energy conservation agency: Portugal – Paper prepared for the agores web site quoted above.
41. For more elaborate investigations of this issue, see Haas et al. (2001). While these authors give some recommendations for an efficient promotion of RES, Lene Hjøllund and Gert Tinggaard Svendsen analyse the actual design of CO_2 taxation in the OECD countries and conclude, 'Strong fiscal incentives drive this policy choice [of CO_2 taxation – jcb and gd] at the expense of environmental concerns because this allows environmental bureaucracies to maximize budgets' (see Hjøllund and Tinggaard Svendsen, 2001).

42. Personal communication from Freiberger Stadtwerke AG, District Heating Division, 31 May 2001.
43. See: Odgaard (nd).
44. See Swedish Energy Agency: 'Renewable energy policies – Sweden', Paper prepared for the agores web site, see above.
45. See NN: 'Belgium' – Paper prepared for the agores web site.
46. This explains why the policy for the promotion of electricity from RES in France is underdeveloped in comparison with the policy concepts of other member states. Some of the French objectives are rather ambitious (for example, 3000 MW of wind power by 2010) but it is not very clear how they will be fulfilled. See the paper produced by ADEME (the French agency for energy and the environment) prepared for the agores website.
47. Although this chapter does not focus on specific technologies for RES – nor on the question why some of them successfully break through while others remain in an infant stage – this is an important issue. In this context, it is obvious that, in many parts of the world, wind power 'has made it' and is a booming industry. See Asmus (2000); Dannemand and Hjuler Jensen (2000); Klinge Jacobsen (2000). For more general views on the role of technology changes in the energy sector, see Aubus (2000); Azar and Dowlatabadi (1999).
48. See, however, various contributions in this volume.
49. The official term is 'Bonn Agreement for the Implementation of the Buenos Aires Plan for Action'.
50. See Ott: (2001). See also: www.wupperinst.org/Seiten/Index.html.

REFERENCES

Albrecht, J. and D. François (2001), 'Voluntary agreements with emissions trading options in climate policy', *European Environment*, Vol. 11, No. 4, pp. 185-96.

Asmus, P. (2000), 'Trends in the wind: lessons from Europe and the U.S. in the development of wind power', *Corporate Environmental Strategy*, Vol. 7, No. 1, pp. 51-61.

Aubus, H. (2000), 'The treatment of technology development in energy models', *International Journal of Global Energy Issues*, Vol. 13, Nos 1/2/3.

Azar, Chr. and H. Dowlatabadi (1999), 'A review of technical change in assessment of climate policy', *Annual Reviews – Energy & the Environment*, Vol. 24, pp. 513-44.

Bahn, O. (2001), 'Combining policy instruments to curb greenhouse gas emissions', *European Environment*, Vol. 11, No. 3, pp. 163-71.

Boyle, G. (2000), 'DREAM-WORLD: a simple model of energy-related carbon emissions in the 20th and 21st centuries', *Energy & Environment*, Vol. 11, No. 6, pp. 573-85.

Dannemand, P. and P. Hjuler Jensen (2000), 'Wind energy today and in the 21st century', *International Journal of Global Energy Issues*, Vol. 13, Nos 1/2/3.

Eurelectric (2000), 'Implementation of the Internal Electricity Market Directive in the EU Member States'–Draft updated 9 March, at: www.Eurelectric.org/Public/Files/IEMDir.pdf.

European Commission (2000), *Green Paper Towards a European Strategy for the Security of Energy Supply*, COM (2000) 769 final of 29 November.

Glasbergen, P. and R. Groenenberg (2001), 'Environmental partnerships in sustainable energy', *European Environment*, Vol. 11, No. 1, pp. 1-13.

Haas, R., N. Wohlgemuth and C. Huber (2001), 'Financial incentives to promote renewable energy systems in European electricity markets: a survey', *International Journal of Global Energy Issues*, Vol. 15, Nos 1/2.

Hjøllund, L. and G. Tinggaard Svendsen (2001), 'Why green taxation?', *Energy &*

Environment, Vol. 12, No. 1, pp. 29-38.
Klinge Jacobsen, H. (2000), 'Energy technology and foreign trade: the case of Denmark', *Energy & Environment*, Vol. 11, No. 1, pp. 93-107.
Kwant, K.W. and S. Wink (2000), 'Renewable energy in the Netherlands', NOVEM, April, to be found on the agores web site at www.agores.org.
Morthorst, P.E. (2000), 'Scenarios for the use of GHG-reduction instruments – How can policy instruments as carbon emission trading and tradable green certificates be used simultaneously to reach a common GHG-reduction target?', *Energy & Environment*, Vol. 11, No. 4, pp. 423-38.
Odgaard, O. (Danish energy agency) (nd), 'Renewable energy in Denmark', paper prepared for the agores web site at www.agores.com.
Ott, H.E. (2001), 'The Bonn Agreement to the Kyoto Protocol - paving the way for ratification', *International Environmental Agreements: Politics, Law and Economics*, Vol. 1, No. 4.
Owen, G. (2000), 'Energy efficiency and energy conservation: policies, programmes and their effectiveness', *Energy & Environment*, Vol. 11, No. 5, pp. 553-64.
Painuly, J.P. (2001), 'Barriers to renewable energy penetration: a framework for analysis', *Renewable Energy*, Vol. 24, pp. 73-89.
Pocklington, D. (2001), 'The UK climate change levy – innovative but flawed', *European Environmental Law Review*, Vol. 10, No. 7, July, pp. 220-27.
Sugiyama, T. (2000), 'Strategic value of carbon recovery and storage technology: political and administrative dimension', *Energy & Environment*, Vol. 11, No. 6, pp. 647-54.
Toman, M.A., R.D. Morgenstern and J. Anderson (1999), 'The economics of "when", flexibility in the design of greenhouse gas abatement policies', *Annual Reviews - Energy & and the Environment*, Vol. 24, pp. 431-60.
Voogt, M., M.G. Boots, G.J. Schaeffer and J.W. Martens (2000), 'Renewable electricity in a liberalised market. The concept of green certificates', *Energy & Environment*, Vol. 11, No. 1, pp. 65-79.
Wiser, R.H. (1999), 'Greening the electricity industry: the dynamic logic of the *Green-e Certification Program*', Corporate Environmental Strategy, Vol. 6, No. 1, Winter, pp. 24-36.
Yates, A.Y. and M. Cronshaw (1999), 'Pollution permit markets with intertemporal trading and asymmetric information', *Journal of Environmental Economics and Management*, Vol. 42, pp. 104-111.

10. Domestic capacity, regional institution and global negotiations: lessons from the Netherlands-EU Kyoto Protocol negotiation

Norichika Kanie

1 INTRODUCTION

What kind of potential does a regional institution have in relation to the global climate change regime-building process? Studies in the field of international relations have shown that the countries that muster domestic support more effectively than other countries in translating international agendas into domestic policy agendas and/or in implementing international agreements may enjoy a bonus in multilateral diplomacy. This bonus is identified as 'leadership through unilateral action' in leadership theory in the field of multilateral negotiation (Young 1991; Underdal 1994). However, this type of unilateral leadership occurs less frequently than expected in actual multilateral negotiations, partly because those countries that have a potential to exert this type of leadership in terms of domestic policy are often a middle power, which is not necessarily influential globally. If this is the case, then can a regional institution influence the degree of potential of a middle power in the actual leadership of a global negotiation process? This chapter tries to tackle this question by drawing lessons from the descriptive case analysis on the Kyoto Protocol negotiation process of the Netherlands-EU.

2 BACKGROUND

International Regime Building and the Potential of Middle-power Diplomacy

International regimes may be recognized as 'intervening variables' that stand between fundamental characteristics of world politics such as the international distribution of power on the one hand and the behaviour of states and non-state

Domestic capacity, regional institution and global negotiation 231

actors on the other.¹ Thus the relative power relations between states are not necessarily fully reflected in the behaviour of the member states within a particular regime. Consequently, some of the so-called 'middle-power' countries, such as the Netherlands, have been able to exert disproportionate influence. The role of 'middle-power' countries in international regime-building processes has gradually come under greater scrutiny.² Studies of small countries, middle power and 'middlepowermanship' have also identified the bridge-building role of middle-power countries such as Australia, Canada, Denmark, Norway and the Netherlands in situations of conflict. Robert Cox has argued that this tendency has been encouraged by the transition from the post-war reconstruction regime to the current form of multilateralism, which increasingly depends upon the principle of power dispersion and social equity (Cox 1989).

According to leadership theory, the mode of leadership in international multilateral negotiation may be categorized as shown in Figure 10.1 (Underdal 1994, p. 185). In a recent study these leadership categories have been further elaborated, but the basic typology is similar. The new typology is indicated in parentheses in Figure 10.1 (Grubb and Gupta 2000).

In this model, the kinds of leadership which small or middle powers may

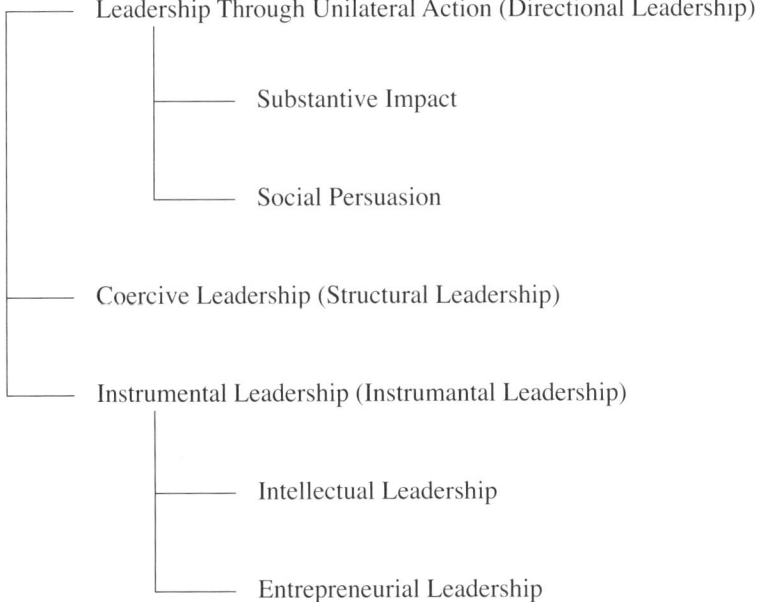

Figure 10.1   Three modes of leadership in international multilateral negotiation

demonstrate are: (i) leadership through unilateral action accompanied by social persuasion; (ii) intellectual leadership as instrumental leadership; and (iii) entrepreneurial leadership as instrumental leadership. In particular, leadership based on unilateral policy action ought to be increasingly important in tackling global issues, given the importance of bottom-up multilateralism in tackling such issues.

While these arguments are convincing, and however great the potential to exert such types of leadership may be, case study evidence in which middle powers in fact exert such leadership have been rare. A case study of the negotiations on the New International Economic Order in the early 1970s has shown that, while Jan Pronk, then Minister for Development Cooperation in the Netherlands, and Thorvald Stoltenberg, then Norway's Under Secretary of State for Development Cooperation, led a coalition of 'like-minded countries' and tried to lead the negotiation by bridging the gap between the north and the south, they failed actually to take a leadership role in November 1975 (Dolman 1979). In order for a middle-power country, or a group of middle-power countries, to exercise leadership in a global negotiation process some kind of 'additional condition' may be necessary. The case of the Netherlands in the Kyoto Protocol negotiation has shown that the European Union (EU) as a regional institution has the potential to become such an 'additional condition'.

This chapter examines the potential of a regional institution in the climate change regime-building process in terms of the interface between domestic politics and international negotiation. The focus of the case study is on the negotiation process on quantified emission limitation and reduction commitments (QELRCs) of the Kyoto Protocol in the Netherlands, the EU and the Kyoto Conference in 1997. Thus the chapter focuses on the period between 1995, when the QELRC negotiation began to make substantive progress under the Berlin Mandate, and 1997, the year of the Kyoto Protocol adoption. The chapter then tries to draw lessons for the possible role of regional institution in the future climate change regime. The study is done through descriptive case analysis on the basis of interviews with stakeholders such as negotiators, civil servants, industrialists, environmentalists and researchers mainly from the Netherlands. Interviews were conducted at various times between 1997 and 1998 with follow-up interviews conducted during COP6 in The Hague in November 2000.

3 THE KYOTO PROTOCOL NEGOTIATION PROCESS

As the European Commission (EC) is a party to the UNFCCC, most of the negotiations are undertaken by the EU Troika, composed of the member states

Domestic capacity, regional institution and global negotiation 233

which are the former, current and next presidents of the Council of Ministers. A middle-power country such as the Netherlands, negotiating as part of the EU, obtains greater negotiating power in global multilateral diplomacy than it would alone. Comparing the situation to the 'like-minded coalition' with Sweden, Norway and so on, built in the 1970s, the Dutch chief negotiator for the Conference of the Parties (COP) of the United Nations Framework Convention on Climate Change (UNFCCC) said:

> 'Like-minded' coalition was a looser arrangement in the 1970s and 1980s than the EU, and certain member states of the EU preferred to work with non-EU member states rather than with the members of the EU. But, that has changed dramatically. If the Netherlands speaks to a few other countries, people may listen politely. But, if the EU speaks, people listen to it seriously. It has more weight.[3]

Although the EU took certain initiatives throughout the UNFCCC process, the negotiation process itself did not make substantive progress until COP2. Some had even doubted whether the Protocol would be adopted by COP3 at Kyoto. For the EU the Protocol negotiation was seen as a great opportunity to take a leadership role in the UNFCCC process (Huber 1997).

Before the Dutch Presidency of the European Council

Since the EC became involved in the multilateral climate change negotiations in the early 1990s, the EU member states and the Commission have realized that burden sharing within the EU was necessary. It was not until the first half of 1997, however, that the EU succeeded in reaching its internal burden-sharing agreement for greenhouse gas (GHG) reduction. The EU burden-sharing arrangements, particularly in preparation for the COP3 Protocol negotiation, were tabled by the Italian presidency during the first half of 1996. The discussion intensified during the second half of 1996 under the Irish presidency. An EU climate change policy workshop 'Towards a European Consensus' was held on 2-3 September 1996 in Dublin. The workshop was attended by negotiators from member states participating in the EU Ad Hoc Group on Climate, officials from the Commission's DGXI (Environment Directorate-General) and Dutch researchers on climate change policy including the Utrecht university research group and professional facilitators (Ringius 1997).

At the workshop DGXI proposed an overall CO_2 reduction target of 10 per cent by 2005 and a mechanism for burden sharing among the member states. Under the proposal, nine member states would reduce their emissions, two would be allowed to increase, and the rest would stabilize their emissions. However, most of the member states objected to the proposal and argued that an emission reduction target of around 5 per cent by 2005 would be more realistic. Although these proposals were rather similar to those agreed in March 1997, the Commission's proposal was thought to lack substantive content. Describing the atmosphere at the Dublin workshop, Ringius wrote:

> [T]he DGXI proposal set a too ambitious common target, appeared to be ill prepared, and was not persuasively prepared in Dublin. For these reasons, especially the lack of substantive content, the proposal failed to build consensus among the member states. Explained an astute Irish negotiator: 'The proposal did not have the necessary intellectual rigour'(Ringius 1997, p. 19).

At the European Environment Council meeting on 15 October 1996 in Luxembourg, most of the time was spent on the three main agendas of the Irish presidency, and agreement on the EU burden sharing was not reached.[4] The presidency called on delegations to work 'actively' on the burden-sharing arrangements 'in order to reach conclusions at the Environment Council in December'.[5] This in a sense reaffirmed the statement of the Council made under the Italian presidency in June 1996, intending a firm commitment of the EU to reach a political conclusion on the strategy for the Protocol negotiation in order to take leadership and give 'substantive guidance for the Protocol negotiations'.[6] The Ministerial Declaration of COP2 had already called for proposals from the parties to be submitted by the fifth meeting of the Ad Hoc Group on the Berlin Mandate (AGBM5) in December to make substantive progress in the negotiation process. For the EU this represented a great opportunity to take a leadership role in the global setting.

However, the results of the Environment Council held in Brussels on 9-10 December 1996 were disappointing. Climate change was among the two main agendas during the discussion held in the evening sessions. However, the Irish presidency's proposal for burden sharing, which included a 5-10 per cent reduction of GHGs by 2005 and a 10-15 per cent reduction by 2010 (relative to 1990), failed to achieve agreement because of disagreements on the method for calculating CO_2 emissions and over the burden-sharing pattern among member states.[7] Therefore, the EU was unable to present a proposal to AGBM5, which had been held over until 13 December, and the issue was deferred for further discussion under the Dutch presidency. Meanwhile, the political pressure to find a solution to the burden-sharing issue was increasing to enable the EU to take a lead in the Protocol negotiation process.

Domestic Political Agreement for the Dutch Presidency

By the time the Netherlands took over the presidency of the European Council during the first half of 1997, it was clear in the Netherlands that, despite its ambitious target and policy measures on climate change, domestic CO_2 emissions had been increasing due mainly to low energy prices and unexpected economic growth. CO_2 emissions in the Netherlands had increased by 7 per cent between 1990 and 1996.[8] This increase led to widespread scepticism regarding the prospects for climate change policy. Yet this scepticism itself did not make the Netherlands reluctant to tackle the burden-sharing issue. There had, of course, been a substantive domestic debate concerning an ambitious CO_2 reduction target. For example, when the Temporary Second Chamber Commission on Climate Change concluded in September 1996 that 'the CO_2 emission reduction in the Netherlands in 2020 must be around 30-40 per cent', the employers' organization, VNO-NCW, strongly objected to it as 'absolutely unfeasible'.[9] The Director of Environmental Affairs at VNO-NCW, who had been lobbying 'at any level' of the policy-making process, however, still felt that the wind was favourable for the environmentalists and said, 'Parliament and society were very much in favour of those environmental targets. We felt that pressure'.[10] Although the 'environment boom' since 1989 when the environment was one of the main issues of the election had gone, the general political atmosphere was still in favour of stricter environmental measures. One example of Dutch interest in environmental issues is that almost 11 per cent of the population were members to either one of the three major environmental non-governmental organizations (ENGOs), WWF (World Wide Fund for Nature) Netherlands, Greenpeace Netherlands and Milieudefensie (Friends of the Earth Netherlands).

The domestic atmosphere, together with the inherent 'international approach' of the Dutch climate change policy, made it politically possible for the Minister for Housing, Spatial Planning and the Environment (VROM), Ms de Boer, to propose an ambitious target for the forthcoming EU negotiations. At the highest level of the decision-making process, many members of the cabinet in the coalition government agreed that it was not possible to realize the ambitious EU-wide target. At the same time, however, no one strongly objected to presenting such a presidency proposal to the Council of Environment Ministers. Ms de Boer said, 'nobody believed it could be realized. Everybody thought, "no, impossible", but said, "if you want to go for it, go."'[11]

Finally, climate change policy was given the highest priority among the five important agendas in the environment area for the Dutch EU presidency. An EU presidency memorandum notes:

> Climate policy plays an important role nationally in the Netherlands as has been regularly emphasised by the Lower House of Parliament. It is clear that we will have to have European and internationally coordinated instruments for a further reduction in CO_2 emissions at national level. Internationally as well as we are being confronted with a major turning point. During the third Conference of Parties to the World Climate Convention (December 1997, Kyoto, Japan) agreements will have

to be made on post-2000 emission reduction. The preparations for this at the European level will have to take place under the Dutch Presidency, which will entail a major effort on the part of the EU president. ... It is clear that the role of the European Union and the influence of the Union on international environmental policy will depend on whether the EU succeeds in deciding on a common position in March 1997.[12]

Dutch Contributions and the EU Burden-sharing Agreement

It was clear from the failure of the Irish presidency that the key to an arrangement on burden sharing was a convincing approach backed up by scientific rigour. The EU Ad Hoc Group on Climate organized a workshop at Zeist, the Netherlands, on 16-17 January 1997. The Dutch presidency presented its strategy of a combination of an EU-wide abatement target and a proposal for an internal burden-sharing mechanism, called the 'Triptych approach'. Dr Bert Metz, then the Dutch chief negotiator on climate change, thought that a sectoral approach would work, but would require a sound scientific basis to be realized. In consequence, a research group at the Utrecht University, led by Professor Kornelis Blok, formulated such an approach. The Utrecht University research group had a good reputation in the field of energy research at the EU level, having produced a report on EU-wide policies and measures in collaboration with WWF Netherlands to be submitted to the EU Ad Hoc Group on Climate. Professor Blok was also an energy policy consultant to the Dutch government. The research group had undertaken several research projects under the governmental research programme NOP-MLK (Nationaal Onderzoek Programma Mondiale Luchtverontreiniging en Klimaatverandering), which managed a budget of approximately 13 billion Dutch guilders over ten years. The group's knowledge bases related to climate change policy were enhanced through the programme-funded projects, and they knew where to go in order to obtain the necessary data by the time Dr Metz contacted them.

The main characteristic of the Triptych approach was its sectoral approach; each member state was divided into three sectors, namely the light domestic sector, the energy-intensive, export sector, and the electricity generation sector. An emission reduction target in each country was set by adding up the potential for emission reductions in each sector. The EU-wide target was set by totalling these figures. This approach was based on the lessons learned from the earlier failure to agree on burden sharing, where agreement had been reached on the total figure for members, but it had proved impossible to allocate reductions among the member states. By using the sectoral approach, the Triptych approach successfully penetrated the 'black boxes' of member states (Gupta et al. 1998; Phylipsen, Bode, et al. 1997).

The Triptych approach gave rise to considerable discussion. Some member

Domestic capacity, regional institution and global negotiation 237

states called for adjustments, and some poorer member states were not satisfied with their quota. Yet the overall response to the mechanical formula of the approach was basically positive; Danish government officials in particular fully supported it. Officials from member states said that the 'approach presented a new, promising angle on the whole question of burden sharing', and 'accelerated the process of differentiation' (Ringius 1997, pp. 25-26).

On 27 January 1997 the Dutch presidency sent a proposal based on the Triptych approach to the member states. The burden-sharing formula was modified and revised based on the discussion held at the workshop, and the emission increases in the poorer member states were supplemented by additional reductions by Finland, Italy and the Netherlands. In an accompanying letter, Minister de Boer pointed out the importance of the member states' contributions to establishing leadership at the forthcoming global negotiation, and asked for further compromise:

> A challenging period is ahead of us and I firmly believe the world is looking to the European Union to play an ambitious leading role in the international negotiating process. That role can never be more than the sum total of the commitments individual Member States are willing to make.[13]

The EU Ad Hoc Group meeting on 17-18 February discussed the proposals, but no conclusion could be reached. The discussion made clear that it would be very difficult to achieve EU-wide 15 per cent reduction and that a target of 10 per cent would be more realistic. The German Minister for the Environment and the European Commission were concerned that setting an EU target below 10 per cent would mean that the EU's position in the global negotiations was not politically credible (Ringius 1997). International ENGOs such as WWF and Friends of the Earth also expressed their anxiety about the situation, but thought that the 15 per cent target was in fact impossible.

As the member states' reduction targets had been reduced in the course of discussion, it was clear that the total reduction for GHG emissions would be approximately 10 per cent relative to 1990, and not 15 per cent as suggested in the proposal. This figure was thought to undermine the EU's ability to take a lead in the global negotiation. Subsequently, the Dutch proposal at the EU Council of Environment Ministers on 3-4 March was revised to an EU-wide target of 12 per cent reduction. The Council seemed unable to agree on the more ambitious target until the Danish Environment Minister suggested keeping the EU-wide target of 15 per cent, but accepting a 10 per cent internal burden sharing for the time being. It was suggested that if the Kyoto agreement were to exceed a reduction of 10 per cent, then the remaining 5 per cent would have been renegotiated after the Kyoto Conference. Thus, the member states 'agreed that 10 per cent would be enough until Kyoto', and so avoided acrimonious discussion about the division of the remaining 5 per cent.[14] (See Table 10.1.)

Table 10.1 Greenhouse gas reduction target in 2010 relative to 1990

	Triptych approach	Presidency proposal	February informal pledges	Council decision
Belgium	-15	-15	-10	-10
Denmark	-15	-25	-15	-25
Germany	-21	-30	-15	-25
Greece	2	5	10?	30
Spain	9	15	15	17
France	-4	-5	-5	0
Ireland	-5	5	10?	15
Italy	-7	-10	-5	-7
Luxembourg	-20	-40	-30	-30
Netherlands	-9	-10	-10	-10
Austria	-5	-25	-15	-25
Portugal	20	25	25?	40
Finland	-8	-10	-5?	0
Sweden	20	5	5	5
UK	-20	-20	-10	-10
Total	-12	-15	-11	-15
Gases	CO_2	CO_2, CH_4, N_2O		CO_2, CH_4, N_2O

Sources: Gupta et al. (1998), Ringius (1997)

The EU proposal of 15 per cent emissions reduction in three GHGs by 2010 relative to 1990 had a strong impact on the AGBM6 negotiation held at Bonn.[15] The proposal facilitated the concrete discussion, and thus led the negotiation as the EU presidency had intended. By proposing the first QELRCs from the Annex I parties the EU, led by the Dutch presidency, contributed to facilitating the global negotiation process.[16]

Dutch Response and Institutional Background

The Council decision of March 1997 required the Netherlands to achieve a 10 per cent reduction by 2010. This was a very tough target for the Netherlands, a country which had introduced climate change policy as early as 1990 and yet nevertheless faced difficulties in reducing the GHGs.

Policy debate on the implementation of the Council decision and the strategy for the Kyoto Protocol negotiations in the Netherlands started in the summer of 1997. As in the case of the National Environmental Policy Plan (NEPP), the debate was held among the four main ministries involved. They were the ministries of VROM, Economic Affairs (EZ), Agriculture, Nature Management and Fisheries (LNV) and Transport, Public Works and Water Management (V&W). The ministries of Development Cooperation and Finance were also invited on occasion. The meeting was chaired by a member of the climate change department of the Ministry of VROM. In addition, Task Force Protocol Negotiations (TPO) were formulated between the ministries of VROM and Foreign Affairs (BuZa), which met in principle every week in order to have more in-depth discussions in preparation for the interdepartmental meetings. The final decision with respect to the final position at international negotiations was taken in the committee chaired by the Ministry of BuZa. It was called the Coordination Committee on International Environmental Issues, and one of the subcommittees was the Subcommittee for Climate Change, which met every two or three weeks and was attended by representatives from all the ministries involved. In short, the issues were discussed and common grounds were found at the committee chaired by the Ministry of VROM; the committee chaired by the Ministry of BuZa finalized the Dutch position for international negotiations.[17]

Domestic coordination with target groups was mainly undertaken by the Ministry of VROM, as is the case of environmental policy in general under NEPP, but other ministries in charge of specific issues were also charged with negotiations with target groups, the groups responsible for GHG emission, in their own areas of responsibility. As for the ENGOs, which do not fit into the framework of target groups, formal talks and exchanges of information were institutionalized as 'DOMILO', at which they met every two months with a director-level civil servant at the Ministry of VROM. There was also 'DOMILO-PLUS', at which they met four times a year with the Minister for VROM. These institutionalized meetings were, however, designed for general lobbying on the environmental policy, and more substantive and effective exchange of information and views regarding climate change was undertaken through informal contacts that were held almost daily. The main channel for this was the international policy department of the Directorate-General for climate change in the Ministry of VROM. With regard to contact and exchange of information, everything went smoothly and satisfactorily for both the ministry and the ENGOs, as Dr Metz saw ENGOs as 'allies' rather than enemies.[18] Furthermore, those who were interested in the issue, including civil servants, NGO representatives and researchers gathered at a policy study group meeting called *bezinngsgroep*, to exchange views and information.

They attended the meeting not as representatives of each organization, but by their personal capacity.[19]

The government sent a letter, generally known as the 'Kyoto letter', to Parliament on 17 October 1997 concerning the Netherlands' negotiating target in Kyoto. Following the projections made in the Fourth National Environmental Outlook (NEO4) presented by RIVM (1987) that the national target of −3 per cent for CO_2 reduction in 2000 would not be met under current policies and that additional policies would be necessary in order to meet the target, the Kyoto letter also concluded that it would be difficult to meet the objective of 10 per cent reduction for CO_2, CH_4 and N_2O in 2010. It also showed two policy options as examples, both of which emphasized the need for an internationally coordinated policy at the EU level.[20]

In general, the industrial sector opposes a strict climate change policy, as it is afraid of its negative impact on international competitiveness. This impediment prevents many countries from tackling climate change. The Netherlands is not exceptional in this sense, but the 'internalization' approach of the NEPP, which internalize the national-level target into policies on target groups, has helped to prevent the industrial sector from undermining an ambitious position in multilateral negotiations.

The industrial sector has introduced long-term agreements (LTAs) on energy efficiency since 1992 and striven for improvement of energy efficiency by signing LTAs with the government. This approach has given the industry a free hand in improving its own energy efficiency in the most cost-effective manner, while the government promised not to introduce additional policy measures as long as it improves energy efficiency in line with the LTA. In addition, the introduction of the Benchmarking Covenant was under discussion for the energy-intensive industry in 1997. This covenant binds the government not to introduce additional policies on condition that the Dutch industry reaches the 'top level in the world', at least in three years, in energy efficiency relative to comparable industry sectors abroad. The government can introduce additional policies only if industry fails to reach the target.[21] This, on the one hand, is a strategy of industry which needs to maintain international competitiveness and to prevent the unilateral introduction of additional policy measures suggested in the Kyoto letter. On the other hand, these agreements have made it clear what policy measures are to be applied to the industry, and prevent it from worrying about the unknown. Mr Chris Dutilh, the coordinator of environmental policy at Unilever and a member of the environment board of VNO-NCW, said:

> [R]egarding the Kyoto target, industry accepted it because we thought it was a political debate. The government manages their business, and we sort out something with our government so that they know what they can expect from us. If they want

Domestic capacity, regional institution and global negotiation 241

to go further, it is their responsibility. It is reasonable to think that if Dutch society wants to do more, because of political motivation, then that should be paid for by Dutch society. The Benchmarking Covenant will mark the bottom-line of the industry. This is what the industry contributes to the government's plan. If they plan more, then society has to pay'.[22]

Interdepartmental deliberations on the Dutch strategy were held twice at the Ministry of BuZa before the Kyoto Conference. Some points were left unsolved, and the final decision was made at the cabinet meeting attended by the Prime Minister, which was held in the last week of November.

Kyoto Conference

After AGBM7 in July–August and AGBM8 in October it became apparent that the EU proposal contained the most ambitious target for the reduction of GHGs. The G77 and China expressed support for the EU proposal at AGBM8. Accordingly, a group of influential ENGOs, Climate Action Network (CAN) had also expressed support for the EU proposal by the time the Kyoto Conference was convened.[23]

The EU Troika negotiated on behalf of the EU at Kyoto. The Troika consisted of the Netherlands, Luxembourg and UK. In practice, however, the Netherlands and the UK led the negotiation because of the small size of the Luxembourg delegation.[24] The chief negotiator for the Netherlands before the arrival of the minister was Dr Bert Metz, the main architect of the Triptych approach and the main negotiator for the Netherlands since COP1. At COP1 he was nominated as an EU negotiator together with British and German delegates by the French presidency.[25] He was well known, and his reputation was one of the best among the EU negotiators. Although the Dutch delegation to COP3 did not include an NGO representative as had been the case at the Earth Summit at Rio, contacts with NGOs were held frequently throughout COP3, in addition to a formal gathering held every morning.[26]

As the EU had the most ambitious target among the Annex I parties at Kyoto, and subsequently won the support of many developing countries and ENGOs, the Dutch governmental and non-governmental network on climate change could work to facilitate the negotiation towards a more ambitious output. For example, a representative of the Netherlands' WWF supported the EU proposal, and tried to convince American and Japanese delegations by citing the intellectual rigour and existing policies and measures for realizing the 15 per cent target of the EU proposal. Likewise, the role of a representative of Friends of the Earth Netherlands at this stage was to convince Norwegian and Australian, rather than Dutch or EU representatives.[27] In fact, after the DOMILO-PLUS meeting before the Kyoto Conference, Mr Teo Wams, the leader of Friends of the Earth Netherlands, took a 'big step' and decided that his organization would basically support the EU position at Kyoto in order to make the Kyoto Conference successful.[28]

It is difficult to measure the impact of these patterns of behaviour and persuasion on the outcome of the negotiations. However, it is clear that both governmental and non-governmental organizations generally supported the EU proposal at this stage, which gave the Dutch government another channel for influencing intergovernmental negotiations, using the network of ENGOs, a branch organization of which has been actively involved in Dutch climate change policy making.

In Kyoto, 38 industrialized countries agreed to reduce six types of greenhouse gases (carbon dioxide, methane, nitrous oxide, HFCs, PFCs and SF_6) by 5 per cent relative to 1990 levels between 2008 and 2012. The shares were 6 per cent for Japan, 7 per cent for the US and 8 per cent for the EU. As Sjostedt argued, it is possible that 'EU diplomacy has contributed to producing the inadequate but still encouraging results obtained in Kyoto' (Sjostedt 1998, p. 227).

THE POTENTIAL OF A REGIONAL INSTITUTION: LESSONS FROM THE CASE OF THE NETHERLANDS-EU KYOTO PROTOCOL NEGOTIATION PROCESS

The policy approach and intellectual rigour of the Netherlands were obviously important factors enabling the EU to adopt ambitious proposals for the Kyoto Protocol negotiation, which then led to the outcome of the Kyoto Conference. As one of the negotiators expressed at one point, the biggest difference between the Irish and the Dutch proposals concerning the EU internal burden-sharing negotiation was the intellectual rigour behind the two proposals. In addition, by proposing the strictest target among the developed countries in the global negotiation, the EU won the support of the ENGOs and G77 and China by the time of the Kyoto Conference. The power of international public opinion contributed to making the American and Japanese GHG reduction targets higher, although the EU also had to make substantive concessions to them (Sjostedt 1998). In any case, the process started with the policy-related intellectual rigour of the Netherlands, which was developed and utilized by the policy-oriented network. In other words, the Dutch research output could be utilized, although different from the original proposal, in the form of the Kyoto Protocol because of the negotiating power device of the EU.

Regarding domestic capacity, the Dutch government successfully eliminated a potential negative influence on its negotiating posture by signing an environmental covenant with the domestic industrial sector. Since industry, most of which is often regarded as being opposed to stricter climate change policies, had negotiated an agreement with the government in advance, which made clear the burden it would have to bear, it had no strong incentive to object to the development of an international agreement at the time of the Kyoto Conference.

Another important domestic capacity is the Netherlands' institutional basis. The institutional basis that made such interaction between domestic, regional and global negotiation possible was the Dutch policy network that had gradually been developed since the beginning of the 1990s. ENGOs such as WWF, Greenpeace and Friends of the Earth which have a global network also have branches in the Netherlands, and people in these organizations form part of the Dutch policy network. Moreover, the fact that the Netherlands was a part of the EU Troika during the Kyoto Protocol negotiation process magnified the influence of the policy network, which was instrumental in linking 'the local and the global' (Princen and Finger 1994).

This case shows the potential of a regional institution as an important device when a middle-power country wants to exert a greater influence in global negotiations. This is especially true in a case such as the UNFCCC in which a regional organization is also entitled to be a party. In such a case a middle power can exert its influence in a regional institution, where its relative influence is greater than in the case of global negotiation, and then negotiate on behalf of the regional institution in the global negotiations. The Dutch case is a good example of this. The Netherlands influenced the EU burden-sharing negotiation process by developing the Triptych approach, which was a product of the intellectual rigour developed though the domestic policy of the Netherlands, and subsequently attended the global negotiation as a part of the EU Troika and negotiated on behalf of the EU. The timing and the power-diffusive nature of the EU institutional setting were obviously important factors for the intellectual capacity of a middle power to actually exert its influence on the regional negotiation process.

In short, the framework of a regional organization may provide additional power to a middle power, which allows it to obtain the necessary conditions of adequate diplomatic skill level and intellectual and institutional capacity to take an international initiative, as Richard Higgott has argued (Higgott 1997, pp. 36-7). There are three factors that made the Dutch success possible.

First, it is clear that intellectual rigour as well as the instrumental (diplomatic) capacity at the national (domestic) level made it possible for the Netherlands to make use of the EU as a device to extend its potential. Without this, a regional organization cannot be used as a device to enhance its diplomatic influence.

Regarding implementation capacity, the Dutch government successfully attributed the failure to achieve its domestic CO_2 reduction target to the failure to reach agreement at the international level. After taking unilateral measures in domestic climate change policy, further domestic policy developments became conditioned upon the development of international agreement. Also, the difficulty of achieving the target facilitated the climate change-related research, and thus enlarged the intellectual output. Therefore, the lack of unilateral implementation capacity was successfully used to link the domestic policies with the international negotiation process.

Second, the coherency of the EU as a coalition within the realm of QELRC negotiation made it possible to be a recognizably important power bloc in the global negotiation process. In spite of the EU's internal incoherency in other issues negotiated in the period leading up to Kyoto Conference, it could manage to create a common EU position on QELRCs under the Dutch presidency.[29] Of course, its economic scale was another important factor that made the EU one of the most important parties in the final stage of the negotiation. However, without coherency (in this case common QELRCs) the EU could not speak with one voice externally. Therefore, I argue that it was the coherency of the EU that made it one of the most important players.

The third important factor is the institutional design of the EU. Because the EU presidency can exercise a fairly large degree of political power both internally and externally, and because the EU Troika can negotiate externally on behalf of the EU, the climate change political/diplomatic potential of the Netherlands was able to be transformed into an EU negotiation. Therefore, the institution design of the regional institution is an important factor for a middle power in order to effectively utilize a regional institution as a negotiating power device.

It may be possible that the role and importance of regional (economic) institutions in the climate change negotiations will grow in the future, given that economic (market) factors are attracting more attention in the climate change regime. This, in return, means that once internal coherency is attained, the framework of a regional institution may provide additional negotiating power for a middle-power country that wants to exert influence in global negotiation but in fact is not able to realize it. A middle power can first exert its influence at a regional level, and then it may be able to negotiate with other countries on behalf of the regional organization at the global level. Taking into account that the interests of developing countries are now being dispersed in climate change negotiations, it can even be a useful option for a developing country to make use of the power of a regional organization. Of course, more environmental consideration and more institutional development are necessary for many regional institutions other than the EU. Nevertheless, regional institutions have a great potential in the future of the climate change regime-building process.

NOTES

1. International regimes are usually defined as 'principles, norms, rules, and decision-making procedures around which actor expectations converge in a given issue-area (Krasner 1983, p. 1)'. Of course, not every scholar completely agrees with this definition. For example, Oran Young (1983) sees international regimes primarily as institutions.
2. See, for example, Cooper et. al. (1993); Cooper (ed.) (1997).
3. Interview with Dr Bert Metz, 20 March 1998. For the 'like-minded' countries, see Dolman (1979), pp. 1–13.
4. Priority of the Irish presidency for the Environment Council was given to (i) the fifth action plan for sustainable development; (ii) the fuel programme for vehicles; and (iii) the strategy for water management.
5. *Europe Daily Bulletin*, No. 6835, 18 October 1996.
6. Press Release: Luxembourg (25 June 1996), No. 8518/96 (Press 188). In the EU the lack of climate change strategy had already been criticized, see *Europe Daily Bulletin*, No. 6821, 28 September 1996.
7. France, Sweden, Spain, Greece, Italy and Portugal recommended a reduction per inhabitant, while the other countries preferred a reduction in absolute value. *Europe Daily Bulletin*, No. 6871, 11 December 1996.
8. Change, July–August 1998.
9. *Het Financieele Dagblad*, 12 September 1996.
10. Interview with Mr Wiel Klerken, Director of Environmental Affairs, VNO-NCW, The Hague, 1 December 1998.
11. Interview with Ms Margaretha de Boer, The Hague, 18 November 1998.
12. VROM DGM/IMZ/96074893.
13. Letter 27 January 1997, cited in Ringius (1997, p. 27).
14. *Europe*, No. 6927, 5 March 1997. See also Ringius (1997).
15. The three conditions for the EU proposal were: (i) same level of reduction by other developed countries; (ii) internationally coordinated policies and measures to be decided at Kyoto; and (iii) joint implementation between Annex I countries.
16. *Earth Negotiations Bulletin*, **12** (45), 10 March 1997.
17. Interviews with Mr Henk Merkus, Ministry of VROM, 13 May 1998 and Mr Andreas te Boekhorst, Ministry of BuZa, 24 February 1998.
18. Interview with Dr Bert Metz, 20 March 1998.
19. This kind of cooperation between governmental and NGOs is often referred to in connection with Dutch political culture. See the following, for example: Visser and Hemerijck (1997); Kickert et al. (eds) (1997); Kalders (1998).
20. Tweede Kamer, Vergaderjaar 1997–1998 (Second Chamber, Session 1997–1998), 24785, No. 4.
21. Ministry of Housing, Spatial Planning and the Environment, Ministry of Economic Affairs, Ministry of Agriculture, Nature Management and Fisheries and Ministry of Transport, Public Works and Water Management (1997).
22. Interview with Mr Chris Dutilh, Rotterdam, 9 November 1998.
23. 'ECO Cop no.3, No.2'; http://www.climatenetwork.org/eco/eco2_cop3html.
24. Interview with Ms Margaretha de Boer, The Hague, 18 November 1998.
25. The French presidency did not formulate the Troika at COP1.
26. According to the interviews, NGO delegations were satisfied with their contact and exchange of information with government delegations during COP3.
27. Interviews with Mr Sible Schone, WWF Netherlands, 24 November 1998, and with Mr Wim Kersten, Milieudefensie (Friends of the Earth Netherlands), 12 March 1998.
28. Interview with Mr Teo Wams, Milieudefensie, 26 October 1998.
29. With regard to the internal EU incoherency and coherency in terms of the Kyoto Protocol negotiation, see Yamin (2000).

REFERENCES

Cooper, A.F. (ed.) (1997), *Niche Diplomacy: Middle Powers after the Cold War*, Basingstoke: Macmillan.
Cooper, A.F., R.A. Higott and K.R. Nossal (1993), *Relocating Middle Powers: Australia and Canada in a Changing World Order*, Melbourne: Melbourne University Press.
Cox, R.W. (1989), 'Middlepowermanship, Japan, and future world order', *International Journal*, **44**, autumn, 823-62.
Dolman, A.J. (1979), 'The like-minded countries and the New International Order: past, present and future prospects', in *Cooperation and Conflict*, **14**, 57-85.
Grubb, M. and J. Gupta, (2000), 'Leadership: theory and methodology', in Gupta and Grubb (eds).
Gupta, J. and M. Grubb (eds), (2000), *Climate Change and European Leadership*, Dordrecht: Kluwer Academic.
Gupta J., C.J. Jepma and K. Blok (1998), 'International climate change policy: coping with differentiation', *Milieu*, **13** (5), 264-74
Higgott, R. (1997), 'Issues, institutions and middle-power diplomacy: action and agendas in the post-Cold-War era', in Cooper (ed.).
Huber, M. (1997), 'Leadership in the European climate policy: innovative policy making in policy networks', Duncan Liefferink and Mikael Skou Andersen (eds), *The Innovation of EU Environmental Policy*, Copenhagen: Scandinavian University Press, pp. 133-55.
Kalders, P.R. (1998), *Besturen op termijn: Tijd grilligheid en trajectmangement in grondwaterbeleid (Administration in the future: time irregularity and route management in groundwater policy)*, Uitgeverij Eburon.
Kickert, W.J.M., Erik-Hans Klijn and J.F.M. Koppenjan (eds) (1997), *Managing Complex Networks: Strategies for the Public Sector*, Beverly Hills, CA: Sage.
Krasner, S.D. (1983), 'Structual causes and regime consequences: regimes as intervening variables', in Krasner (ed.), *International Regimes*, Ithaca, NY and London: Cornell University Press, pp. 1-21.
Ministry of Housing, Spatial Planning and the Environment (VROM), Ministry of Economic Affairs (EZ), Ministry of Agriculture, Nature Management and Fisheries (LNV) and Ministry of Transport, Public Works and Water Management (V&W) (1997), *Policy Document on Environment and Economy: Towards a Sustainable Economy*, June.
Phylipsen, G.J.M., J.W. Bode, K. Blok, H. Merkus and B. Metz (1997), 'A Triptych sectoral approach to burden sharing; GHG emissions in the European bubble', unpublished paper, Ministry of VROM.
Princen, T. and M. Finger (1994), *Environmental NGOs in World Politics: Linking the Local and the Global*, London: Routledge.
Ringius, L. (1997), *Differentiation, Leaders and Fairness: Negotiating Climate Commitments in the European Community*, CICERO Report, Centre for International Climate and Environmental Research, Oslo.
RIVM (National Institute for Health and the Environment) (1987), *Concern for Tomorrow: A National Environmental Survey 1985-2010*, Ultrecht.
Sjostedt G. (1998), 'The EU negotiates climate change: external performance and internal structural change', *Cooperation and Conflict*, **33** (3), 227-56.
Underdal, A. (1994), 'Leadership theory: rediscovering the art of management', in Zartman (ed.), pp. 178-97.

Visser, J. and A. Hemerijck (1997), *A Dutch Miracle: Job Growth, Welfare Reform and Corporatism in the Netherlands*, Amsterdam: Amsterdam University Press.

Yamin, F. (2000), 'The role of the EU in climate negotiations', in Gupta and Grubb (eds), pp. 7-66.

Young, O.R. (1983), 'Regime dynamics: the rise and fall of international regimes', in S.D. Krasner (ed.), *International Regimes*, Ithaca, NY and London: Cornell University Press, pp. 93-113.

Young, O.R. (1991), 'Political leadership and regime formation: on the development of institutions in international society', *International Organization*, **45**, 3: 281-308.

Zartman, I.W. (ed.) (1994), *International Multilateral Negotiation: Approaches to the Management of Complexity*, London: Jossey-Bass.

11. Global environmental change regimes: impact assessment on the basis of an extended GTAP model

Shunli Wang, Peter Nijkamp and Onno Kuik

1 SETTING THE SCENE

Global environmental change policy is increasingly looking for sustainability strategies that take into account the transboundary nature of environmental externalities. As a consequence, we observe an increasing popularity of international environmental agreements. The implementation of such agreements is however, fraught with many difficulties, as in an international context there are always (absolute or relative) winners and losers. This situation leads to complex negotiation procedures which have to find a balance between sound economic principles, environmental sustainability requirements and political acceptability conditions. Nevertheless, the awareness has grown that a coordinated environmental policy is of critical importance to ameliorate the impacts of environmental decay on the natural eco-system as well as on the socio-economic system. The reasons for this growing recognition of the need for a strong transboundary policy stem from the following factors: (i) pollution sources are multiple and difficult to identify; (ii) transaction costs for coping with individual pollutants may be very high; (iii) in an international context, it is almost impossible to create a negotiation platform among all polluters and all victims; (iv) the space-time interactions are impossible to map; and (v) the individual costs of coping with a distinct pollution case in the light of global environmental effects are hard to assess.

Several economic instruments have been set up to offer decision support for international environmental agreements, such as game strategies (see, for example, Carraro 1999) or institutional procedures. A bottleneck has always been how to deal with burden sharing of environmental policies. Such equity issues may also explain differences in the acceptance rate of involvement in global environmental policy agreements by individual countries. The history of long-range transboundary air pollution (LRTAP) policies offers many illuminating learning effects on the caveats in such international agreement processes (Nijkamp and Castells, 2001). In 1979 a first protocol to LRTAP, under the auspices of the UN Economic Commission for Europe, was signed with 33 signatory parties and ever since various additional protocols have come into place. Also the European Commission is playing an increasingly active role in environmental negotiation platforms, as can be seen in recent discussions centred around the Kyoto Protocol.

In this context, there is also a rising need for solid scientific information which may support intertemporal decision making in the environmental field. Integrated assessment models, policy scenario studies and general equilibrium models are examples of scientific tools aimed at understanding the transboundary complexity of global environmental issues and related policies. Nevertheless, it is clear that many hurdles still have to be taken before a satisfactory policy support system can be developed.

This chapter forms one of the steps by applying a general equilibrium model to the global environmental change regime, defined as institutes and institutions for controlling global environmental change – where institutes refer to formal relationships such as organizations and legislation concerning climate change policies and where institutions indicate informal relationships such as principles, norms and values in regard to the behaviour of actors in case of climate change issues (Haas 1975; Krasner 1983).

More specifically, this chapter focuses on possible future regimes under the Kyoto Protocol. In 1992, some countries – now widely labelled Annex I countries – voluntarily agreed to restrict their greenhouse gas (GHG) emissions to a certain threshold amount under the United Nations Framework Convention on Climate Change (UNFCCC). This agreement was an important step in the setting up of a global environmental change policy regime.

A provisional contour of this international policy regime is visible after the achievement of the Kyoto Protocol in which Annex I countries agreed on legally binding emission restrictions in the 2008–12 period. In this protocol, three policy mechanisms, namely, emissions trading (ET), joint implementation (JI) and the clean development mechanism (CDM), are introduced in order to have a 'cost-effective' way of implementing the GHG emission restriction. Currently, after the failure of the world community to achieve an agreement during the Sixth Conference of Parties (COP6) in the year 2000 and the likely scenario of the United States' non-participation, it becomes clear that the global environmental regime is still fragile. For analysing policy impacts of global climate policies, this fragility of the climate regime leads us to consider policy impacts of various possibilities of a future climate policy regime.

In order to illustrate the impacts of various regimes, this chapter tries to answer the following research questions: what are the likely impacts of some

Kyoto regimes (in particular CDM) for the world as a whole as well as for the individual participating world regions or countries?

For reasons of simplicity, this chapter compares the impacts of four possible future regimes to the non-action regime, that is, the business as usual (BaU). The BaU refers to the situation where no change occurs and is interchangebly used to refer to the base regime where no policy instruments are introduced. The four possible future regimes are set up in two dimensions. In these four policy regimes, the agreed Kyoto emission restrictions are imposed for the participating countries. The difference between the four regimes is that along the one dimension, we have a division between two regimes, one in which the USA participates as an Annex I country and another regime in which the USA does not participate; along the second dimension, we have to distinguish the use or non-use of the cost-effective instrument of CDM.

The first dimension is introduced, as the possible non-participation of the USA will definitely make a difference for the macroeconomic variables as well as for the global level of CO_2 emissions. This influence is not only important because of the USA's large share in global economic activity and related global carbon emissions, but also because of the political influence and the domino effect it might have on the global climate change negotiation. The danger is, of course, that the USA non-participation regime may become a precursor to other future regimes. The uncertainties and difficulties which prevailed previous to the Berlin negotiations, which is a follow-up of the COP6 negotiation in The Hague, may underwrite this political implication. However, the USA non-participation regime may deeply illustrate the economic and political consequences of a decision of a single country to withdraw from an international agreement.

The second dimension introduced here, namely, the quantification of the specific impact of CDM, is widely discussed, since domestic emission restriction is a costly regime. It is often argued that the Kyoto Protocol instruments may be more cost-effective. This chapter will provide a tentative quantitative assessment of this instrument, as (i) there are still many arguments for and against the cost effectiveness of CDM (see, for example, Begg et al. 2001), and (ii) international emissions trading (IET) is already widely discussed in the literature (for modelling experiments see, for example, Parry and Williams 1999; and for policy impact analysis, see, for example, Jensen and Rasmussen 1999; Bollen et al. 2000; Zhang 2001a, b).

We like to emphasize the tentative character of our quantitative assessment, as the results largely depend on specific assumptions to be discussed in Section 2 and further specified in the introductory parts of Sections 3 and 4. Three important simplifying assumptions influencing the cost effectiveness of the CDM instrument are: (i) the rate of carbon emission reduction of 20 per cent; (ii) the amount of CDM investments from the Annex I countries; and (iii) the determination of the baseline. We shall illustrate this restriction briefly by presenting two different institutional settings for the baseline of CDM.

Against the background of the previous introductory remarks, the organization of the chapter is now as follows. Section 2 deals with methodological issues on the analysis carried out in this chapter. Section 3 presents an analysis for the regimes without CDM, while Section 4 presents the results for the regimes with CDM. Section 5 offers some concluding remarks.

2 METHODOLOGICAL ISSUES

Introduction

The findings in this chapter are based on insights and analysis from recent simulations with the so-called GTAP-CDM model, a static global applied general equilibrium model that is extended with characteristics of CDM, such as technological progress and investments from the Annex I countries in the non-Annex I countries in order to achieve the Kyoto reduction target on emissions.

The simulations in this chapter are based on aggregated data related to the year 1995 from the GTAP-4E database, collected by the Global Trade Analysis Project team as well as the energy volume data from the International Energy Agency (IEA) transformed for the GTAP-E model by Truong (1999). We have chosen not to project the data to the year 2010, since such a projection requires several assumptions about future developments, which are still uncertain. In other words, in the trade-off between (i) a less 'intuitive' approach based on historical data, and (ii) a more 'intuitive' approach for the year 2010 data with many uncertainties about the 'validity' and 'reliability' of the projected data, we have opted for the first approach. The aggregated data refer to five world regions, that is, the USA, the EU, Economies in Transition (EIT), Rest of Annex I regions (RAX), and Rest of the World (ROW). Furthermore, the model employs 12 types of commodities, that is, five primary input factors, five energy products (coal, petroleum, gas, oil and electricity), one capital good and one other output.

Our choice for the use of GTAP is partly driven by the fact that the GTAP team also provides a fully documented, publicly available global database as well as software for simulation purposes. Furthermore, GTAP is a proper model for analysing the impacts of climate change regimes related to the Kyoto agreements and mechanisms, as those are related to the medium term, that is, the 2008-12 period (see also Kremers et al. 2001 for other reasons). Of course, the choice for a static versus a dynamic model is a trade-off between taking the real-world complexities into account versus neglecting long-run behaviour and effects. In the dynamic context, complexities relating to technological change, for example, the timing issue (Grubb 1997; de Groot 2000) and the related uncertainties (Carraro and Hourcade 1998) play an important role. However, the use of a dynamic model, like WorldScan for example, would lead to a problematic interpretation of the short-run impact (see, for example, Bollen et al. 2000). The static model reduces these complexities, but could add information on regional differences and trade effects in the analysis. In this context, the emission reduction target should be interpreted as if it is (i) voluntarily agreed, and (ii) the optimal solution from a dynamic optimization process which takes the complex interaction between the economic and the ecological systems into account.

The Model

GTAP-CDM is based on GTAP-E, which is a specific application of the base GTAP model to energy substitution, as developed by Truong (1999). GTAP-CDM is innovative, because it provides a way to endogenize technological change in a general equilibrium model (see, for example, Carraro and Hourcade 1998). The base GTAP model is a static multiregion, multisector applied general equilibrium model developed by the Global Trade Analysis Project team. In the GTAP model, the world economy is divided into a number of regions. In each of the regions, consumers are described by an aggregated regional household sector. Its income is allocated to consumption, public expenditure and savings. The share of consumption and public expenditure makes it possible to perform a demand for goods which are produced by the production sector (either domestic or foreign). The savings are used for investments in capital goods.

On the consumers' side, GTAP-E has reformulated the demand function by adding a carbon tax for consumers' expenditure as well as public expenditure on final goods which will emit carbon gases when used, for example, petroleum and gasoline. On the producers' side, the reformulation by GTAP is more complicated. We shall briefly describe the production structure of the model, because this is relevant for the formulation of CDM. For an overview of the GTAP base model, we refer to Hertel and Tsigas (1998); for an overview of relevant changes and the complete GTAP-E model, we refer to Truong (1999).

On the production side, GTAP-E uses the GTAP formulation for the output of an industry j in region r (denoted as $qo(j,r)$) which is modelled in a nested structure. The first-level nest in the production function has a Leontief form and is produced by two input factors, namely, a composite intermediates nest ($qf(i,j,r)$) and a value-added nest ($qva(i,j,r)$); a graphical representation is

Global environmental change regimes 253

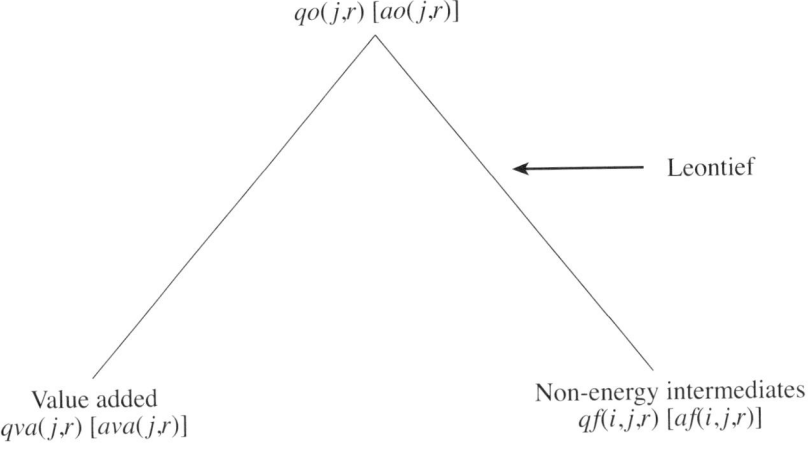

Source: Truong (1999).

Figure 11.1   First level of production structure for output of industry j *in region* r *in GTAP-E*

given in Figure 11.1. In contrast to the base GTAP model, the composite intermediate nest in GTAP-E consists of non-energy commodities. The energy commodities are transferred to the capital–energy composite in the value-added nest.

The use of a Leontief form for the first level is, as Hertel and Tsigas (1998) stated, based on the assumption of separability in production. This assumption implies that the elasticity of substitution between any individual primary factor, on the one hand, and intermediate inputs, on the other hand, is equal to zero. Hertel and Tsigas stated that, because of the Leontief form in the first nest and the assumption of constant return to scale, the mix of intermediate inputs is independent of the prices of primary factors.

The subsequent nests are, independently from each other, in turn characterized by constant elasticity of substitution (CES) production functions; see Figure 11.2. The CES function has the property that the substitution elasticities ($esub(i)$) between all the input factors within the function are constant (see Hertel and Tsigas 1998).

Furthermore, GTAP-E follows the base GTAP model in the use of the Armington assumption in the composite intermediates nest in order to allow for intra-industrial trade. This means that the commodities are assumed to be different according to the location where they are produced. In other words, the CES function has as intermediate production factors: (i) domestic inputs (qfd) and (ii) foreign inputs (qfm), where the foreign inputs are in turn assumed to be composed of foreign inputs from the individual foreign regions (qxs).

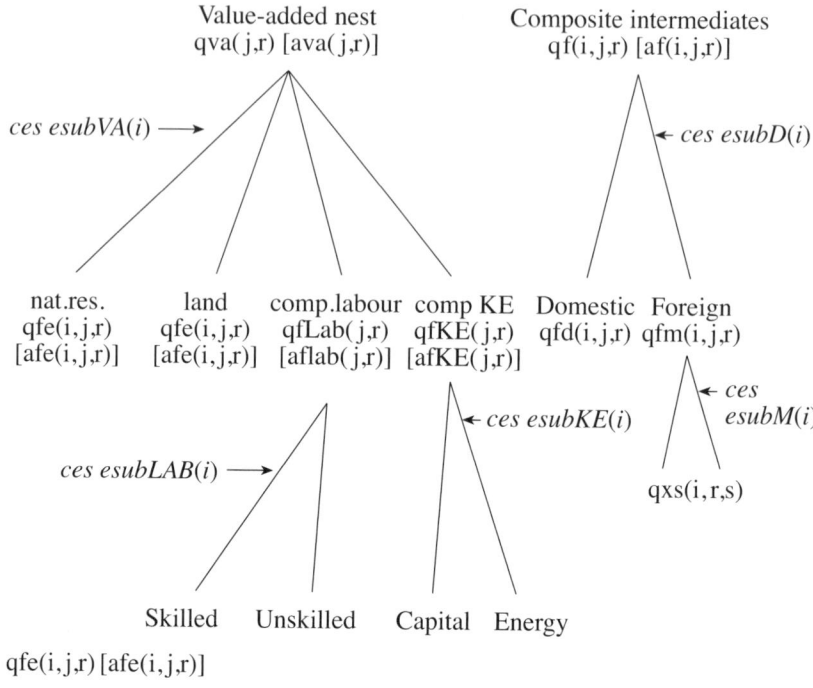

Source: Truong (1999).

Figure 11.2 The value added nest and composite intermediates nest in GTAP-E

Two relevant substitution parameters are related herewith, that is, *esubD(i)* and *esubM(i)*. The first is the elasticity of substitution between domestic versus imports of all agents in all regions and the second is the elasticity of substitution defined for imports from different regions.

The value-added nest is in GTAP produced by a CES production function from the endowments: land, composite labour (*comp.labour*, which is a CES function of the skilled and unskilled labour) and capital. In GTAP-E, natural resources (*nat.res.*) form a new endowment in the production of the value-added nest. Furthermore, Truong (1999) merged the energy composite with the capital endowment as part of input factors for the value added. The capital-energy composite (*comp KE*) has a CES form with capital and energy as inputs; see Figure 11.3 for a graphical representation. The energy nest is further produced by a multilevel structure of electric and non-electric energy. The non-electric nest is composed of coal and non-coal inputs. the non-coal nest is composed of the inputs: gas, oil and petroleum products. It is interesting to note that the inputs in the energy nest also meet the Armington assumption.

From a conceptual point of view, this means that Truong assumes that energy inputs are part of the endowment commodities which are owned by producers. The advantage of this formulation is, as Truong stated, that this formulation allows for: (i) substitution between the fuels; (ii) substitution between energy and capital in the energy-capital composite nest; and (iii) substitution between the energy-capital composite nest and other factors (see Troung 1999: 33). In this way, GTAP-E allows capital and energy to depend on the model parameters, substitutes or complements.

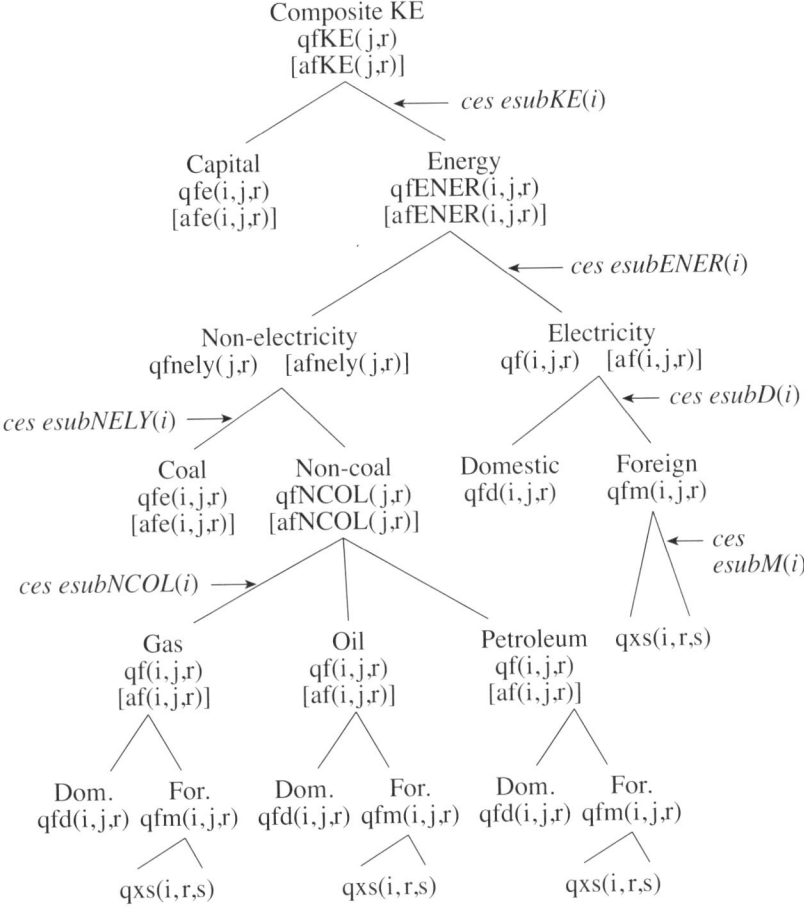

Source: Truong (1999).

Figure 11.3 The production of energy in GTAP-E

GTAP-CDM tries to offer a framework for endogenizing technological progress for the non-Annex I countries; this in contrast to GTAP and GTAP-E, where technological progress, which in Figure 11.3 is shown in brackets, is exogenously given for each nest. The main problem associated with endogenizing technological progress in a general equilibrium framework is the lack of data. As collecting data needs time, GTAP-CDM used a proxy for the function of technological progress in order to offer some *ex ante* prognoses of the cost effectiveness of CDM. This proxy of the function of technological progress uses the main idea of endogenous growth theory on technological progress. More precisely, it is based on the idea that the R&D sector uses the input factors that would otherwise be used in the production sectors to produce its own product, namely, technological progress. Then, at a given moment where the economy is in equilibrium, the level of technology in that economy may be approximated by the existing level of input factors in the production sector. Moreover, the marginal productivity of the input factors in the research and development (R&D) sector may also be determined by the marginal productivity of the input factors in the production sector.

Thus, the incorporation of CDM in GTAP-CDM has the following characteristics: (i) Annex I countries levy energy taxes to raise funds for CDM investments; (ii) non-Annex I countries receive these investments and will hence face increasing technological progress in their production processes; and (iii) as a result of technological progress in the non-Annex I countries, Annex I countries will acquire a certain amount of emission credits.

The Regimes

The base regime is the BaU situation, whereby no policy change, namely, carbon emission restrictions, would be introduced. In this regime, the CO_2 emissions would be unrestricted, although not unlimited, as the level of economic activity is determined by the available endowments. In this regime, there is no shadow price for carbon emissions, and countries would, from an economic point of view, not exploit CDM activities, as there is no reward for it. The base regime forms our benchmark on which the impact of other regimes will be applied and presented.

For our analysis, we work with two policy dimensions to construct distinct regime possibilities. The first dimension relates to the participation of the USA and the second dimension relates to the implementation of CDM. Regime possibilities in which the USA participate are denoted as USA_P and USA_P+CDM. Along the second dimension, regimes USA_P+CDM and USA_NP+CDM includes CDM. Table 11.1 summarizes the regimes which will be analysed in this chapter.

In regimes USA_P and USA_P+CDM, the participating Annex I regions

Table 11.1   Regime possibilities under analysis

	No emission constraint	Emission constraint	Emission constraint and CDM
US participation	base regime (=BaU)	USA_P	USA_P+CDM
US non-participation		USA_NP	USA_NP+CDM

are the USA, the EU and RAX; the non-participating region is EIT, and the non-Annex I countries are formed by ROW. In these regimes, the participating Annex I regions are reducing 20 per cent of their 1995 CO_2 emissions, while the non-Annex I countries and the non-participating Annex I countries would act under a BaU strategy. The amount of 20 per cent has been chosen for the reason of comparability with the Truong (1999) results for the GTAP-E model.

In these regimes, the reduction targets are achieved by each individual participating Annex I region through a tax on energy products (oil, coal, petroleum, gas and electricity). The reason for not including EIT as participating Annex I countries in our analysis is that (i) this region would not exceed their emission targets, and (ii) this region would not undertake CDM activities, because (a) these countries are themselves in transition and (b) there are no incentives to undertake CDM activities, because the shadow price for emission reductions in these countries is negligible. In regimes USA_NP and USA_NP+CDM, non-Annex I countries are still the rest of the world; the non-participating countries are EIT and the USA.

3   EFFECTS WITHOUT CDM

Introduction

In regimes USA_P and USA_NP, the carbon tax is differentiated according to the CO_2 emissions of the energy products. Given the CO_2 emission coefficient in GTAP-E, which is given in Table 11.2, the differentiated carbon tax means that the tax on coal is higher than that on gas. These coefficients are purely technical relationships in terms of CO_2 emissions of the underlying energy products. The reason for crude oil to have a zero CO_2 emission coefficient is that crude oil is mainly used as a material input into petroleum refining; thus it is a 'feedstock' rather than an energy input. For electricity, the zero coefficient is used in order to avoid double-counting, as electricity is produced from other primary fuels which have non-zero CO_2 emission coefficients.

Table 11.2 CO_2 emission content in GTAP-E model

	Energy products				
	Coal	Petroleum	Crude oil	Gas	Electricity
CO_2 coefficient	3.8107	2.7638	0	1.8844	0

Note: In tonnes of CO_2 per tonne of oil equivalent (toe).

Source: Truong (1999, p. 47).

The carbon tax would in the first instance lead to a higher price for the energy products, so that demand decreases. A secondary effect of the differentiated carbon tax is that there will be some switching in the demand for energy products, because energy products with a higher carbon content (coal, for example) become relatively more expensive as compared to energy products with a lower carbon content (for example, gas). Furthermore, in a multisector, multiregion economy where the economic activities are interrelated, there are also other kinds of secondary effects. The price differences between the energy products would also result in switching in import demand for products produced with these energy products. Together with changes in the world markets of energy products, this would lead to carbon leakage. An implication of carbon leakage is that the final world level of emission reductions is less than the aggregate target set by the Annex I countries.

All these direct and second-order effects also play a role in the USA_NP regime, where the United States does not join the Annex I group. The US decision would affect the final allocation of the economic resources as well as the extent of the world level of emission reductions.

Impact on Emission Reductions and Carbon Tax

Table 11.3 shows the emission changes in comparison to the BaU scenario for USA_P and USA_NP. The carbon tax of USA_P corresponds to Truong's (1999) calculation, where a tax of US$21.1 per ton of carbon emissions is found for the EU in an eight-country simulation. Carbon tax from GTAP-E/CDM is low compared to other models because of the incorporation of energy substitution in this model (for a discussion, see Truong 1999). Table 11.3 also shows that the United States has the lowest carbon tax among the Annex I countries. This result corresponds to the above-mentioned study of Truong (1999) and Bollen et al. (2000). The results for the Rest of Annex I

Table 11.3 Carbon emission target and carbon tax for USA_P and USA_NP (relative to BaU in 1995)

	Regime USA_P: Annex I incl. USA			Regime USA_NP: Annex I excl. USA		
	Target CO_2 emission %	Actual CO_2 emission %	Carbon tax US$/t$CO_2$	Target CO_2 emission %	Actual CO_2 emission %	Carbon tax US$/t$CO_2$
USA	−20	−20	14.57	none	1.54	0
EU	−20	−20	22.13	−20	−20	20.79
EIT	none	3.39	0	none	2.58	0
RAX	−20	−20	22.88	−20	−20	21.46
ROW	none	3.18	0	none	2.28	0
World		−8.08	7.85		−3.31	4.27

Source: Own GTAP-E/CDM calculations.

countries differ slightly. This may be due to a different aggregation between the three studies. In Table 11.3, we see that RAX countries have the highest carbon tax. This corresponds to the results of Truong (1999) for Japan, although in Bollen et al. (2000) Japan has almost the lowest carbon tax among the Annex I countries except for the economies in transition (=Eastern Europe and Former Soviet Union).

Interesting in Table 11.3 is the fact that the carbon tax for USA_NP differs only slightly from USA_P for the EU and the RAX countries. The reason for this small difference is that in our calculations each region still has to meet the target within the region. The lower carbon tax, however, is probably related to a lower economic activity rather than more efficiency within both regions. In USA_NP, we see that, in spite of high reduction targets in the EU and RAX, the total world emission is only reduced by 3.3 per cent. Thus, there is some carbon leakage. This indicates that a part of CO_2 emission reduction is cancelled out by extra CO_2 emissions by firms in the USA and other non-participating countries.

Other Economic Impacts

Although the emphasis in the literature on the impact of emission restriction is on the carbon tax and the related marginal reduction costs (see, for example, special issue of *Energy Journal* 1999), it is expected that the economic impacts of the various regimes also encompass other macroeconomic variables. In this chapter, we shall present the estimates of the impact on terms of trade, trade balance in ratio to real GDP instead of real GTAP, real GDP, GDP in value terms and real capital goods. These are indicators which are partly used to measure the 'international competitiveness' of the countries. The relative importance of each of the indicators however, depends on the decision makers. For example from a theoretic economic point of view, the real GDP is more important as it measures the amount of goods which are consumed, while from the consumers' point of view the value of GDP may be more important as it also accounts for the relative price changes. From a trade point of view on the other hand, the terms of trade or the trade balance are more attractive as indicators than GDP. Finally, the indicator of 'real capital goods' is used for long-run economic development.

In Table 11.4, we find that the terms of trade for USA_P are positive for the participating Annex I regions. This result suggests a plea for participation, as these countries receive relatively more for the same amount of their exports. However, by looking at the terms of trade for USA_NP, we conclude that the USA is even better off by non-participation. In both regimes, the indicator for trade balance in relation to real GDP suggests that the participating countries would export more than they would import.

Table 11.4 Indicators for international competitiveness for USA_P and USA_NP (relative to BaU in 1995)

	Regime USA_P: Annex I incl. USA					Regime USA_NP: Annex I excl. USA				
	Terms of trade %	Trade balance/real GDP %	Real GDP %	Value GDP %	Real capital goods %	Terms of trade %	Trade balance/real GDP %	Real GDP %	Value GDP %	Real capital goods %
USA	0.13	0.14	−0.17	0.53	−0.88	0.42	−0.32	0.01	0.62	1.87
EU	0.09	0.30	−0.21	0.32	−1.56	−0.10	0.51	−0.37	−0.16	−2.74
EIT	−0.07	−0.60	0.03	0.46	3.09	−0.04	−0.45	0.01	0.16	2.27
RAX	0.17	0.02	−0.22	0.53	−0.09	0.00	0.12	−0.23	0.22	−0.50
ROW	−0.26	−0.54	0.09	0.49	2.15	−0.04	−0.41	0.06	0.37	1.66

Source: Own GTAP-CDM calculations.

The indicator 'real percentage change in capital goods' may encourage the USA not to participate in Annex I, as the participating regions are faced with a decrease in the amount of capital goods. This also holds for the indicator 'percentage change in terms of real GDP', as all participating countries are confronted with a decrease in the amount of GDP. Non-participation of the USA would have some impact on the EU, as in USA_NP, the EU would be confronted with a real GDP loss of 0.37 per cent, whereas in USA_P, this is only 0.21 per cent. Even in terms of value of GDP, USA_NP gives a loss of 0.16 per cent for the EU. The indicator 'value of GDP', however, emphasizes the USA's incentive for non-participation, as USA_NP gives a rise of 0.62 per cent of GDP in value terms for the USA.

4 IMPLICATIONS OF CDM

The clean development mechanism is one of the important instruments in the Kyoto Protocol that aims at contributing in a cost-effective way towards the achievement of the amount of emissions reduction, as voluntarily agreed by Annex I countries under the Kyoto Protocol.

The Formulation of CDM

A basic concept of CDM is that Annex I countries are allowed to invest in projects which achieve sustainable development in non-Annex I countries, for which, in return, Annex I countries receive some amount of 'certified emission reductions' (CERs) which can be subtracted from the voluntarily agreed amount of reduction to be achieved by the country under consideration.

In the analysis carried in this section, investments from Annex I countries for CDM activities are provided by a 5 per cent tax on all intermediate energy products for production of traded commodities in participating Annex I countries. The allocation of these investments to sectors in non-Annex I countries are exogenously determined according to the CO_2 emissions share of the specific sector in the total CO_2 emissions of all sectors in all non-Annex I countries. The amount of investment allocated in the specific sector in non-Annex I countries determines in turn the rate of technological progress in this specific sector.

For the determination of the amount of CERs however, a standard method is not at hand yet. This is a result of uncertainties regarding the practical form of CDM that should be negotiated after the Kyoto Protocol. One of the issues in the negotiations is how to set the baseline and how to calculate (i) the part of emission due to more efficient technologies as a result of investments from Annex I and (ii) the part of emission which would be the case if there were no investments from Annex I countries (see, for example, Begg et al. 2001). Like the introduction of carbon tax, secondary effects would lead to a reallocation of economic resources in a general equilibrium setting.

For reasons of simplicity, CERs in the simulations performed in this chapter are determined solely by emission reductions that are accreditable to technological progress in non-Annex I countries within a single year. However, because GTAP-CDM is a static model, while the technological progress due to CDM investments also has a sustainable character, we shall also perform an analysis for the baseline calculation for CERs which is 10 and 20 times the emissions reduction in the non-Annex I countries within a year. With regard to these numbers, if we look at the economic lifetime of investments, a factor of 10 or 20 is conservative for capturing the sustainable character of CDM investments in a static model, as it represents a period of 8.5 years and 15 years, respectively, at an annual interest rate of 4 per cent. These numbers partly represent the still uncertain institutional setting, as a high baseline (Baseline C) will positively affect the cost effectiveness of the CDM instrument, while a low baseline (Baseline A) requires a relatively high investment from the Annex I countries to achieve the same amount of CERs.

Results for Emission Reductions and Carbon Tax

The contribution of technological progress as a result of CDM activities to CERs attributed to Annex I countries is presented in Table 11.5, which also shows other results of regimes USA_P+CDM and USA_NP+CDM with regard to emission reductions and the corresponding carbon tax. In this table, emission targets from participating Annex I countries are achieved by a non-CDM part and a CDM part. The non-CDM part is the result of carbon tax. In Table 11.5, we see that this part is lower than the target; this leads to a lower carbon tax per tonne CO_2 for participating countries. The non-CDM part of non-Annex I countries is calculated from the demand for energy products by these countries, as if technological progress had not taken place. The CDM part is calculated from the remaining part that can be ascribed to technological progress as a result of CDM activities by Annex I countries. Actual emissions in non-Annex I countries are the eventually perceivable emissions from these countries, while the non-CDM part is calculated from the notion 'as if' no CDM and thus technological progress has occurred. The similarities and the differences between the regimes if CDM is taken into account will be analysed below through decomposition of price and emission components.

Decomposition of Emission Components in Non-Annex I Countries

Figure 11.4 gives a comparison of carbon taxes for participating countries

Table 11.5 Emission reductions and carbon tax for CDM regimes (relative to BaU in 1995)

	USA_P+CDM: CDM and Annex I incl. USA					USA_NP+CDM: CDM and Annex I excl. USA				
	Target reduction %	Non-CDM part %	CDM part CER %	Actual emission %	Carbon tax $/tCO$_2$	Target reduction %	Non-CDM part %	CDM part CER %	Actual emission %	Carbon tax $/tCO$_2$
USA	−20	−19.76	−0.24	−19.76	14.25	no	2.02	0	2.02	0.00
EU	−20	−19.28	−0.62	−19.28	19.78	−20	−19.18	−0.82	−19.18	18.07
EIT	no	3.59	0	3.59	0.00	no	2.70	0	2.70	0.00
RAX	−20	−19.47	−0.53	−19.47	21.81	−20	−19.39	−0.61	−19.39	19.98
ROW	no	3.32	0.66	2.66	0.00	no	2.36	0.55	1.81	0.00
World		−7.78		−8.02	7.41		−2.97		−3.17	3.83

Source: Own GTAP-E/CDM calculations.

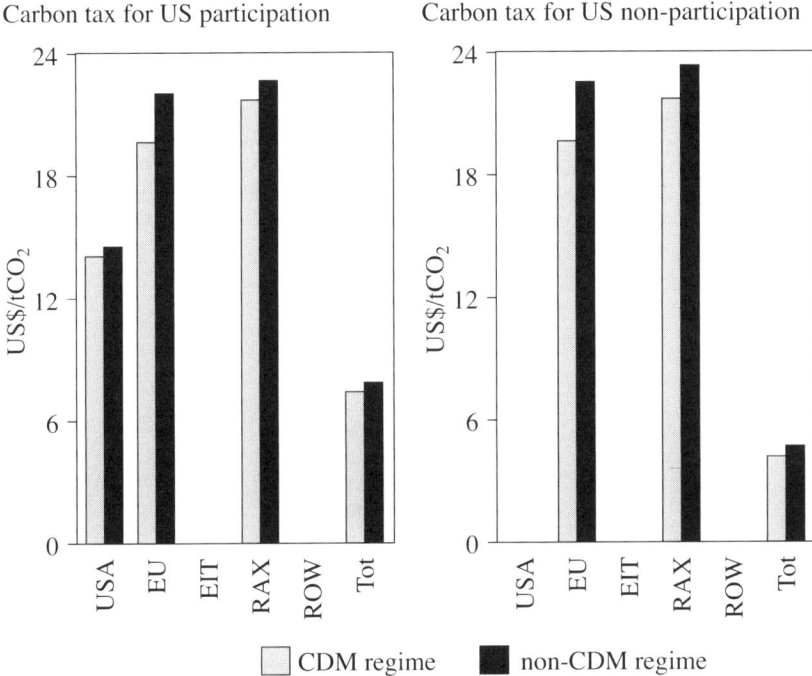

Figure 11.4 Impact of carbon tax as a result of CDM

between the four regimes under analysis. It shows that CDM would result in lower carbon tax. This result is the strongest for the EU, which means the EU would benefit most from CDM activities.

A comparison between Tables 11.3 and 11.5 also gives information on substitution effects which result in carbon leakage. Figure 11.5 breaks down the effect of CDM by comparing carbon leakage to ROW in regimes USA_P and USA_P+CDM. In USA_P, emissions in ROW would grow by 3.18 per cent, while in USA_P+CDM, this growth would be 2.66 per cent.

The reduced effect of carbon leakage could be attributed to (i) a substitution effect as a result of reallocation of economic resources by CDM activities from Annex I countries; this is (3.32%−3.18% =) 0.04 per cent of the emission; (ii) an efficiency effect as a result of technological progress from CDM activities, this is (2.66%−3.32% =) −0.66 per cent of the emissions. In this simulation, there still remains a net substitution effect as a consequence of a price effect caused by carbon tax on energy products, which is 2.66 per cent of the ROW's emissions. In this figure, we see that carbon leakage may be reduced by CDM investments.

Finally, a comparison between Tables 11.3 and 11.5 shows that the final

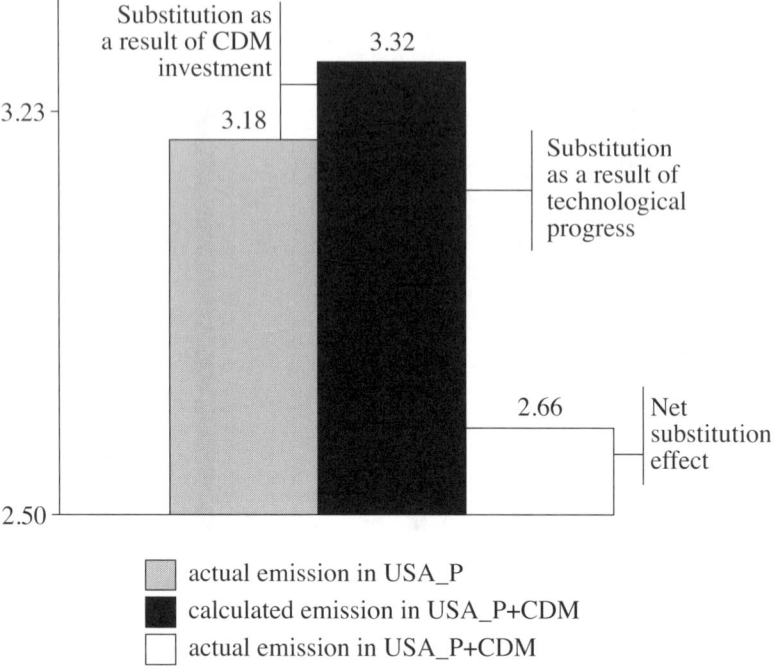

Figure 11.5 Decomposition of carbon leakage for ROW by regimes USA_P and USA_P+CDM

impact on the total reduction for the world as a whole depends on the relationships between the above-mentioned substitution effects. In the case where the USA is part of participating Annex I countries, the world emissions reduction as a result of CDM is slightly higher than in the non-CDM case: –8.02 per cent (USA_P+CDM) versus –8.08 per cent (USA_P). In the case of USA non-participation, however, the world emissions reduction in CDM case is –3.17 per cent, while in the non-CDM case it is –3.31 per cent. Thus, at the world level, we have the so-called rebound effect (Howarth 1997) which indicates that a more energy-efficient production process would lead to an increased demand for energy products for which the demand effect is stronger than the technology effect. This results in an increase in emissions instead of an expected decrease as a result of higher energy efficiency in the production of the output level.

A Comparison of Economic Effects

The economic impacts of undertaking of CDM activities are shown in Table

11.6. In this table, the previous result that participating Annex I countries would be faced with a loss in real GDP still holds. This also applies to the result for the terms of trade and the trade balance: the USA is better off in the case of a non-participation policy. A comparison between Tables 11.4 and 11.6 shows that for the USA, an implementation of CDM will enlarge the impact for the macroeconomic variables under consideration in case of a non-participation strategy.

For the convenience of comparison, we have subdivided the results from CDM regimes by those from the non-CDM regimes. The results from this division are given in Figures 11.6 and 11.7. In both figures, a value above 0 indicates that the impacts on the macroeconomic indicators are in the same direction for both the CDM and the non-CDM cases. A value above 1 indicates that CDM would magnify the impacts (either positive or negative). A value below zero indicate some adverse effects, that is, there is a change in sign. Furthermore, the letter P in the bar indicates the positiveness of the value for the indicator in the CDM case. Clearly, the value for the terms of trade for the USA in Figure 11.6 is positive in the CDM case, and it is around three times as high as the non-CDM case.

Figure 11.6 shows the impact of CDM in proportion to the non-CDM case for the US participation case. In this figure, we see that most variables are above zero. This indicates that there are not so many adverse effects, if the CDM instrument is applied. This holds for both the USA and ROW. However, this indicates that the impacts of emission restrictions would be enlarged by the CDM instrument, for example, the real GDP and the real capital goods for EU become more negative, while the trade balance becomes more positive.

For the EU, there is a minor adverse effect for the terms of trade. By looking at the relevant numbers in Tables 11.4 and 11.6, we see that for the EU, the terms of trade in the non-CDM case are positive (+0.09 per cent), while in the CDM case, they are negative (–0.08 per cent). For the EIT on the other hand, we observe clearly terms of trade gains from CDM. They will change from –0.07 per cent in non-CDM case to +0.09 per cent in the CDM case. For RAX, there are losses in trade balance (from +0.02 per cent to –0.12 per cent), but there are also gains in the terms of real amount of capital goods (from –0.09 per cent to +0.45 per cent).

Figure 11.7 shows the impact of CDM in proportion to the non-CDM case for a US non-participation strategy. In this figure, we see that the adverse effect only applies to the terms of trade for the EIT. Instead of a loss of –0.04 per cent, it gains 0.12 per cent. Furthermore, CDM magnifies the effects, if the USA decides to adopt a non-participation position. For the USA and the EU, the relative magnitude is above 1. This also holds for indicators other than the terms of trade for the EIT and the ROW. This indicates that for these regions,

Table 11.6 Macroeconomic indicators for CDM regimes (relative to BaU in 1995)

	Regime USA_P+CDM: CDM and Annex I incl. USA					Regime USA_NP+CDM: CDM and Annex I excl. USA				
	Terms of trade %	Trade balance %	Real GDP %	Value GDP %	Real capital goods %	Terms of trade %	Trade balance %	Real GDP %	Value GDP %	Real capital goods %
USA	0.42	0.02	−0.22	1.04	−0.22	0.89	−0.66	0.02	1.26	3.83
EU	−0.08	0.94	−0.76	−0.09	−5.02	−0.33	1.25	−0.97	−0.79	−6.72
EIT	0.09	−1.11	0.06	0.90	5.72	0.12	−0.89	0.04	0.44	4.60
RAX	0.38	−0.12	−0.28	1.02	0.45	0.14	0.04	−0.30	0.55	−0.20
ROW	−0.29	−1.07	0.34	1.19	4.54	−0.01	−0.87	0.26	0.96	3.77

Source: Own GTAP-CDM calculations.

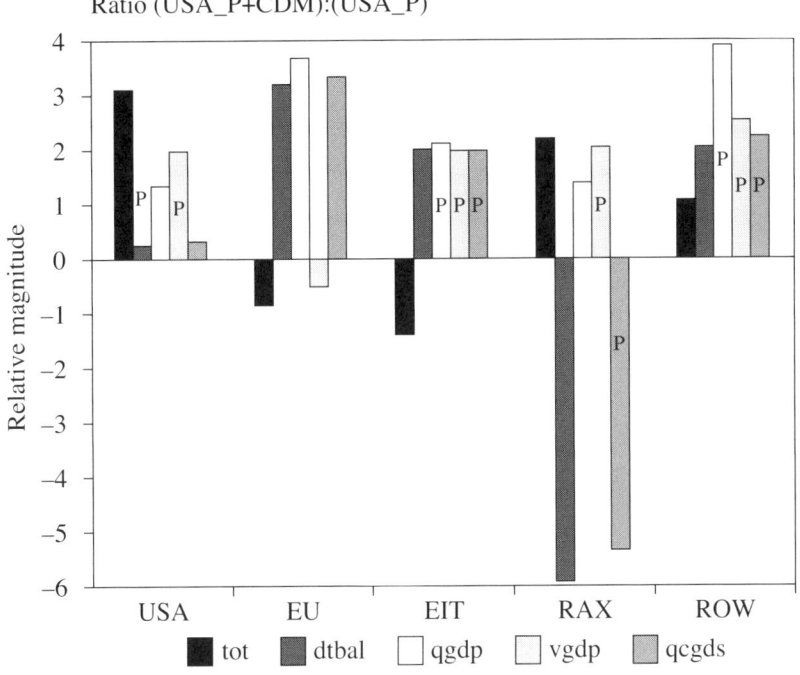

Figure 11.6 Relative magnitude of CDM effect with regard to US participation regimes

CDM would be more costly if compliance with the emission target is costly, as CDM would probably strengthen the impacts of policies aiming at emission reductions. Only for RAX, there seems to be some trade-off between the relative gain from the capital goods sector (instead of –0.50 per cent loss, the CDM case gives –0.20 per cent loss) against relative loss from trade balance (from 0.12 per cent to 0.04 per cent) and a loss of real GDP (from –0.23 per cent to –0.30 per cent in the CDM case).

In the next subsection, the robustness of this conclusion will be tested by showing how the impact of CDM would be altered if another design for CDM is chosen, that is, another calculation method for the baseline is applied.

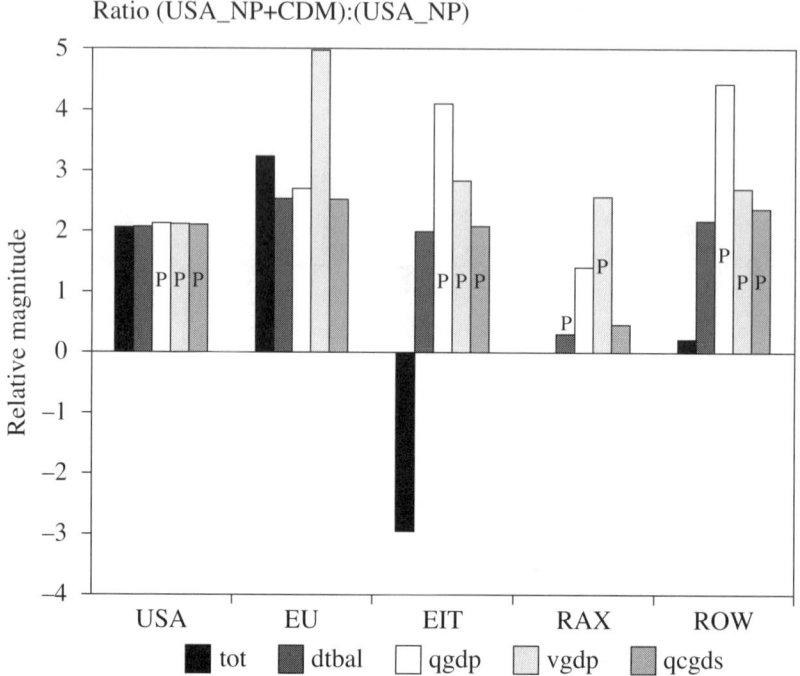

Note: For explanatory terms, see Figure 11.6.

Figure 11.7 Relative magnitude of CDM effect with regard to US non-participation case

The Issue of Baselines

The calculation of baselines forms one of the problems for CDM investments. At the project level, this is shown in Chapter 8 in this book by de Leeuw and van Ierland. In a static general equilibrium model where emission reductions due to CDM investment could easily be spotted, the problem is how to value the future emission reductions as a result of current CDM investments.

As is pointed out, a standard method is not at hand yet. Therefore, we shall illustrate this by showing for the US participation regimes how the indicators as discussed in the previous sections will be affected by a different calculation of the baseline, that is, Baseline B for a case where CERs are worth 10 times as much as in the single-period case (Baseline A); and Baseline C where CERs are worth 20 times as much as in the single-period case. Table 11.7 shows the results for the US participation case for Baseline B and Baseline C.

As a result of higher CERs for the same CDM investments, we see that the non-CDM part of the emission reduction is accordingly lower. This results in a lower carbon tax, which is quite standard. More interesting is that for the EU, if CERs are worth 20 times as much as in the single-period case, the CDM part is nearly 75 per cent of the target. This indicates that for the EU, CDM is a cheaper instrument than for the USA. The carbon tax for the EU could be lowered to 3.81 US$/tCO$_2$ in comparison to 22.13 US$/tCO$_2$ in the non-CDM case, while for the USA, the difference is only (14.57−9.79=) 4.78 US$/tCO$_2$. RAX falls between the EU and the USA.

From the macroeconomic indicators, we see that, except for a minor change in the trade balance and real capital goods for the USA, and the value of GDP for the EU and real capital goods for RAX, all other values for Baselines B and C tend to be lower than Baseline A. This indicates that the magnifying effects as observed in Section 4, above will be diminished if the baseline is set more in favour for the Annex I countries. An intuitive explanation for this result is that, considering the lower carbon tax, the economic distortion as a result of emission restrictions is cancelled out by CDM activities. The impact of this process is bigger as the baseline for CDM is set higher.

5 CONCLUSION

In this chapter, we have analysed the impacts of possible future global environmental change regimes. For this purpose, we constructed four regimes based on the dimensions of (i) the implementation of emission restrictions with or without the Kyoto 'CDM' instrument; and (ii) the decision of the USA to be either participating or not. Of course, the future regime is still fragile and the categorization of possible future regimes may be done along many dimensions.

One of these possible future regimes that has not been analysed in this chapter is - as we have already mentioned - the pessimistic, but real, possibility of the domino effect of the USA's non-participation. An optimistic future regime that we have not analysed in this chapter is that it is also possible that the process which has been set in motion, for example, to prepare activities implemented jointly (a precursor for the CDM projects) and CDM, the negotiations with the Umbrella group and so on, will have its own momentum. The industries and businesses in the USA that have already invested considerable time and manpower may wish to continue with 'implementation without ratification', such that technological progress in the non-Annex I countries will be faster. In this sense, we may have a situation whereby the total CDM investments in the US non-participation regime may be higher. The conclusion on the impact of CDM however would, as would be

Table 11.7 Emission reductions, carbon tax and macroeconomic indicators for different assumptions on baseline calculations under USA_P+CDM (relative to BaU in 1995)

	USA_P+CDM Target reduction %	Non-CDM part %	CDM part CER %	Actual emission %	Carbon tax $/tCO$_2$	terms of trade %	Trade balance %	real GDP %	value GDP %	real capital goods %
Baseline B										
USA	-20	-17.53	-2.47	-17.53	12.08	0.30	0.09	-0.18	0.82	-0.59
EU	-20	-12.75	-7.25	-12.75	10.54	-0.02	0.70	-0.58	0.00	-3.77
EIT	no	2.89	0	2.89	0.00	0.06	-0.91	0.05	0.77	4.71
RAX	-20	-14.66	-5.34	-14.66	14.78	0.30	-0.09	-0.21	0.84	0.33
ROW	no	2.86	0.57	2.19	0.00	-0.28	-0.89	0.32	1.01	3.80
World		-6.08		-6.32	5.31					
Baseline C										
USA	-20	-15.02	-4.98	-15.02	9.79	0.22	0.12	-0.15	0.65	-0.77
EU	-20	-5.37	-15.63	-5.37	3.81	0.00	0.54	-0.47	0.01	-2.94
EIT	no	2.08	0	2.08	0.00	0.06	-0.75	0.05	0.67	3.85
RAX	-20	-9.23	-10.77	-9.23	8.43	0.26	-0.08	-0.14	0.70	0.34
ROW	no	2.20	0.68	1.52	0.00	-0.25	-0.73	0.30	0.86	3.18
World		-4.21		-4.45	3.42					

expected, not change very much. Clearly, this analysis needs much future work. In addition, one of the important issues related to technological change is the intertemporal trade-off (leading to the so-called first-comers, late-comers dilemma) (see also Nijkamp and Castells 2001).

From the regimes that are analysed in this chapter, we sum up the following main conclusions. Firstly, in a multiregion, multisector general equilibrium model, we find that price and other substitution effects may result in carbon leakage, since carbon taxes in the Annex I countries raise the relative prices of their products. This leads to more demand for the products from the countries that do not introduce carbon taxes. Second, the US participation case shows that technological progress as a result of CDM would reduce this leakage. However, a rebound effect may show up at a world level. Third, the simulation results show that carbon tax is the lowest for the USA in the US participation case for 20 per cent of emission reductions through all participating countries. This means that, marginally speaking, the USA is less affected by reduction than the other participating countries. In the US non-participation case, the carbon tax for other countries does not change drastically. Fourth, the simulation results show that CDM increases the impact for a large part of the macroeconomic variables under consideration. Some adverse effects appear to occur for the EU, EIT and RAX. However, as is shown in Section 4, this impact depends on the method for baseline calculation. As a consequence, this implies that the institutional arrangement on the design of CDM affects the macroeconomic results. Fifth, as long as the institutional arrangements are not effectively implemented, an important element for the climate change negotiations is the result which may be deduced from the US non-participation regimes. From a single country's perspective, it may be possible that, when only the environmental costs are taken into account (for example, because the environmental benefits are transboundary), countries would have an incentive not to participate in Annex I. This result needs to be confirmed by future research for non-participation of other countries and under different assumptions concerning the reduction percentages, baselines and the calculation of environmental benefits. Finally, as is to be expected, the simulations show that the actual emission reductions at the world level are less if the USA decides not to participate.

The result that world emissions are higher when a country decides on non-participation combined with the result that countries have an incentive not to participate when only the costs are taken into account confirms the ultimate problem of environmental externalities: the prisoner's dilemma. It is well known from earlier theoretical and empirical works on environmental economics that internalizing environmental externalities is welfare improving for the world as a whole. The voluntarily agreed commitment on GHG emission of the Annex I countries in the United Nations Framework

Convention on Climate Change may be regarded as an attempt to internalize environmental externalities caused by carbon emission.

In the case of an externality, the equilibrium where an optimal amount of 'emission constraint' is effectively imposed, is also the optimal one. In a static model, every deviation from this 'optimal' emission constraint should be regarded as non-optimal, unless a trade-off between an extra amount of emission and economic variables related to long-term variables (for example, savings, technology) is explicitly specified. This implies that, if an emission constraint is optimal, the business-as-usual will be non-optimal. Therefore, the impacts associated with emission constraints and CDM activities are actually welfare improving instead of costly.

The prisoner's dilemma also shows that the welfare-improving implication of effective emission constraints does, however, not rule out that there are some incentives for a single country not to participate. The US non-participation case might be interpreted in this way. Therefore, future analysis should point out whether, after taking the environmental benefits into account, it is still beneficial for the USA not to participate or that the advantage will also show up for other countries if they decide not to participate. In that case, rationality may persuade the negotiating parties that free riding, that is, not taking the transboundary environmental benefits into account, would result in a non-optimal, that is, destructive, outcome.

REFERENCES

Begg, K.G., T. Jackson and S. Parkinson (2001), 'Beyond joint implementation – designing flexibility into global climate policy', *Energy Policy*, **29**, 17-27.

Bollen, J., T. Manders and P. Tang (2000), 'Winners and losers of Kyoto', October, 1-41, RIVM (National Institute for Health and the Environment) and CPB (Netherlands Bureau for Policy Analysis).

Carraro, C. (1999), 'Environmental conflict, bargaining and cooperation', in J.C.J.M. van den Bergh (ed.), *Handbook of Environmental and Resource Economics*, Cheltenham, UK and Northampton, MA, USA: Edward Elgar, pp 261-471.

Carraro, C. and J.C. Hourcade (1998), 'Climate modelling and policy strategies: the role of technical change and uncertainty', in: *Energy Economics*, **20**, 463-71.

de Groot, H.C.F. (2000), 'Note on the optimal timing of abatement activities: an application of the DICE-model to the debate on "when flexibility"', CPB, The Hague.

Energy Journal (1999), Special Issue: 'The costs of the Kyoto Protocol: a multi-model evaluation', International Association for Energy Economics.

Grubb, M. (1997), 'Technologies, energy systems and the timing of CO_2 emissions abatement', *Energy Policy*, **25**, 159-72.

Haas, E.B. (1975), 'Is there a hole in the whole? Knowledge, technology, interdependence and the construction of international regimes', *International Organization*, **29**, 827-76.

Hertel, Thomas W. (1998), *Global Trade Analysis: Modeling and Applications*, Cambridge: Cambridge University Press.
Hertel, Thomas W. and Marinos Tsigas (1998), 'Structure of GTAP', in Hertel (ed.), *Global Trade Analysis: Modeling and Applications*, Cambridge: Cambridge University Press, pp. 9-71.
Howarth, R.B. (1997), 'Energy efficiency and economic growth', *Contemporary Economic Policy*, **15**, October, 1-9.
Jensen, J. and T.N. Rasmussen (2000), 'Allocation of CO_2 emission permits: a general equilibrium analysis of policy instruments', *Journal of Environmental Economics and Management*, **40**, 111-36.
Krasner, S.D. (1983), *International Regimes*, Ithaca, NY: Cornell University Press.
Kremers, H., P. Nijkamp and S. Wang (2001), 'A comparison of computable general equilibrium models for analysing climate change policies', Research paper, Department of Spatial Economics, Free University, Amsterdam.
Nijkamp, P. and N. Castells (2001), 'Transboundary environmental problems in the European Union', Research paper, Department of Spatial Economics, Free University, Amsterdam.
Parry, I.W.H. and R.C. Williams III (1999), 'A second-best evaluation of eight policy instruments to reduce carbon emissions', *Resource and Energy Economics*, **21**, 347-73.
Truong, T.P. (1999), 'GTAP-E: incorporating energy substitution into the GTAP model', GTAP Technical Paper no. 16, Purdue.
Zhang, Z.X. (2001a), 'An assessment of the EU proposal for ceilings on the use of Kyoto flexibility mechanisms', *Ecological Economics*, **37**, 53-69.
Zhang, Z.X. (2001b), 'The liability rules under international GHG emissions trading', *Energy Policy*, **29**, 501-8.

PART II

After Kyoto

12. The multi-sector convergence approach to global burden sharing of greenhouse gas reductions

Jos J.C. Bruggink

1 SCOPE OF STUDY

Meaningful Participation of Developing Countries Remains a Contentious Topic

The UN Framework Convention on Climate Change (UNFCCC) recognizes the problems of burden sharing by noting that countries have common but differentiated responsibilities. Since then, international negotiations have been dominated by discussions about fairness in solving problems of climate change related to setting greenhouse gas emission targets or sharing abatement and adaptation costs. The first round of this political battle has taken place around setting emission reduction targets for the developed countries (Annex I signatories). The second round involving potential commitments from developing countries (non-Annex I signatories) is likely to be much more complex. The vast differences between the average developing and developed country with respect to historical cumulative emissions, present income levels and future adverse impacts lead to difficult ethical questions. Moreover, some Annex I countries, in particular the USA, threaten to withdraw from the Kyoto Protocol if no provisions are made for meaningful participation of non-Annex I countries. This makes the issue of global burden sharing a highly contentious and politically decisive bottleneck in international climate change negotiations.

Global Burden-sharing Rules as Negotiation Tools

In this chapter we shall outline the role of principles of distributive justice in establishing global burden-sharing rules, to summarize lessons learned from international negotiations between Annex I countries so far, and to present a new, pragmatic approach to global burden sharing. Global burden-sharing rules are considered effective tools to structure the dialogue on global participation and global solidarity in solving problems of climate change. Of course, simple rules will never solve the complicated issue of global burden sharing, but they can illustrate the consequences of normative choices quantitatively and thus provide a road map towards common ground.

2 PRINCIPLES OF DISTRIBUTIVE JUSTICE IN INTERNATIONAL CLIMATE CHANGE NEGOTIATIONS

Four Basic Principles of Distributive Justice

Climate change actions involve costs and benefits that are distributed across nations in different and uncertain ways. Presumably, international negotiations on the distribution of costs and benefits are based not only on pure self-interest, but also on principles of distributive justice, if only because pure self-interest would lead to deadlock and result in an outcome detrimental to all involved. There are, however, many principles of distributive justice that can be considered for deriving burden-sharing formulas in the case of climate change actions (den Elzen et al., 1999; Ringius et al., 2000; Metz, 2000; Rose et al., 1998). It is important to note that the term 'burden sharing' is ambiguous since it may describe the sharing of initial emission reduction targets or the sharing of ultimate financial costs. In this chapter we refer primarily to the distribution of initial emission reduction targets, although we also reflect on the distribution of costs where appropriate. We shall not, however, discuss burden sharing in the sense of either who pays for initial adaptation measures or who suffers ultimate financial damages of adverse impacts. To illustrate the range of principles of distributive justice, the four that are commonly used are:

- *Historical contribution to the problem* Countries that are the cause of the problem should pay to solve it. The guilty should pay for their sins. Thus this is often referred to as the guilt principle or the polluter-pays principle. It involves assigning the highest targets to countries with the highest cumulative historical share in greenhouse gas emissions. This is the essence of the so-called 'Brazilian proposal'. Not easily measurable when sinks are included.
- *Ability to pay for the solution* Countries that can afford the economic burden should shoulder it. The wealthy should pay for the action; hence often called the capacity-to-pay principle. Emission reduction targets should reflect GDP levels in one way or another. This is a basic reason why the Kyoto Protocol distinguishes Annex I countries from non-Annex I countries.
- *Equality of rights* Everybody has the right to an equal share of allowable greenhouse gas emissions. The rich should pay the poor if they wish to use more than their fair share of the global commons (equality principle). This principle sounds eminently fair if equally unrealistic for immediate action. Often used as a distant target to strive

for and prominent in burden-sharing rules based on the contraction-and-convergence approach.
- *Historic claims* Present levels of emissions constitute a fair initial distribution of emission rights. This principle of grandfathering is often invoked as a convenient starting point for assigning emission rights, but is highly unfair to new entrants, *in casu* the developing countries, and therefore not very appealing from an ethical perspective. Sometimes named the sovereignty principle. Perhaps useful for pragmatic reasons in the short term, but not helpful as a guiding principle for the long run.

All these are known as allocation-based principles. They address the allocation of either activity obligations to reduce global pollution loads (first two) or ownership rights of global carrying capacity (last two). In addition to this type of justice one can also formulate norms reflecting the fairness of the process of setting allocation rules and the institutional arrangements for doing so. Such process-based principles do not refer to the rules and outcome of the allocation, but to the procedures and institutions necessary for allocation and are not discussed in this chapter.

The Problem of Moral Ambiguity

It is tempting to concentrate the discussion on burden sharing on the problem of how to rank or weight the various ethical principles in order to decide on a theoretically optimal formula for burden sharing. This search is bound to be fruitless because this kind of multi-criteria approach involves finding some other equally normative principle for ranking or weighing principles which will simply reopen the justice discussion on another level with dire consequences for transparency. Moral ambiguity is a fact of life in a world of divergent interests and contextual norms. Ethical considerations are an absolutely necessary ingredient for establishing a menu of possible burden-sharing rules, but they are far from sufficient to establish a politically feasible solution. It is therefore important to search for rules that combine different principles transparently and pragmatically. This requires that principles can be translated into quantitative formulas containing variables, which can be estimated on the basis of unquestionable empirical observations or widely shared visions on the future. Such formulas must be developed from a persuasive, not a coercive perspective.

Pragmatic Focus on Initial Reduction Targets and Emission Rights

In choosing between principles, two basic issues stand out. First: should we apply rules with respect to initial reduction targets or with respect to ultimate reduction costs? Second: are we to prefer rules concerning obligations or rules concerning rights? With respect to the first question, it is undoubtedly easier to apply rules to reduction targets rather than cost. There are too many uncertainties associated with the expected costs of reduction, because they depend to a large extent on fuel-price changes and technology improvements that are notoriously difficult to predict for the long run. Moreover, costs are expressed in money values and are thus likely to evoke further debate on the marginal value of money in different countries. The utility or welfare associated with one dollar in the Indian countryside tends to be vastly different from that of the same dollar in an American suburb. With respect to the second question, there are some basic problems associated with principles depending on obligations. The guilt principle is weakened because the guilty countries can argue that they were unaware of the consequences of their action at the time of perpetration. The capacity principle is weakened because it will substantially reduce the meaningful participation of rich countries, which have never been inclined to follow this principle in other areas of international negotiation. We therefore conclude that the principles of distributive justice most appropriate in the area of international climate change regimes should be based on rights rather than obligations and should concern initial distribution of targets rather than ultimate distribution of costs.

3 LESSONS LEARNED FROM INTERNATIONAL CLIMATE CHANGE NEGOTIATIONS IN THE PAST

Situation at the Time of the Kyoto Conference of the Parties

Following the establishment of the 1992 UNFCCC, equity issues remained a hotly debated topic where principled positions mattered more than practical outcomes. However, the explosive nature of the issue had already been defused, by dividing nations into two separate groups (Annex I and non-Annex I) right from the start. Exemption from any near-term burden is of course very effective in avoiding controversy. In the negotiations for the 1997 Kyoto Protocol the discussion thus shifted easily from equity between developing and developed nations to equity among developed nations. Non-Annex I countries now became more important from the point of view of efficiency than from that of equity. Annex I countries had agreed to shoulder the initial burden. Now they sought to lighten that burden by implementing targets cost-effectively and the clean development mechanism (CDM) seemed one way to achieve this. Moreover, differentiation between reduction targets of Annex I countries was bound to lean on efficiency considerations just as much as equity considerations, because levels of income, a crucial indicator of equity, were no longer substantially different.

The Relation between Equity and Efficiency Considerations

Since 1997 the discussion on burden sharing has become muddled, because equity considerations are not clearly distinguished from efficiency considerations. Certainly, it makes sense to take abatement measures where the marginal costs of abatement are lowest, but this does not automatically imply that the country where such costs are lower should shoulder a higher burden and vice versa. Indeed, the basic idea behind the drive for using flexible instruments is based on the assumption that nations invest abroad to fulfil their targets. If targets are distributed in such a way that marginal costs of abatement are similar everywhere without foreign investments, there would be no need for flexible instruments. This situation would of course produce a very cost-effective solution, but there is no guarantee at all that this solution would be considered very fair by all involved.

Cost Effectiveness, Carbon Leakages and Dynamic Efficiency

Additional complications arise because economic concepts of efficiency can be invoked on many different levels to defend one position or another. In its simplest form it just refers to cost effectiveness. Once the level of an activity is determined in one way or another (in the case of climate change the required emission reductions to reach an admissible level of concentration) it is desirable to achieve that level with the least resources possible. Unfortunately, this may be difficult if the consequences lead to side-effects that actually worsen the intended positive results. Competitive pressures worldwide may eventually lead to substantial shifts in the distribution patterns of carbon-intensive activities (so-called 'carbon leakages'), particularly if some countries are exempt from taking any climate change actions at all. Finally, what ultimately counts is not so much the static neoclassical efficiencies of a competitive economy from economics textbooks, but the Schumpeterian dynamic efficiencies based on innovation cycles which drive life-style evolution and technology development towards less carbon-intensive societies. Economies that succeed in bringing about the required changes in patterns of consumption and production may ultimately be most successful in a world that takes climate change seriously. In the international negotiation process following Kyoto all these aspects of efficiency (cost effectiveness, carbon leakages, dynamic efficiency) have influenced discussions on burden sharing. The first lesson to be learned from this muddled past is that focusing on principles of distributive justice only is not likely to be an effective route to agreement. Considerations of efficiency and competitiveness must play an additional role.

Horizontal versus Vertical Negotiations

An important observation to make in this respect concerns the dual fronts on which international negotiators are engaged in political battles (Gupta, 1997). Horizontal negotiations between different countries take place at the international level, but vertical negotiations between governments and domestic interest groups take place at the national level. What is considered politically fair at the international level must be economically feasible on the national level. It is no use defending the high moral ground in international horizontal negotiations without being able to sell the consequences to domestic interest groups in vertical negotiations. In many ways, the strategy of national interest groups is to look for reasons to shift potential burdens domestically rather than internationally on the basis of economic arguments that have more to do with efficiency and competitiveness than with equity and justice. After all, multinational companies are likely to appeal to their chances of national survival while the transport lobby is likely to point out their high marginal costs of abatement. Their arguments would be difficult to base on their poor financial disposition with respect to their counterparts in the developing world.

The Success of the Triptych Approach

The second lesson to be learned from the past concerns the fact that the use of formal burden-sharing rules can be an important tool in helping international negotiators to evaluate their own position and interests relative to others when setting national targets. They form an excellent vehicle for persuasion and communication. This has been clearly demonstrated in the case of differentiation of the European Kyoto target among member states (van Harmelen et al., 2001; Ringius, 1999). The burden-sharing tool used in this case is the so-called Triptych model developed by Utrecht University (Phylipsen et al., 1998; Groenenberg et al., 2000). The model is based on a sectoral approach to target setting, which allows the application of different principles for each sector. Three sectors are considered: power generation, the energy-intensive industry and the domestic sector (including households, transportation and remaining production sectors). Only the domestic sector is subject to the equality principle of distributive justice (per capita convergence). The principle for the energy-intensive industry is based on a norm for efficiency improvement (historical rights). The principle for power generation is based on the national generation mix (not actually an equity argument, but based on considerations of acceptable marginal costs of abatement). Reduction targets were based on baseline reference scenarios that allowed for higher growth in lower-income countries. Calculations with the model turned out to be very valuable in reaching agreement. The success of the model was due to the eclectic approach with respect to different sectors and the allowance made for specific national circumstances. It judiciously combined elements of equity principles with realistic considerations regarding cost effectiveness and carbon leakage.

THE MULTI-SECTOR CONVERGENCE APPROACH

General Features

The multi-sector convergence approach was developed in a joint study by CICERO (Centre for International Climate and Environmental Reseach, Oslo) and ECN (Energy Research Centre of the Netherlands, Petten), (Jansen et al., 2001; Sijm et al., 2001). It was developed specifically for global burden sharing where the participation of non-Annex I countries forms the key issue. Its main features are based on a mix of general principles of fairness, sectoral target levels and country-specific allowances. It combines the ethical elements of the contraction-and-convergence approach with the sectoral detailing of the Triptych approach. In addition to covering both developed and developing countries it also includes the complete basket of greenhouse gases agreed upon in Kyoto except the small share (2 per cent) of fluorocarbons.

Evolutionary Development from Grandfathering to Equality

Initial short-term commitments are based on expected greenhouse gas emissions by sector in a near-term reference year (e.g. 2010). Ultimate long-term commitments are based on equal per capita emission norms per sector in a long-term reference year (e.g. 2100). This amounts to a gradual transition from grandfathering to equality on the basis of sectoral emission rights. Non-binding sectoral emission rights are aggregated to binding national emission rights leaving scope for flexibility in national negotiations on the vertical level. These rules are similar to the well-known set-up of the contraction-and-convergence approach as developed and promoted by the Global Commons Institute. Reduction rates are, however, not determined on the national level, but on the sectoral level. This is a basic feature of the Triptych approach. The number of sectors in the multi-sector convergence approach is, however, higher to account for the much greater divergences between developed and developing nations and the higher number of greenhouse gases. There are five sectors where emissions are primarily related to energy use and two sectors where emissions are not related to energy use. This level of detail will not lead to substantial data problems. The sectors distinguished include:

- power generation;
- industry;
- services;
- households;
- transportation;
- agriculture; and
- waste treatment.

Advantages of Sectoral Allocation

The introduction of sectoral allocations implies that the trajectories for national emission reductions are strongly dependent on the relative importance of the seven key sectors in national emission balances as mentioned above. For instance, a relatively high national emission contribution from the transport sector will lead to relatively modest per capita emission reduction targets, because presumably the potential for continuous emission reductions in the transport sector is relatively low. The required annual rate of change for this sector is assumed to be zero in Table 12.1, below. Alternatively, a relatively high national emission contribution from the waste sector will lead to relatively elevated per capita emission reduction targets, because presumably the potential for continuous emission reductions in the waste sector is relatively high. The required annual rate of change for this sector is assumed to be –1.1 per cent in the table. The sectoral approach to target calculations thus automatically takes account of both the sectoral composition of different national emission balances and the global technical potential for emission reductions per sector. In addition, the sectoral targets are not considered binding, which means that they are only used to calculate a binding national target as the sum of the non-binding sectoral targets. This leaves individual countries considerable freedom to make their own national choices regarding actual sectoral emission reduction policies, because from the perspective of national implementation opportunities and bottlenecks they may prefer to divert from the sectoral reduction norms as agreed upon internationally.

Allowance Factors and Commitment Thresholds

Setting per capita emission targets at the sectoral level takes due account of the specific sectoral conditions per country, thus acknowledging emission needs based on sectoral structure and development patterns. Yet, there may be physical circumstances at the sectoral level which set countries apart on the basis of unchangeable factors that are generally considered legitimate reasons for variation in reduction rates between countries: in the power sector this could be the renewable resource base; in the household sector the number of heating degree days; and in the transportation sector the population density. In addition to these unchangeable factors, there may be social circumstances shared by groups of countries that set them apart on the basis of economic factors that are only changeable in the long term. Economies that are experiencing a transition period towards a market-oriented economy are an example and have already obtained a special status in the Kyoto Protocol. If these differentiating economic circumstances are such that actual commitments are in fact not desirable or feasible as is the case for most developing countries, it would be unwise to establish allowance factors that would simply result in negating the commitments. In that case, agreements on thresholds for joining the list of committed nations are more to the point. Such commitment thresholds should determine which combination of per capita sectoral emission levels would justify the setting of targets and how long the preceding notification and following adjustment periods should be.

Numerical Illustration

A step-by-step numerical case study can illustrate the basic features of the multi-sector convergence (MSC) approach.

- We use 2010 as base year. Emission data per sector and per capita for Annex I countries are determined assuming that the Kyoto targets are realized and that the sectoral composition of emissions remains unchanged. Emission data for non-Annex I countries are based on one of the latest International Panel on Climate Change (IPCC) scenarios (SRES A1). Average global per capita emissions for each sector in 2010 can be calculated from these figures (Table 12.1, column 1).
- We assume that each sector is characterized by an annual global rate of carbon efficiency improvement (Table 12.1, column 2). Applying these improvement rates for the next 90 years to 2100 results in average global sectoral emission standards for 2100 (Table 12.1, column 3).
- At each point in time countries are divided into two groups: those with national per capita emissions above the global average for the corresponding year and those with national per capita emissions below this average. For the first group we assume a geometric convergence from their 2010 sectoral per capita emissions to the final per capita sectoral allowance in 2100. For the second group we assume a specific annual growth rate of emissions (3 per cent) until they have reached the

Table 12.1 Input data for MSC burden-sharing rule

MSC sector	tC/capita 2010	Global emissions 2010 (GtC)	tC/capita annual rate of change (%)	tC/capita 2100	Global emissions 2100 (GtC)
Power	0.28	1.946	−0.6	0.16	1.495
Industry	0.57	3.911	−0.7	0.25	2.288
Transport	0.21	1.451	−0.0	0.21	1.916
Households	0.10	0.655	−0.9	0.05	0.467
Services	0.09	0.634	−0.7	0.24	0.445
Agriculture	0.24	1.682	−0.0	0.24	2.222
Waste	0.06	0.421	−1.1	0.02	0.205
Total	1.55	10.711		0.98	9.038

global average. Then they are allowed an adjustment period of 15 years after which they are subject to the convergence conditions of the first group of countries. Figure 12.1 presents the calculated emission trajectories for four major global emitters: the USA, the EU15, India and China.

Interpretation of Case Study

The case study demonstrates some key points of contention, which are likely

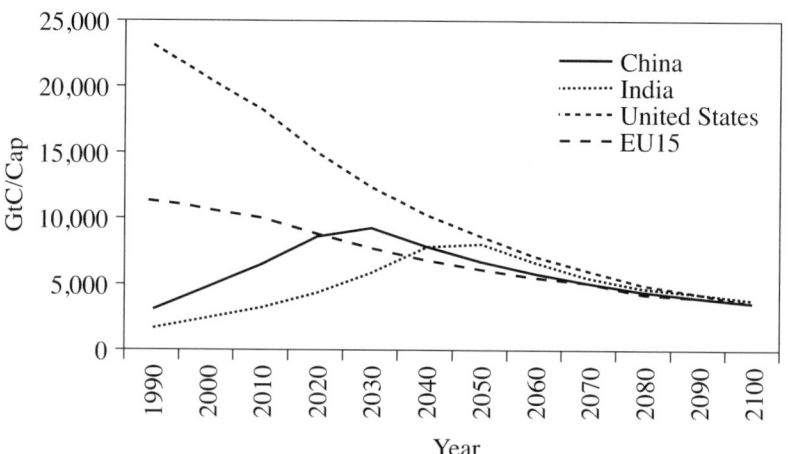

Figure 12.1 Per capita total emission allowances

to dominate global burden-sharing discussions. First and not surprisingly, it suggests much harsher reduction rates for the USA than for the EU, while allowing a reasonable and differentiated growth trajectory for developing countries. India and China even exceed the EU in per capita emissions in the second quarter of the century. Second, most of the action takes place in the first half of the century. By 2050 national trajectories already show remarkable similarities. It should be noted that the assumptions about ultimate per capita norms for sectoral emissions determine to a large extent the concentration of greenhouse gases by the end of the century, which in this example far exceeds the present concentration. Of course, this pattern just illustrates how to start the dialogue on meaningful participation of developing countries when negotiating global burden sharing. Further dialogue no doubt will concentrate on three major issues: allowance factors, emission credit trading and land-use impacts. Allowance factors could reduce the relative burden of the USA in a way acceptable to others. Emission credit trading may substantially reduce the costs of emission reduction for the USA, if developing countries are in a position to remain substantially below their assigned targets as countries in Central and Eastern Europe are at present. Emissions trading would thus effectively result in a global income redistribution trajectory coupled to successful climate change policy. Finally, land-use changes are not included in the present model and may also improve the relative burden of the USA (and a number of developing countries) when introduced in an environmentally sound way.

5 CONCLUSION

The Persuasive Function of the Multi-Sector Convergence Approach

The most challenging task of any scheme for global burden sharing is to entice the participation of developing countries while not alienating the developed countries already committed. The evolutionary approach from grandfathering to equalization over the greater part of a century in the multi-sector convergence approach is intended to provide a road map towards such incremental regime evolution. This evolutionary feature will allow a persuasive function in horizontal negotiations between nations at very different levels of economic development. On the other hand the sectoral orientation will be important for vertical negotiations within nations between governments and sectoral interest groups. Combining these two features in a hybrid approach will provide international negotiators with an effective communication tool on both the international and national levels. The negotiated determination of sectoral emission standards, national allowance factors for physical and economic circumstances and thresholds linked to adjustment periods provides significant flexibility for compromise.

Equity Issues Crucial for Establishing Global Climate Policy Regimes

The debate on the Kyoto Protocol has focused strongly on the division of targets among Annex I countries and the implementation of flexible instruments. Efficiency issues and cost effectiveness have become the central themes in this debate. Nevertheless, any conceivable extension of the Kyoto Protocol is bound to refocus on equity issues. Successful global climate change regimes must necessarily involve not only the question of how cheap solutions could possibly be, but also how fair. Questions of efficiency are intricately linked with questions of equity. This basic observation implies that the search for promising global climate change policy approaches should focus on finding ways to introduce issue linkages. Issue linkages refer to the phenomenon that international negotiators try to advance their joint interests on multiple issues at the same time, thus creating more room for mutual consensus than would be possible on any single issue. The CDM option is an example of issue linkage in the sense that it combines the immediate climate change goals of developed countries with the immediate development goals of developing countries. Meaningful participation of developing countries necessitates the linkage of global income redistribution and global climate change abatement. Formal approaches to describe the consequences of such linkages quantitatively and to analyse questions of burden sharing from the perspective of basic emission needs may help to improve the quality of the required dialogue between developed and developing nations.

REFERENCES

den Elzen M., M. Berk, M. Schaeffer, J. Olivier, C. Hendriks and B. Metz (1999), 'The Brazilian proposal and other options for international burden sharing: an evaluation of methodological and policy aspects using the FAIR model', Dutch National Research Programme on Global Air Pollution and Climate Change, RIVM (National Institute for Health and the Environment) report 728001011, Bilthoven.

Groenenberg, H., D. Phylipsen and K. Blok (2000), 'Differentiating the burden world wide - global burden differentiation of GHG emissions reductions based on the Triptych approach: a preliminary assessment', Department of Science, Technology and Society, Utrecht University, Utrecht.

Gupta, J. (1997), 'The climate change convention and developing countries: from conflict to consensus?', PhD dissertation, Amsterdam.

Jansen, J., A. Torvanger, L. Ringius, A. Underdal, C. Battjes, F. Ormel, J. Slijm, C. Volkers and R. Ybema (2001), 'Sharing the burden of greenhouse gas mitigation', Dutch National Research Programme on Global Air Pollution and Climate Change, Bilthoven.

Metz, B. (2000), 'International equity in climate change policy', *Integrated Assessment*, **1**, 111-26.

Phylipsen, G.J.M., J.M. Bode, K. Blok, H. Merkus and B. Metz (1998), 'A triptych sectoral approach to burden sharing: greenhouse gas emissions in the European bubble', *Energy Policy*, **26**, 929-43.

Ringius, R. (1999), 'Differentiation, leaders and fairness: negotiating climate commitments in the European Community', *International Negotiation*, **4**, 133-66.

Ringius, L., A. Torvanger and A. Underdal (2000), 'Burden differentiation: fairness principles and proposals', CICERO WP-1999:13/ ECN-C-00-011, Centre for International Climate and Environmental Research, Oslo, February.

Rose, A., B. Stevens, J. Edmonds and M. Wise (1998), 'International equity and differentiation in global warming policy', *Environmental & Resource Economics*, **12**, 25-51.

Sijm, J., J. Jansen and A. Torvanger (2001), 'Differentiation of mitigation commitments: the multi-sector convergence approach', *Climate Policy*, **1** (4), 481-97.

van Harmelen, A.K., M.J. Boomsma, R.H.J. Korenramp, M. Andersson, A. Mol and H.B. Diepenmaat (2001), 'Building bridges, building dikes: evaluation of the Dutch policy making on climate change', Dutch National Research Programme on Global Air Pollution and Climate Change, Bilthoven.

13. The Dutch energy transition and its institutional problems: report from a stakeholder assessment

Matthijs Hisschemöller

1 SCOPE AND FOCUS OF THE CHAPTER

Climate and energy policies for the decades to come face a serious dilemma. On the one hand, government intervention in most countries has decreased during the last decades and is supposed to decrease further as a consequence of the liberalization of energy markets. On the other hand, the transition towards a sustainable energy system, which implies a drastic reduction of CO_2 and other greenhouse gases in the decades to come, may require a huge government involvement, nationally as well as at the EU level. These observations appear quite irreconcilable. Does this mean that there is an institutional deficit for dealing with the issue of climate change? Should we think of new institutional forms to replace the traditional enforcement mechanisms of government? Or should we look at institutional arrangements in a new way?

This chapter investigates these questions, thereby presenting and analysing some results from the Integrated Environmental Assessment project Climate OptiOns for the Long term (COOL).[1] This project, carried out in the period from 1999 to 2001, identified long-term climate policy strategies for the Netherlands in a European and global context. It addressed the question: how can considerable greenhouse gas emission reductions up to –80 per cent for the Netherlands and Western Europe be realized by 2050? The project did *not* address the question as to whether such reductions might be necessary or desirable.[2]

The project took a participatory approach, as it wanted to take into account and assess the value of different views and perspectives and encourage learning by both scientists and practitioners. At the national, European and global levels, dialogue groups were formed to assess the strengths and weaknesses of policy options for reducing greenhouse gas emissions. These dialogue groups included participants from government agencies, environmental and consumer non-governmental organizations (NGOs), business and agriculture.[3] All participants took part on a personal basis. In the National Dialogue, the main focus of this chapter, the dialogue took place in four groups, each dealing with one sector: industry/energy, housing, transport and agriculture. The project team's main task was to organize, structure and facilitate the dialogue. In the tradition of participatory integrated environmental assessment (Hisschemöller et al., 2001) one of COOL's major tasks was to identify, confront and synthesize state-of-the-art knowledge from various science disciplines and practical knowledge from stakeholders who represent different views and interests related to the subject.[4] Hence, the dialogue groups were supported by scientific support units whose main task it was to provide the dialogue participants with the state-of-the-art scientific information on options for greenhouse gas emission reductions.[5]

Many of the topics discussed relate to technological options to address the climate change issue. The dialogue analysed opportunities and barriers for *inter alia* a large-scale introduction of energy efficiency (all sectors), biomass (especially for industry/energy, transport, and agriculture), CO_2 removal and storage and hydrogen (industry/energy and transport), onshore and offshore wind (housing, industry/energy and agriculture) and solar (housing, industry/energy). In this exercise, the dialogue identified many issues related to the institutional aspects of emission reductions as well as different views and dilemmas with respect to the institutional trajectories most effective to obtain −80 per cent emission reductions by 2050.

Yet, one of the project's main conclusions is that the institutional aspects of the transition towards a low emissions economy need further investigation. Therefore, this chapter is far from conclusive in the sense that it does not provide straightforward recommendations on what institutions are better fit for realizing drastic emission reductions than others. It draws conclusions, though, with respect to the kind of questions that appear relevant. The outcomes of the dialogue indicate that the discussions on the most effective or desirable technologies and institutions do not stand apart. They are linked to one another in the sense that a preference for a certain technology (for example, CO_2 removal and storage or biomass) often coincides with a preference for a specific kind of institution (for example, a market for tradable permits, a benchmarking agreement between government and industry or some kind of regulatory system).[6]

Section 2 presents three trajectories for managing the transition towards a low emissions economy that were discussed within the dialogue groups. Then, Section 3 compares and discusses these trajectories, thereby focusing on the main results from the COOL dialogue across the dialogue groups. This section identifies critical assumptions with respect to the institutions for governing the energy transition and discusses whether these assumptions are conflicting or complementary. Section 4 presents the chapter's main conclusions.

2 THREE TRAJECTORIES TOWARDS A LOW EMISSIONS ECONOMY

This section describes three possible institutional trajectories - or scenarios - towards a low emissions economy: 'regulation', 'emissions trading' and 'shared responsibility'. The dialogue groups discussed what institutional form would be preferable to support the energy transition in their specific sectors.[7]

Regulation

An emissions reduction up to 80 per cent by 2050 can be effectively realized by means of progressive standard setting and long term standards for specific sectors. In the COOL National Dialogue, arguments in support of these forms of regulation have been put forward for the housing and transport sectors.

Dutch policy to improve the energy efficiency of buildings is effectively implemented through progressive standard setting. Initially, the Dutch government used a system of detailed standards, which prescribed how a building (using specific technologies or materials) should meet certain requirements. Since 1995, this system of detailed regulation has been replaced by a system of regulation related to output. Pivotal for the success of this system has been the consensus among parties in the sector on a methodology for measuring the energy efficiency of buildings. The government regularly adjusts the standards with reference to the latest technologies.[8] Standards for new buildings are immediately enforceable. Standards for existing buildings are implemented on a voluntary basis, using subsidies as an incentive.

Progressive standard setting has a positive record in the housing and construction sector. The dialogue group is confident that significant emission reductions in the future may be realized using this regime. Although progressive standard setting is intended to follow technology development rather than to force further innovations, in practice the system seems to work both ways. Since companies know for certain that the energy standards are increased regularly, they tend to shift from a reactive to a proactive attitude. For the housing sector, it is also important to note that the Dutch national government tends to be independent of policy developments elsewhere or at the EU level.

Regulation may thus, as the housing sector shows, relate to progressive development of standards which are enforced and implemented in the *short term*; another possibility is to establish *long-term* standards. This has been proposed for the transport sector as an effective way to shift from conventional fossil fuels to CO_2 neutral fossil fuels. The major advantages of this instrument are that the burden on administration is comparatively low and it prevents unfair competition, which makes it acceptable for business.[9] The assumption which underlies a long-term standards regime is that companies, once they are legally bound, will work to provide technological solutions, because they may expect their competitors to do the same. The key condition for successful implementation of long-term standards is that business is given adequate time to accommodate them. Regulation must take into account the investment cycle characteristic for the sector, for example, the replacement of vehicles, normally once every five to ten years. Hence, governments could require that within a 15–20-year period all vehicles are adjusted to use CO_2-neutral fuels only. In the case of clean fuels for the transport sector, the EU has considered the minimal level to develop and implement long-term standards.

In general, regulatory regimes are considered appealing, as they have an immediate impact on technology development and the adaptation of technological innovations. Regulatory regimes can be called 'technology forcing'. This is especially the case when regulation implies a government choice to promote specific technologies (wind, solar) and push them into the market. But even if regulation is not intended to prescribe the application of technologies in particular and leave it to the private sector to find its own solutions in a competitive way, as in the examples of progressive and long-term standards, policy may in a more indirect way set the agenda for technological alternatives. A second general characteristic of regulation is that it may imply financial mechanisms, such as subsidies or taxation. The National Dialogue groups for both the housing and transport sectors proposed taxation as an effective means for internalizing environmental costs in addition to regulation.[10]

Emissions Trading

It is widely expected that an emission-trading regime at a global, European and national scale is emerging and will become reality in the near future. According to participants in all dialogue groups, trading is probably the most sophisticated mechanism to foster a low emission economy in the context of a liberalizing energy sector. Others, although willing to acknowledge the current trend towards emission-trading regimes, were sceptical with respect to its advantages. In fact, emissions trading may turn out to become a highly controversial instrument, but this will be further explained in Section 3. This section aims at presenting the core assumptions in its support:[11]

- The polluter pays for his or her entire emissions.[12] Under a regulatory

regime, polluters only pay the costs of adjusting to more stringent standards.
- Another distinction with regulation is that society sets environmental targets but private parties are free in their choice of how to cope with these.
- Compared to other market instruments, especially ecotaxation, trading has the advantage that private parties are free to decide how to spend their money.
- Investments in emission reductions may be financed by selling emission rights. If reductions in the own company or sector are too expensive, the money can be invested in reductions elsewhere.

Basically, the procedure may be as follows: private parties, such as energy companies, are entitled to a certain quantity of emission rights for a certain period of time (five or ten years). Governments may initially provide the rights for free or through an auction. According to economic theory, parties will start trading emission rights if they need more or less than they initially acquired. Supply and demand determine the market price for CO_2. If the (marginal) costs of emissions reduction outweigh market price, parties will try to sell their rights. This observation relates to two additional arguments with respect to the effectiveness and efficiency of trading. First, private parties are always in search of opportunities to maximize profits. The opportunity to sell emission rights provides them with an incentive to look for ways to reduce their emissions in a cost-effective manner. Second, the incentive to reduce emissions implies a continuous stimulus to search for technological innovation.

As the total number of allocated rights, which is related to Mt CO_2, gradually decreases over time, the CO_2 market price increases, which enables private parties to implement reduction options that were not cost-effective before.

At this point in time a trading system is emerging worldwide. Emissions trading has been embraced by many in the COOL dialogue, especially because of its claim to provide a level playing field and at the same time contribute to cost-effective solutions. However, a well-functioning emission-trading regime may still be far away, at a global level in particular. Emissions trading must cope with some major difficulties that it cannot avoid.

The prerequisites for trading must be present
Markets imply competition, both on the supply and the demand sides. However, it is in the interest of private parties to limit competition as much as possible. Limitation of competition disturbs the working of any market. This problem can be overcome if a large number of parties is entitled to trade

emission rights. The more parties present and the more diverse their interests, the lower is the risk that parties will be able to frustrate the market by hidden agreements and alliances. From this perspective, a per capita allocation of emission rights might be the best way to create a well-functioning market.

The role of government
A market is often considered an institutional arrangement, which limits the role of government to maintain the rule of law. This is, however, different in the case of tradable emission rights. To start with, government must create a market. Then, regular interventions are required to keep the market working in such a way that reduction targets are met. There seem to be three problems in particular. First, a cap must be established and periodically revised in a downward direction. Second, parties must be entitled to emission rights. Third, emission rights must be distributed among the parties. It seems rather obvious that these matters may cause political controversy, both at the start of the process and later.

The cap The cap must correspond to international obligations and will be revised accordingly. If private parties see that the regime works satisfactorily and are confident that future reductions will be realized, they are unlikely to bother much about cap adjustments. If a substantial number of parties fail to reduce according to schedule, then political turbulence will arise. If the views diverge too much or this happens too frequently, the regime itself may come under serious strain. It is hard to predict how to deal with a heavy decline in social support for a trading regime.

Identification of parties Who will be provided with emission rights initially? Who will be entitled to buy and sell them? Is it only the major emitters, such as the energy producers, or do we also think about consumers? Assuming that a trading regime will be installed worldwide, will the identification of parties in various countries be made in the same way? How can the exclusion of a sector or group be legitimized? Is it desirable to distinguish between parties who are active on the world market (the so-called 'exposed' sector) and parties who are active at a national or local scale (the so-called 'sheltered' sector)? The identification of entitled parties is a complex political choice which will largely influence the further process. Some dialogue participants showed a strong preference to allocate emission rights on an individual basis. This is expected to empower consumers, which might accelerate the realization of emission reductions.

Two further issues are especially important if the regime is to be installed on a global scale. First, such a regime might function properly to the extent that governments maintain and enforce environmental laws, which is a well-known problem in the Netherlands today. Furthermore, a trading regime works to the extent that it is able to resist corruption and fraud. Many countries do not yet have a well-functioning market, let alone a market for emissions trading. These observations may lead to the conclusion that building the institutions for emissions trading on a global scale will still take some time. It should also be noted, however, that markets for CO_2 are already functioning, but not in a transparent way and with little government control.

Shared Responsibility

So far, this section has described one regime which is widely practised (regulation) and one regime of which much is expected, but which is still in its infancy (emissions trading). To encourage critical reflection, one trajectory is added here: an emissions reduction up to 80 per cent by 2050 can be realized by private parties who cooperate to achieve self-established emission targets on a voluntary basis. Initially, government does nothing. Shared responsibility was not discussed within the dialogue groups, but played a role in the discussion across these groups. It articulates important elements from the discussion that are not or are barely dealt with in regulation and emissions trading.

The notion of the government taking a wait-and-see position does not seem to make much sense, especially since it is bound to implement the climate change treaty and its protocols, which include binding quantitative commitments. As will be argued below, under a shared responsibility scenario, a social demand for government intervention is likely to emerge; government policies may even evolve into forms of regulation or emissions trading. The relevance of this scenario as a thought experiment is that, in contrast to regulation and trading, it takes a bottom-up approach. It does not assume, as these policy scenarios do, that socio-economic actors are basically unwilling or unable to voluntarily engage in environmental change.[13] Why is it so important to question this assumption? If the assumption is wrong, and this might be the case more often than one may expect, policies provoke resistance not because of their goals but because of their style.

Shared responsibility starts with two basic assumptions. First, private parties are willing to develop a climate policy on a voluntary basis, because there is broad social agreement that significant emission reductions are needed. Parties are committed to a 'fair share' in total reduction up to 80 per cent by 2050, provided that current knowledge on the seriousness of climate change risks is confirmed. Second, at the same time, parties are uncertain as to whether they will be able to realize significant reductions. At least, many of them do not feel confident that they can handle the matter by themselves.

Innovations are to be realized through cross-fertilization among and within sectors of the economy and different disciplines.

For a good understanding of this trajectory it is important to note that two further assumptions, often inferred from those above, are left out here, because they are unnecessary and even confusing. First, it would be naive and unwarranted to infer that at a given point in time all social actors would be simultaneously working to reduce emissions. It is sufficient if there is a particular group that has an influence on society as a whole and the power to involve laggards. What we then see is not yet a regime but a process of regime formation; those who participate do not know yet where this will end and what the regime will look like. Second, it would be easy to make the inference that, when parties seek for cooperation and exchange of knowledge, 'shared responsibility' is just another term for the Dutch *polder* model of consultation and accommodation. It is easily assumed that this trajectory resembles this model of (to a large extent) self-managing policy networks that consist of many actors, government being just one. This model is highly fashionable today, especially in the emerging policy science of transitions and transition management (Rotmans et al., 2001). But this is not the implication of shared responsibility, because this scenario includes the assumption that, in the course of socio-economic interactions, a demand for government intervention may emerge.

The starting point in shared responsibility is that parties decide to take their own. The first and necessary step is for each of them to develop insights with respect to emissions and opportunities for emission reductions. What emerges is a (virtual) market for the exchange of knowledge, this concept being broadly defined as technological, behavioural and process knowledge. A second step is probably to find an *accounting system* for greenhouse gas emissions. This enables each party (including many companies and individuals) to make its own emissions visible and to calculate its share in the total emissions. Then, parties develop programmes to reduce their emissions accordingly. Where a party is dependent on the behaviour of other parties (for example, energy distribution companies are dependent on energy producers as well as consumers) they encourage the others to do the same.

What specific tasks for government can be identified if it is asserted that government responds to a societal demand for climate policies? The role of government will initially be limited to facilitation and support. Government may assist with respect to the development of an accounting system which enables private parties to map their greenhouse gas emissions, shapes conditions for a transparent knowledge infrastructure (see van der Sluijs et al., 2001), and negotiates with private parties who have an interest in binding agreements with government. Then, government may be asked to address existing practices which constitute major barriers for private actors to reduce their emissions. In this respect, the dialogue groups have pointed to many examples, including the present knowledge infrastructure, incentives from the Kyoto Protocol to burn rather than to recycle wood and a policy bias in favour of complex high-tech solutions rather than cheap and simple solutions to be realized by integrated design. Taking away barriers may imply that existing regulation is revised or abandoned. It can also generate a demand for new regulation, especially to internalize environmental costs. Thus, societal initiatives to promote significant emission reductions may provoke strong pressures on government to address laggards.

It might be speculated that, after a while, other, more classic types of intervention will be asked for in order to assist private parties with the social demand for climate policy. Private parties who seriously try to realize a long-term reduction target up to 80 per cent may express a joint interest in monitoring their efforts. They may ask government to address free-riding. Since planning for a period up to 50 years is impossible, they might ask government to formulate interim targets for a period of, say, 7 years. They may further be interested in translating these interim targets into binding agreements for each sector or company. The main reason for doing this is to balance the various efforts.

For both government and the private parties, there are roughly speaking two possible strategies to accomplish this. First, for parties facing similar problems with respect to reducing emissions, such as the producers of motor vehicles, government could impose long-term targets on them, as this chapter pointed out under Regulation.

The second strategy is to set a price for CO_2. This strategy may apply to dissimilar parties who face quite different opportunities and obstacles in realizing emission reduction targets. Perhaps parties will agree to what might be called a 'voluntary ecotax'. This tax is not paid to government but invested in the emission reduction programme developed by the party itself. Given the context of the mainstream debate on institutions and instruments, a voluntary tax seems a contradiction, but such a regime might very well be consistent with the shared responsibility trajectory. In addition, it is conceivable that such a regime, as a consequence of societal demand, evolves into something like a regime for emissions trading. If private actors realize that the output of their programme falls short, they may prefer to transfer their investment money to parties able to realize substantially more reductions.

To conclude, although government does not initiate climate policies in shared responsibility, as in regulation and emissions trading, this does not imply that (national and European) government is absent. On the contrary, the role of government remains extremely important. The most important distinction with the two other scenarios is that, in shared responsibility, climate policies are not aimed at making private parties do things that they would not have done otherwise, but to facilitate the realization of emission reductions that parties had already decided upon themselves. The lesson here is that policies are much more effective to the extent that stakeholders have the opportunity to initiate a policy regime in their own interest.

A second lesson from this exercise is that the needs for policy and policy regimes change over time. What starts as a bottom-up voluntary process may end up as a kind of regulatory regime, which may again evolve into an emission-trading regime over time. So, trajectories or regimes that are, at first glimpse, based on conflicting assumptions may turn out to be complementary if a longer time frame is taken into account. Those who are resistant to government steering at a certain point in time, may articulate a demand for steering at another point in time. The shared responsibility trajectory is, therefore, not at all equivalent to policy networking in the Dutch *polder* model. Yet, this trajectory cannot be considered as some alternative institution; it primarily helps to see existing institutions in a new light.

3 CONFLICTING STAKEHOLDER ASSUMPTIONS LINKING TECHNOLOGY AND INSTITUTIONAL FORM

This section explores the issues that may be critical for addressing the most challenging question for the work to be done in the near future: this question is, what kind of institutional trajectory would enable the Netherlands to realize emission reductions up to 80 per cent by 2050? Apparently, this is not one of the trajectories discussed above. Although many expect regulation and emissions trading to be part of the institutional solution, there are many uncertainties and doubts as to whether these would produce the intended outcomes. Taking a cross-sectoral approach, one may expect conflict between institutional approaches at the sector level and it is still highly uncertain what an integrated institutional strategy should look like. The National Dialogue and the other COOL dialogues do not provide straightforward answers to this question but they provide some direction where the answer can be found.

The National Dialogue results per sector are summarized in Table 13.1.[14] Note that the dialogue has used the concept 'trajectories' in a technological rather than an institutional sense. In brief the results for the sectors can be characterized as follows:

- *Housing* Reductions up to 80 per cent likely, no major technological innovations needed.
- *Industry and energy* Reductions up to 80 per cent uncertain but possible. Results may largely depend on the efficiency improvement to

Table 13.1 Findings for the four sectors in 2050

	Sector			
	Housing	Industry and energy	Agriculture	Transport
Trajectories identified by the dialogue groups	Existing buildings: sustainable	Clean fossil: CO_2 storage, H_2, efficiency	Primary sector Energy and materials	Drawback of demand for transportation
	New buildings: sustainable	Sustainable energy system: Biomass, wind, solar, high efficiency	Sinks	Efficiency – modal shifts Clean fuels
		Hybrid: CO_2, H_2, sustainable		
Emission reduction	80–90%	50–100%	100–150%	??–100%

be realized in industry and the social acceptance of large-scale CO_2 removal and storage;

- *Agriculture* Reductions up to 80 per cent likely, additional reductions in other sectors possible.
- *Transport* Reductions up to 80 per cent likely, however largely dependent on the availability of clean fuels, which in turn will be dependent either on the large-scale availability of biomass or on the availability and acceptance of clean fossil fuels.

These results, quantified by the scientific support team, provide a quite optimistic picture of opportunities for emission reductions which is to a large extent compatible with the conclusions in the IPCC Third Assessment Report (2001) and a report by the group Reconsidering Energy Policy (Bezinningsgroep Energuiebeleid, 2000): technologies are available against reasonable cost, but the main obstacles relate to institutional factors.[15]

However, a closer look reveals that dialogue participants were less optimistic than this general observation suggests. To understand how the participants felt, one must clearly distinguish between the general picture and the picture with respect to the various crucial options. Many agreed with the technological feasibility of –80 per cent in general but were sceptical with respect to the feasibility of the crucial reduction options: CO_2 removal and storage, renewables, biomass and efficiency. Large-scale removal of CO_2 met with scepticism because of the risks associated with it for the long term.[16] At the same time, it is acknowledged that renewables (solar and wind in particular) are by and large unable to satisfy the Dutch energy needs for the coming decades.[17] If climate risks turn out to be as serious as expected by the IPCC, there might be a willingness to accept CO_2 removal and storage. Nevertheless, it is seen by many as a necessity for the time being, only to be applied in the period of transition to a sustainable energy system.

The biomass option is greeted with scepticism mainly for two reasons. On the one hand it is doubtful whether biomass will be available in the quantities needed without leading to social disruptions in the countries of production (in Africa, Latin America and Eastern Europe). On the other hand, sustainable use of biomass is considered necessary, which implies starting with high-level use (for example, building materials) and then slowly moving downwards (pulp and paper) to finish in the oven or in the car. Dialogue participants were sceptical as to whether institutional arrangements can be created to safeguard sustainable production and use of biomass.

Efficiency is considered of utmost importance, too. Some, however, questioned the physical possibility of maintaining the efficiency curve for the decades to come. Besides, it was widely felt that an improvement in efficiency is likely to slow down emissions but, generally, efficiency will be unlikely to reduce emissions, given the expected economic growth.

Given these considerations, it is no surprise that the dialogue groups, asked to identify immediate actions to be taken by government, business, science and consumers, and in spite of their prevailing general view that significant reductions are possible using state-of-the-art technology, stressed the need for research and development (R&D). Moreover, they stress the need for cooperation within and across sectors. In particular, the reports of two discussion groups from business are interesting in this respect.[18] One discussion group (industry, big companies) reported that it is of primary importance to enhance 'new arrangements for research to enable a contribution for the long term that exceeds the interests of the individual companies'. Private and public institutions must be involved in these arrangements, which must be based on the principle of 'non-competitive exchange of knowledge'. At the same time, this group argued for experiments with emissions trading. Another group (industry, including representatives from small business) stressed the need for finding an arrangement for assisting smaller companies in establishing new products in the market. Traditionally, this role of 'lead customer' has been taken by government, but the government's role seems to be declining. At the same time, this group strongly argued for emissions trading. It was widely acknowledged though that fundamental research is being neglected by both business and government. This is considered to be an issue of major concern.

Reflecting upon the outcomes of the discussion, it would seem that the participants' views underlie different assumptions with respect to climate policy regimes which cannot immediately be explained by diverging preferences for regulation, emissions trading or shared responsibility. Basically, different views on the institutions needed for realizing an energy transition seem to underlie two dimensions, which reflect the following questions:

- Are specific knowledge and the technology needed for the energy transition available?
- How should emission reduction options be developed and/or implemented?

Using a simple dichotomy, the answer to the first question may be Yes or Not yet. The answer to the second question may be: by competition, or, by cooperation. Four different views can be traced to these dimensions, which link preferences with respect to the institutional features, competition or cooperation, to considerations with respect to the availability of knowledge and technology. These views are summarized in Table 13.2.

The table articulates, in an ideal typical sense, the 'extreme positions' with

Table 13.2 Views on institutions needed for realizing the transition towards –80 per cent greenhouse gas emissions by 2050

Technology available?	How to realize the option?	
	Through cooperation	Through competition
Not yet	R&D through cross-sectoral cooperation. Technology choice and R&D infrastructure D	Long-term standard setting for specific sectors or technologies technologies B
Yes	C Support innovations that are still weak in the market (create lead customers)	A Acceptance by market parties and consumers

respect to both questions. Obviously, reality is more complex than the table suggests. As the COOL dialogue already indicates, the question as to whether technology is available may be surrounded by many physical, social, economic and political uncertainties. With respect to the institutional features, it should be acknowledged that competition and cooperation are not by definition exclusive categories. Reality is also more complex, because Table 13.2 leaves out dimensions that may also have a bearing on the institutional debate.[19] But even if the oversimplification of the table is ignored, the complexities that emerge are huge. Before going into these, the distinct positions in the cells of the typology are addressed; each cell articulates a solution for a specific problem.

Cell A articulates the view that competitive technologies, including CO_2 removal and storage, are available. The major problem addressed in cell A is how to force these technologies into the market. Setting a price for CO_2 is supposed to facilitate the acceptance of CO_2-neutral technologies by companies and consumers. Cell B articulates both a problem and a solution that are different from those in cell A. It is assumed that, for specific sectors such as the car industry, CO_2-neutral technologies are not yet fully developed, but there are serious options with high potential. The problem is how to encourage companies to fully develop and implement these options. The solution is found in long-term targets that are imposed on business. Cell C articulates a set of problems that arise when technologies are available but there is no market for them or there are market deficiencies. One problem situation that may be referred to here is that the product is not competitive, such as solar technology.[20] Another situation might be a nationally oriented sector which hardly allows for competition from abroad, such as the Dutch construction sector.[21] Yet another situation occurs when a certain product needs early movers before it can become competitive.[22] Regimes that provide some sort of immediate regulation, like progressive standard setting, are capable of dealing with many of these problems. Cell D articulates a strategy, which provides technological innovation by public-private mega investments. Precedents are the Apollo and Manhattan projects in the US or the European Space programme. These kinds of programme have led to significant technological innovations in the twentieth century. The key problem is the choice of the 'best' technologies. Government does not have a good record here, but nor has the market. Most likely, government is the only party capable of organizing and implementing mega investments to foster breakthroughs.

If it is possible to distinguish four positions, which both articulate a problem as well as a corresponding institutional solution, and provided that the dialogue has produced more than contingent results in this respect, what makes these results so complex? The answer to this question is that as yet we do not know how the positions in Table 13.2 relate to one another. The COOL project only suggests that they can be conflicting or complementary.

The cells of the typology can be related to *conflicting* assumptions with respect to the same problem situation. This can be illustrated by the discussion on efficiency. Some of the dialogue participants argued that in the last 50 years, efficiency technologies were developed and implemented under the pressure of energy crises. Their forecast is that under pressure industry will succeed in realizing further efficiency measures. Others argued that efficiency needs further technological breakthroughs, which can only be realized by cooperation between industries (non-competitive exchange of knowledge). In this case, assumptions may turn out to be in conflict.

However, the views articulated in the cells of the typology can also be related to different problems. What the problem looks like may then be dependent on the specific sector that one has in mind. As regards the availability of technologies and knowledge, if the question is addressed from the perspective of the housing sector, the answer most likely is that the Netherlands can realize –80 per cent using state-of-the-art technology. An additional assumption is that consumers are prepared to pay for the image of renewables, but there is still a price difference with fossil fuels. If one addresses this question from the perspective of the industry or transport sectors, the answer is likely to be that major technological breakthroughs have to be realized. Theoretically speaking, knowledge may be available, but it is expected to take decades or more to realize applications. In this situation, assumptions may turn out to be *complementary* rather than conflicting.

So, the consistency between the different positions in the typology may vary according to the situation. As the illustrations above indicate, the level of analysis may be relevant here. The conclusions with respect to efficiency 'in general' will be different from those with respect to efficiency at the sector level and again if specific technologies are taken into account. Yet another important variable is time. The development of promising options is unequally distributed over time. Options and even sectors may be confined in different phases of the energy transition at one point in time. Table 13.2 can be understood as a process model: an option may be developed in cell D, be introduced into the market through C and adopted through A. At that very moment, however, another relevant option can benefit from C or B. If the level of analysis and the factor time are taken into account, it becomes clear that we have no idea whether options are consistent.

Therefore, even if we start with an oversimplified typology, linking availability of knowledge to either competition or cooperation, the complexities are huge. This becomes especially evident when the two major characteristics of transitions are taken into account, as defined by Rotmans et al. (2001): transition processes take a long time and are characterized by unequal developments at different levels of society. It is of course impossible to design institutions in such a way that all options blossom and this is not the purpose of transition theory or management. What must be concluded, however, is that institutional choice by definition implies a choice of technology. This issue cannot be avoided by proclaiming that technology choice is left to the market, since the choice for the market already implies that some selection of technologies takes place. Nor can one escape from the issue by proclaiming that as many as possible options will be taken into account for as long as possible. Even then, choices with respect to institutions and technologies are unavoidable.

It might be agreed that the choices to be made must be based on careful deliberation rather than coincidence. However, we simply do not yet know how different institutions may affect the development and implementation of innovations over time. Given the above reflections on the results of the COOL dialogue, Table 13.2 may be no more and no less than a valuable starting point for asking the 'right' questions, especially with respect to the availability of knowledge and the need for institutions to foster competition and/or cooperation.

4 CONCLUSIONS

This chapter started by pointing to the observation that there is an institutional deficit in dealing with the issue of climate change. Hence, should we think of new institutional forms to replace the traditional enforcement mechanisms of government? Or should we look at institutional arrangements in a new way? This chapter shows that there is a need for both but stresses the latter. The most challenging question for long-term climate policy and for the energy transition appears to be how to deal with the issue of technological and institutional choice over time.

The first conclusion from this chapter is that evaluative judgements and preferences with respect to institutional trajectories such as regulation, emissions trading or shared responsibility, (implicitly) articulate a view on technological innovation. Also, judgements and preferences with respect to technological options imply a need for specific institutional arrangements. This conclusion fits in with the increasing attention to the relationship between technology and institutions, but is not generally accepted yet, especially when it comes to scientific practice.

After all, to compare and integrate options related to technology and institutions, requires an interdisciplinary approach. Remarkably, this is by and large not the case in the Netherlands today. Many economists argue for market solutions without being interested in technology. Policy scientists argue about transition management as process management without having a sincere interest in substance. Technologists are fascinated by technology and much less by the environmental problems these technologies might help to solve. A successful investigation into the roles of government and private actors and the institutions to accompany them in the process of transition must, first of all, include the various disciplines such as policy sciences, economics, law, and technology sciences, and enable scientists who are used to telling their own story, to learn from one another.

Second, this chapter points to two dimensions that deserve special attention in further exploring the energy transition. The investigation should take account of insights, expectations and conflicting views with respect to the question: do we have the technology available? At the same time, and not as a separate (in which case, abstract) exercise, it will ask: should the institutions in support of the transition foster a bias towards competition or towards cooperation and, more specifically, by whom, when and how? Addressing these questions will raise some difficulties because of the complexity of the issue, especially if its multilevel and long-term character is taken into account.

The third conclusion, implicated by the COOL dialogue and reflections in this chapter, is a methodological device that may turn out to be critical. Looking into institutions in a new way requires that, alongside an interdisciplinary approach, the views and the knowledge of stakeholders are taken into account. There are plenty of scientific studies into technological and institutional effectiveness. Most of them do not and cannot address the issues from the floor. The relevance of stakeholder participation has been justified in several ways. The COOL dialogue points to one in particular: practical knowledge from stakeholders from business, government and environmental NGOs is indispensable for an integrated assessment of this complex issue. Relying on stakeholder knowledge and preferences may provide a short-cut in addressing the many complexities of the energy transition. Stakeholder-scientist interaction, may therefore turn out to be a more effective approach than a purely scientific one. Perhaps surprisingly, it may prove to be the only approach.

NOTES

1. The COOL project was initiated by the Dutch National Research Programme on Global Air Pollution and Climate Change. The project was carried out by Wageningen University, the National Institute for Health and the Environment (RIVM) and the Institute for Environmental Studies at the Vrije Universiteit Amsterdam. The author thanks the many participants and team members who shared their insights. Without them this chapter would not have been possible. I would especially like to thank Marleen van de Kerkhof, Willemijn Tuinstra and Magnus Andersson. I am indebted to Harmen Verbruggen and Onno Kuik for their comments and suggestions with respect to Section 2, Emissions Trading.

The Dutch energy transition 309

2. For a full report on the results and recommendations from the COOL project, see Berk et al. (2001a), for the COOL National Dialogue, Hisschemöller (2001), Hisschemöller and van der Kerkhof (2001), for the European Dialogue, Andersson et al. (2001) and for the Global Dialogue, Berk et al. (2001b).
3. Among them Shell, DSM, Corus, National Capital Bank, AkzoNobel, Essent, Gasunie, Greenpeace, Church and World, Ministry of Environment, Ministry of Economic Affairs (dialogue group Industry/Energy), Nuon, Wilma, Siemens, De Clerque Planontwikkeling, DuBo, Klimaatverbond (city of Tilburg), Global Action Plan, Ministry of Housing (dialogue group Housing), Nederland Distributieland, Van Gend & Loos, CTE-Rotterdam, Railnet, Friends of the Earth, Dutch Federation of Labour Unions FNV, Ministry of Transport (dialogue group Transport), LTO-Nederland, Cosun Suikerunie, Rabobank, Dutch Federation of Country Women, WWF (dialogue group Agriculture).
4. This chapter does not address issues related to dialogue process and the lessons drawn in this respect. For the process evaluation of COOL, see Hisschemöller and Mol (eds) (2001), van de Kerkhof and Hisschemöller (2001) and van de Kerkhof et al., forthcoming.
5. Scientific support was mainly provided by Utrecht University, the National Institute for Health and the Environment (RIVM), the National Energy Research Center (ECN) and Ecofys (see Faaij et al., 1999; Treffers et al., 2001).
6. (New) institutional theory defines institutions as formal and informal rules of the game that shape actors' behaviour. It thereby often distinguishes between institutions and organizations (for example, Young, 1999). This chapter follows this definition in so far it stresses the importance of the implicit and non-organization bound institutions (like markets), but it does not exclude organizations from the concept, as they may also impose rules of conduct and behaviour to individuals and groups.
7. This section is largely based on Hisschemöller: 'De bestuurlijke aspecten van lange termijn klimaatbeleid; Notitie ten behoeve van de integratieworkshop' (The governance aspects of long-term climate policy. Note for the integration workshop), Paper discussed at the COOL National Dialogue Integration Workshop, 1 and 2 March, 2001.
8. The method for measuring, EPN, and the number to express a building's achievement, EPC.
9. The transport group refers to positive experiences with this policy instrument in the case of reducing aeroplane annoyance.
10. For the Dutch housing sector this is congruent with current practice, where the construction sector is bound to progressive standard setting, the consumers pay ecotax and receive subsidies for purchasing the most energy-efficient durables.
11. There are many forms of emission-trading regimes as with regulation. Discussions with respect to this instrument tend to focus on technical details, which however are beyond the scope of this chapter. Elaborate analyses of the technical aspects of emissions trading are found elsewhere in this volume.
12. This claim, which also applies to ecotaxation, can be considered a moral critique on existing regulatory practice. One may argue that the claim is unjustified when the polluter acquires emission rights for free, but this is not relevant. Through an emission-trading regime, all emissions that a company or individual produces will represent a money value that has to be paid for at some point in time.
13. Shared responsibility shows features from the voluntary agreements approach, characteristic for Dutch environmental policy. The *benchmarking agreement* between government and industry holds that Dutch industry will play its part in the reductions under the Kyoto Protocol by remaining a world leader in energy efficiency. In exchange government will not impose taxes or other measures on industry for the years to come. For *shared responsibility*, the point of departure is, however, not a voluntary agreement between private parties and national government, but initiatives by private parties themselves. Yet, a benchmarking agreement may be the outcome of a process in which private actors seek cooperation.
14. For a more detailed discussion, see Hisschemöller (2001).
15. The National Dialogue recognized that cost effectiveness is one of the most important criteria in assessing the feasibility of options for emission reductions. However, costs as such do not play a major role in the long-term assessment, since it is impossible to predict the cost development of options over a 50-year period.

16. Theoretically, because of geological reactions CO_2 could escape into the atmosphere, causing death by suffocation for those in its immediate vicinity. Besides, it is doubtful whether governments are capable of monitoring large CO_2 stores for an infinite period of time.
17. Renewables may fulfil the needs of the housing sector, provided that consumers are willing to pay for the positive image of these technologies.
18. At the concluding Integration Workshop of the National Dialogue, Doorn, March 2001.
19. One of these is certainly the expected role of consumers, alongside government, industry and environmental NGOs, in the energy transition.
20. Hence, the plea for a protected area for renewables in a liberalized energy market.
21. Timber construction is common in the US but still rare in the Netherlands.
22. New products suffer from teething problems, the heat pump being an example. This product appears to be widely adopted in Sweden and the US, but the experiences in the Netherlands have been discouraging.

REFERENCES

Andersson, M., W. Tuinstra and A.P.J. Mol (2001), *Climate OptiOns for the Long term (COOL). European Dialogue*, Wageningen University, NOP (Dutch national research programme on global air pollution and climate change) report 410.400.117, Bilthoven.

Berk, M.M., M. Hisschemöller, T. Mol, L. Hordijk and B. Metz (2001a), *Strategieen voor lange termijn klimaatbeleid. De resultaten van het COOL-project* (Strategies for long-term climate policy. The outcomes of the COOL project), Uitgeverij Programmabureau NOP, Bilthoven.

Berk, M., J.G. van Minnen, B. Metz and W. Moomaw (2001b), *Climate OptiOns for the Long term: COOL Global Dialogue*, RIVM, NOP report 410.200.418, Bilthoven.

Bezinningsgroep Energuiebeleid (2000), *Klimaatprobleem: Oplossing in zicht* (Climate problem: solution within reach), Bezinningsgroep Energiebeleid, Delft.

Faaij, A., S. Bos, J. Spakman, D.J. Treffers, C. Battjes, R. Folkert, E. Drissen, C. Hendriks and J. Oude Lohuis (1999), *Beelden van de toekomst. Twee visies op de Nederlandse energievoorziening ten behoeve van de Nationale Dialoog* (Two future images of the Dutch energy system, prepared for the National Dialogue), Department of Science, Technology and Society, Utrecht University, Utrecht.

Hisschemöller, M. (2001), *The National Dialogue (COOL). Results and recommendations* (Dutch version: Resultaten en aanbevelingen van de Nationale Dialoog), Instituut voor Milieuvraagstukken, Amsterdam,

Hisschemöller, M. and A.P.J. Mol (eds) (2001), *Evaluating the COOL-dialogue*, Wageningen University and NOP Report 410 200 120, Bilthoven.

Hisschemöller, M., R.S.J. Tol and P. Vellinga (2001), 'The relevance of participatory approaches in integrated environmental assessment', *Integrated Assessment*, **2** (2), 57–72.

Hisschemöller, M. and M.F. van de Kerkhof (eds) (2001), *Climate OptiOns for the Long term - Nationale Dialoog. Deel B - Eindrapport*, Amsterdam: Institute for Environmental Studies E (Ext. r. no. 01/05) and NOP, Bilthoven.

Intergovernmental Panel on Climate Change (IPCC) (2001), *IPCC Third Assessment Report*, Cambridge, UK: Cambridge University Press.

Rotmans, J.R., M. van Asselt, F. Geels, G. Verbong and K. Molendijk (2001), *Transities en Transitiemanagement. De casus van een emissiearmeenergie voorziening* (Transitions and transition management. The case of a low emission energy system), ICIS, Maastricht.

Treffers, D.J., A. Faaij, J. Spakman and A. Seebregts (2001), 'Exploring the possibilities for setting up sustainable energy systems in the long term; two visions on the Dutch energy system in 2050', Department of Science, Technology and Society, Utrecht University, Utrecht.

van de Kerkhof, M. and M. Hisschemöller (2001), 'Stakeholder participation as a process of policy oriented learning. Methodological experiences from the Dutch project Climate OptiOns for the Long term', Paper presented at the IHDP (International Human Dimensions Programme of Global Environmental Change) Open Science Meeting, 6-8 October, Rio de Janeiro.

van de Kerkhof, M., M. Hisschemöller and M. Spaniersberg (forthcoming, 2003), 'Shaping diversity in participatory foresight studies. Experiences with interactive backcasting in a stakeholder assessment on long-term policy in the Netherlands', *Greener Management International*, Special Issue on Foresighting and Innovative Approaches to Sustainable Development Planning, January.

van der Sluijs, J.P., M. Hisschemöller, J. de Baer and P. Kloprogge (2001), *Climate Risk Assessment: Evaluation of Approaches*, Synthesis Report, Utrecht University, Programmabureau NOP, Bilthoven.

Young, O.R. (1999), *Institutional Dimensions of Global Environmental Change*, IDGEC Science Plan, IHDP Report Series, Bonn.

14. Modulating dynamics in transport for climate protection
René Kemp and Ellen Moors[1]

1 INTRODUCTION

This chapter delineates a modulation approach for reducing greenhouse gas emissions from transport. In transport many instruments are available to reduce greenhouse gas emissions:

- land-use planning;
- pricing/taxation (road tolls, emission and vehicle purchase taxes based on fuel consumption, congestion pricing, scrapping bonuses);
- infrastructure/mode management (investment in transit information systems and light rail, Park and Ride schemes, bus lanes, cycle priority and road space);
- standards (for emissions, fuel quality and targets for alternative fuels);
- freight management and general economic policies (ecological tax reform, tradable mobility credits).[2]

The instruments may promote the use of climate-friendly technology and behavioural changes that confer climate benefits, besides other benefits. Examples of such technologies and behavioural changes are: low-emission vehicles, public (collective) transport, traffic management technology (consisting of driver information systems, cruise control, vehicle guidance (platooning)), teleworking and teleshopping, local recreation patterns, low mobility life styles and intermodal travel).[3]

Climate protection may be an additional reason for the introduction of such instruments but it is unlikely to be a driver. Climate protection is a minor concern within transport policy; it coincides with the policy goal of energy saving but that is not an important target for national transport authorities.[4] In transport the overriding concerns are: accessibility, safety, environmental protection, noise reduction and 'liveability' of neighbourhoods. In the absence of a commitment to reduce greenhouse gas (GHG) emissions from transport, climate protection has to ride piggyback on other issues.

Much has been written on the instruments available for making transport more sustainable (for instance, Verhoef 1994; Banister et al. 2000; Feitelson et al. 2001). We shall not repeat this. Rather than talking about the choice of instruments we shall focus on modes of governance and process management (modulation) as mechanisms for altering the dynamics in climate-friendly directions. The reason for this is twofold: first, for instruments to be effective they have to fit the context in which they are applied, and second, joining in with *ongoing dynamics* is often easier than forcing changes.

Thus, in terms of the intervention measures, the focus will be on *modes of governance* and *intervention approaches*, not on instruments. By modes of governance we mean the relationship between business and government, in the case of environmental issues: the reliance on self-regulation, regulation, economic incentives, covenants (agreements), the involvement of private actors in policy-making processes, and the enforcement styles (the extent to which regulations are enforced upon business by private or public enforcement agencies).[5] The mode of governance has to do with how policies are made and enforced, the standard operating procedures within policy subsystems, who decides, on what basis, the devolution of responsibilities, and policy principles such as the polluter-pays principle, the precautionary principle, the reliance on concepts of best available technology as the basis for permits and so on.[6] Climate protection policy can be pursued under the principles of internalizing external costs and the precautionary principle.

The mode of governance is related to *policy belief systems* (see Sabatier and Jenkins-Smith 1993) about the appropriate role of actors, appropriate intervention strategies and to rights and obligations (such as the right of freedom to travel and the obligation of automobile manufacturers to produce cars that are safe and clean) although there may be a gap between the dominant policy beliefs and actual policies.

By *intervention approaches* we mean the type of interventions: command and control, agreements, economic incentives and the use of policy models such as transition management.

The reason for focusing on policy approaches and wider governance aspects is that they are not considered subordinate to the instrument choice issue. Processes of agenda building and anticipation for guiding investment and imagination and the use of experiments are considered to be just as important as the actual policies for achieving GHG reductions through climate-friendly investments (say a CO_2 tax). One still needs traditional policy instruments that change the economic frame conditions but these should be complemented by policies aimed at learning and institutional change.

Dealing with climate change is never a simple choice between policy instruments: for example, taxes versus standards or tradable permits. The dynamics of sociotechnical change have to be taken into account and steering cannot be more than modulation of ongoing dynamics. Without that, policy interventions are likely to be futile or counterproductive. Technical change is typically cumulative and patterned, embedded in sociotechnical regimes and shaped by the sociotechnical landscape that surrounds it, but there are also niches for 'variations', all with their own type of development (Rip and Kemp 1998; Geels 2001).[7] These developments have their own dynamics which should be taken into account. Policy should be concerned with dynamics and issues of learning and embeddedness rather than the abstract choice of instruments. One way to influence dynamics for instance is through the creation of niches for promising technologies. This is the tenet of strategic niche management (Kemp et al. 1998a and 1998b; Weber et al. 1999; Weber and Dorda 1999; Hoogma et al. 2002): the creation and management of niches for new technologies or new configurations. The aim is not so much to push them into use, but to develop them and create paths in a mutual way.

Accordingly, this chapter does not point out specific instruments or tools to achieve climate change goals, but instead focuses on modes of governance and intervention approaches for modulation, the outcomes of which are not predetermined or predefined. The focus is on 'de facto' modes of governance: the alignments and dynamic interaction patterns, which have developed within the transport domain and which at the same time direct the further developments within the domain.

This chapter focuses on the lessons learned for governance from the dynamics of sociotechnical change in the transport and mobility domain. Section 2 describes the governance arrangements within the transport and mobility domain and the actor positions and their strategies towards the climate change issue. Section 3 describes the Dutch transport technology policy. Section 4 analyses the patterns and change mechanisms in the transport and mobility domain. The final section offers suggestions for modulation policies for reducing GHG emissions in transport. It outlines a new model for public policy to work towards sustainability, that is, transition management.

2 GOVERNANCE ARRANGEMENTS IN TRANSPORT AND MOBILITY AND RESPONSES TO THE CLIMATE CHANGE ISSUE

The policy arena in the field of transport and mobility consists of a divergent set of entities, both in terms of constituent actors as well as in terms of the issues they deal with. There are local, regional, national and supra-national policy levels, each of which has its own set of policy tools to stimulate specific developments in the public interest. Local levels focus on issues closest to home, such as local air quality or congested city centres, while higher policy levels look at issues with a wider scope, such as GHG emissions, energy security or the negative effects of congested roads on the economy. Thus, the government authorities do not function as a homogeneous entity that pursues well-defined goals. The approaches at the different levels are often not used in synergy to attain these goals, and policies at various levels even tend to work against one another, so that the effect of these policies is often limited.

As there is much more consensus in society on the need to tackle air pollution than to tackle GHG emissions – the impacts of which are still considered uncertain and controversial – the former has been dealt with quite effectively over the past decades, while progress in the domain of the latter has been very difficult and limited (as shown by the recent failure of the Sixth Conference of the Parties (COP6) conference in November 2000 in The Hague). In general, the main approach adopted to limit pollutant emissions is command and control (direct regulation) by the responsible policy level. The formulation of such policies is the outcome of a complicated process, in which various actors and different policy cultures play a role, restricting the policy makers' room for manoeuvring. As a result, strong regulation of vehicle emissions, for instance, came into force in the US much sooner than in the EU (Elzen et al., 1996; Hoogma 2000).

The transport and mobility domain is characterized by a *corporatist arrangement*, in which peak associations play an important role.[8] In the Netherlands the peak associations are: the ANWB (representing the interests of car users), the RAI and the BOVAG (representing car dealers and garages), VVN (the freight transport companies), KNV and 'Nederland Distributieland' (Holland International Distribution Council) for motorized transport and the NS, VSN and the large town public transport companies (especially GVA and RET). The ministry of traffic and transport consults with these associations and will accommodate their interests. Disabled people, cyclists, pedestrians and local actors are weakly represented in the closed policy network which as far as transport planning is concerned is dominated by transport engineers. Environmentalists are not very active in the transport domain, although the environmentalist group Milieudefensie (Friends of the Earth) is quite active with regard to Schiphol, the main airport.

In the sociotechnical transport regime, the government accepted the tasks of organizing the building and maintenance of the road infrastructure, of reducing emissions and noise, and of supporting public transport through subsidies. These policies have been insufficient for dealing with the growth of transport and of making public transport more attractive to users.

For dealing with emissions, the government relied upon the use of technical fixes, such as the catalytic converter and use of lead-free petrol. Car traffic flows and volumes are not really controlled; driving speed is controlled but for reasons of safety, not energy consumption. The freedom to drive is accepted by state transport authorities, as is people's perceived need for mobility. Only within cities and towns is the freedom to drive curtailed, through the use of car-free zones and one-way streets. At the national level, in politics and the ministry of traffic, the freedom of mobility is seen as an entailed right. Within the current prevailing policy belief system, changes in technology and behaviour are seen as alternative ways for dealing with transport problems that are pursued in parallel. The same is true for private and public transport, which are seen as separate, rather than symbiotic. Transport authorities have not yet embraced the new perspective of integrated mobility (*ketenmobiliteit*), which tries to integrate public and private transport; they are still locked into the old idea of a *modal shift* as a strategy for dealing with congestion and pollution.

One explanation for this is that there is no perspective of sustainable transport that is guiding decision makers, which makes it hard to put into place policies for a technological regime shift. Existing roles of private and public actors and preferences of transport users are very much accepted and accommodated. No attempt is made to alter mobility patterns. Vehicles have become much more clean but there has been no regime shift in the form of integrated mobility and collective use of means of transport or to automatic vehicle guidance. In this chapter we shall outline how public policy could be used to achieve a regime shift. We shall argue that in order to make transport more sustainable, the current mode of governance has to change. This will not be easy and implies a great deal of change. It will imply a new role for state transport authorities, which should engage in sociotechnical alignment policies, aligning policies and policy goals to visions of sustainability, and the use of process management. For this the policy process should be broadened and sustainable transport should be made a societal goal for which societal support and resources are mobilized. Government authorities should act as an alignment actor and facilitator of change rather than as a sponsor or regulator. This does not render obsolete the use of subsidies and regulation policies but implies that these should be undertaken as part of a broader approach, which aims at the modulation of dynamics to sustainability goals.

A modulation approach exploits windows of opportunity and aims for mutually reinforcing dynamics. It does not start from goals but from problems, patterns, strategic games, existing structures and expectations (Rip and Kemp 1998; Kemp 2000). A modulation approach seeks to *modulate* ongoing dynamics into socially beneficial directions – here the direction of climate change protection. For this it is important to have an idea of the relevant trends, regime rules, dynamic games and expectations of key actors within the domain of transport and mobility. This should serve as a precursor for both policy and government–business–society interactions and actual policies. Ideas about sustainable transport and opportunities for socially beneficial change through innovation should thus be assessed. Which transport technology will be used in 5 or 25 years' time depends on the investments in alternative transport technology today. That is, it depends upon the research and development (R&D) efforts, investment plans and technological expectations of automobile companies and suppliers of alternative power sources, the resource dependencies and the support and control policies that are 'in the pipeline' or are being considered by public authorities.

It is thus important to have an idea of the R&D strategies of car manufacturers, the network dependencies, public-private relationships in the transport regime and technology policies. This will help to identify possibilities for intervention.

Car Manufacturers' R&D Strategies, Competencies and Expectations

In order to cope with the challenge of limiting exhaust emission and also with congestion problems, car manufacturers and their suppliers have committed themselves to important R&D programmes for reducing emissions and vehicle guidance and information systems. They are engaged in a strategic game to produce the 'green car of the future'. Research is done both at the company level and within cooperative organizations such as EUCAR in Europe and USCAR and PNGV (Partnership for a New Generation of Vehicle) in the United States.

It is important to note that current competencies of car manufacturers are based on the mass production of vehicles. They are good at marginally improving their products, based on their accumulated knowledge, routines and production methods. The dominant technology strategy of major car manufacturers, therefore, is to improve the internal combustion engine (ICE) and to introduce advanced transport telematics technologies within the car in order to cut car emissions and reduce congestion. They are ready and willing to modify ICE vehicles in order to operate on different fuels (petrol, LPG (liquified petroleum gas) or (compressed natural gas) CNG), even to couple the ICE with an electric motor (hybrid electric vehicles) but they are unwilling to do away with ICE. That is, they are averse to making a change to a completely different propulsion technique, such as an electric one based on the use of batteries, not just because the knowledge base is different but also for fear that consumers will not opt to buy vehicles with a range of only about 100 km.

Another important element is expectations about the future green car. There is a consensus that hybrid electric vehicles will be able to solve most problems, without great sacrifices in terms of user benefit; performance characteristics will be equivalent to current petrol cars while decreasing by a large extent polluting emissions (thus easier to sell) and without the need to achieve a significant reduction in weight (which means no change in materials used) (Kemp and Simon 2001).

As to the real green car (in terms of performance), car manufacturers very much regard fuel cell electric vehicles (FCEVs) as the long-term 'sustainable' solution. There is a clear acceleration of R&D on this technology (the recent cooperation between Daimler-Benz and Shell is a good example) but only prototypes are on the road today and it seems not unrealistic to expect FCEVs to be commercialized before 2005. The negative consequence of this is that other types of vehicles are given less attention.

Technology development in the dominant car-based transport regime is strongly internationally oriented, and mainly takes place in the large multinational companies, such as Toyota, Honda, General Motors (GM), Ford, Daimler Chrysler and Volkswagen. These companies expect to dominate the market with new technologies and vehicle concepts.

Response of the Automobile Industry to Climate Change Issues

The transport sector plays a very important role in successful efforts to address climate change, as a major emitter of CO_2 (the Dutch transportation sector is responsible for about 18 per cent of the national CO_2 emissions) and because of its capacity to deal with the problem by developing cleaner driving technologies or by shifting towards multimode transport systems.

Until 1998 there was a clear distinction between European and US-based automobile companies with regard to their attitude towards climate change issues. US-based companies strongly opposed mandatory emission controls, they publicly challenged the scientific basis for action and pointed to the high economic cost of controls. The environmental efforts of US companies were focussed on reduction of NO_x and hydrocarbons to meet California air quality regulations and anticipated LEV (low emission vehicle) and ULEV (ultra low emission vehicle) standards for conventional (non-CO_2) emissions.

European automobile companies being more cooperative did not challenge Intergovernmental Panel on Climate Change (IPCC) reports and in July 1999 entered a voluntary agreement with the EU to reduce average CO_2 emissions about 25 per cent in the 1999-2008 period. European technological investments emphasized diesel and small lightweight cars as short to medium-term emission reduction approaches. Furthermore, European companies expected consumers to change their expectations concerning vehicle usage and the role of private vehicles in transportation networks.

The position of US companies has changed quite dramatically, following the signing of the Kyoto Protocol and increasing competitive pressures for producing green cars, even though the market for such cars has not yet materialized. While remaining opposed to the Kyoto Protocol, they moderated

Modulating dynamics in transport for climate protection 319

their position on climate science and the economic consequences of controls. Ford and GM left the Global Climate Coalition and invested in a range of new technologies. This resulted in a process of convergence between European and US automobile companies in a strategy of accommodation in order to exert influence on the international climate policy.

Important differences, however, remain. US companies are still not planning the mass production of low-weight vehicles with high fuel efficiency. European companies are quite explicit in adopting uniform environmental standards globally, while US companies are not. In general, US companies are strong advocates of voluntary and market-based flexible mechanisms and remain opposed to the Kyoto Protocol, while European companies have been more accepting of mandatory emission controls (van der Woerd et al. 2000).

The Evolution of Governance Arrangements in the Transport Regime

In the transport and mobility domain the public governance arrangements had a corporatist and technocratic character for a long time. Governmental institutions interfered with the professional and social networks involved in the transport regime. In the early history of automobiles, for example, the US government officials were lobbied for road-building projects by a large network of institutions. This network was successful in inducing government to undertake massive road-building projects that strengthened the technological regime. And the 'highway lobby' is still recognized today as one of the most powerful interest groups in US fiscal policy (Unruh 2000).

National security has been the basis for government involvement in the construction of many technological systems including highways. During the Second World War, for example, security issues led the US government to invest over $30 billion in the domestic automobile industry, and then, after the war, to assist in rebuilding the auto industries as part of the post-war reconstruction effort (Nester 1997). Accordingly, since the 1940s, *security issues* could be regarded as an ideograph (guiding principle) for the transport regime and its governance arrangements.

Since the 1970s, *environmental issues* and since the Brundtland Report in 1987 (WCED 1987), the notion of *sustainability* became the main ideographs in transport governance arrangements and later on the ideograph of sustainability was broadened to include issues of *liveability* and especially the *congestion issue*.[9]

Since the mid-1980s alternative modes of transport gained widespread attention. This was due to the increasing problems of air pollution in urban areas, the stronger environmental movement and the growing interest in the issue of global climate change, which received more and more interest in public debates. In some countries, an increasingly popular anti-nuclear power movement and an active quest for an energy supply system based on regenerative energy sources also strongly influenced the political environment. On the manufacturing side, in the 1980s *lean production* became an important ideograph, leading to increased transport all over the world.

The Promotion of Electric Vehicles as a Possible Solution to Environmental Problems

Electric vehicles (EVs) have received a great deal of attention. Several routes for forcing their development were attempted (Hoogma 2000):

- a top-down approach where policy makers tried to enforce the car industry to develop and introduce EVs;
- an endogenous, cooperative approach of the automobile industry working together with government;
- a bottom-up approach of concerned citizens engaged in the support and use of alternative modes of transport (such as car sharing and lightweight vehicles).

For some time, roughly the 1980-95 period, battery EVs were considered the most promising avenue towards a more sustainable transport future. The bottom-up developments in Switzerland, the prototypes of the automobile industry and the zero-emission mandate in California generated expectations about the marketability of electric vehicles among a variety of actors, which led them to also become involved in initiatives to develop electric cars and to put them on the market (Hoogma 2000).

The *top-down approach* was most strongly put into practice through the Californian Zero Emission Vehicle (ZEV) mandate, through which the California Air Resources Board (CARB) attempted to force manufacturers to develop cleaner cars. Much of the Californian interest in electric vehicles stems from the failure to meet federal and state air-quality standards in many (especially urban) areas, but economic reasons were important too: California sees the production of EVs and the needed components as an important future industrial activity for local military industry seeking civilian markets and local new innovative firms. With the ZEV mandate, the Californian government wanted to kill two birds with one stone. The mandate led to a bandwagon of initiatives by various actors, both inside and outside the United States. Electric utilities, governments and corporations freed funds to support the development of vehicles and components, mainly to the benefit of small R&D companies. They also adopted experimental and pre-series EVs in their fleets in order to stimulate the development of a market, and invested in recharging facilities. Two states in the US (New York and Massachusetts) followed the example of California (example function) and adopted the mandate too.

The *endogenous, cooperative approach* of the automobile industry can be seen as a reaction to the public criticisms against the automobile. The industry wanted to recoup public trust and pre-empt regulation. Its contribution was seen as crucial for getting high-quality EVs mass produced at an affordable price. In reality, the electric cars developed by the car industry were often very expensive because the industry oriented its activities towards the performance characteristics of existing petrol cars. The industry (perhaps with the exception of Renault and PSA in France) never really believed in EVs but when GM presented a prototype of the impact electric vehicle at the January 1990 Los Angeles Auto Show, EV technology seemed a reality and it encouraged CARB to include them in the regulation it was preparing for adoption in September of the same year.[10] The automobile manufacturers disliked the mandate and started a year-long controversy over it. They mobilized public opinion against EVs. At the same time, however, there was a feeling of inevitability regarding their introduction. The car manufacturers announced that they would put their prototypes on the market, but only in small numbers while they were awaiting better batteries. By doing so they were keeping faith with the new CARB regulations.

The *bottom-up approach* was followed by a wide variety of outsiders from the automobile industry. The most interesting, but so far not very successful, experiences were gained in those countries with no major indigenous automobile industry, namely Denmark, Switzerland and Norway. The experiments produced lightweight, specially designed urban cars. The largest per capita markets for EVs are in France and Switzerland. In France, they were promoted by the traditional car manufacturers, Renault and PSA Peugeot-Citroen collaborating with EDF (the French national electricity company) and local municipalities (for example, in La Rochelle). In Switzerland EV development was undertaken by a number of small and medium-sized enterprises, university institutes, interested individuals and spin-off firms (founded by students from technical universities) which built up a support and innovation network for the development and diffusion of so-called lightweight electric vehicles. These endeavours with lightweight EVs found strong support from the ever-expanding anti-nuclear power movement of that time, although some people within this movement were against all types of motorized vehicles, including electric ones (for example, Mom 19997; Kirsch 2000).

Public R&D Programmes

The European Commission has been heavily promoting R&D programmes that aim to improve car emissions and traffic congestion. The EU spent 256 million ecus on transport R&D in the 4th Framework Programme. Examples are the 'Car of Tomorrow' programme, and the project on automatic traffic systems. In the US, FAST-TRAC (Faster and Safer Travel through Traffic Routing and Advanced Controls) is one of the largest R&D programmes on transport. In Japan, the main focus is on the deployment of advanced traffic management systems and the development and marketing of automobile navigation systems as a platform for in-vehicle information. Transport telematics R&D has largely been supported under the auspices of ATIS (the Advanced Travel Information System) (Kemp and Simon 2001).

On the whole, government policies were supportive for LEVs but in some cases they also created barriers. For instance, the very strict safety requirements in the Japanese natural gas law drove up the price of on-board gas cylinders and refuelling stations to five times the level of other countries, until the laws were changed in the mid-1990s. The Californian ZEV legislation has strongly stimulated the development of electric vehicles but long discouraged the development of hybrid EVs, although the latter may be cleaner if the emissions by electricity production plants are taken into account (Hoogma 2000).

The implementation of catalytic converter technology provides a good example of the interrelation between firm strategy and government policy in shaping the technological development of the car and its subsystems.The need for more immediate compliance with catalytic converter legislation in the US, Japan and Europe has meant that resources were devoted to the refinement of this technology and the additional adjustments that it necessitated (such as the production and widespread availability of lead-free petrol at filling stations). This legislation has thus had the effect of promoting the option with firms devoting resources to the incremental improvement of the catalytic converter. It has not, however, precluded consideration of radical more long-term solutions. The increasing policy emphasis on the reduction of levels of carbon dioxide emissions is bringing consideration of such technological options to the fore. In response to expected future legislation on carbon dioxide emissions, some companies also initiated some work towards the development of lean burn fuel-efficient engines and complex engine management systems, with the future aim of combining this technology with that of catalytic conversion.

Environmental concerns thus had the dual effect of narrowing the more immediate technological agenda of firms, forcing companies to concentrate efforts on the catalytic converter trajectory away from their previous heavy investments in lean burn technology, and broadening its longer-term horizons to encompass a range of more radical future alternative technologies.

In the UK, but also in France and Italy, companies invested substantial resources in lean burn engine technology in the 1970s and 1980s, in the belief that this was the path to emission reductions. The UK government encouraged this technological solution. The EU regulation, however, necessitated a shift away from this technology to that of catalytic conversion, for which the company was able to draw on its US export experience. In the US, the petroleum companies had been in the beginning unwilling to be part of this network, but they were persuaded by GM and the US government. In addition GM carried out aggressive regulatory actions. What the above discussion shows is that for the most part, governments have taken a *reactive approach*, relying heavily upon the solutions offered by car manufacturers, which consisted of the use of technical fixes. The Dutch government relied upon the use of solutions developed elsewhere, adopting a strategy of creative follower.

3 THE DUTCH TRANSPORT TECHNOLOGY POLICY

Contrary to the impression gained from reading government publications, there is no integrated technology policy in the Dutch transport and mobility domain. Some governmental programmes are aiming at behavioural changes in transport and stimulating particular technologies, but technology and behavioural change are seen as separate routes. Policy does not actively promote processes of co-evolution in which people are encouraged to reconsider their mobility needs and ways of satisfying these (personal communication of Elzen) and work towards alternative systems for transport.

The Dutch transport technology policy consists of:

- infrastructure projects (railways, bridges, tunnels, highways, airports);
- the use of traffic management technology aimed at disciplining users of transport and controlling traffic streams through traffic lights, physical measures to slow down traffic (silent policemen) and, in the future, road pricing and toll roads;
- programmes for the introduction of clean and silent vehicles, such as SSZ (*stil, schoon enzuinig verkeer in stedelijk gebied*) and DEMO of the Netherlands Agency for Energy and the Environment (NOVEM).

The Dutch approach tries to solve certain mobility problems (for example, congestion) incrementally. No innovative technology policy is carried out in the Netherlands, focusing on potential new transport and mobility regimes in the future. The only programme that is oriented towards innovative solutions is '*Wegen naar de toekomst*' (roads to the future) which involves a series of experiments with novel concepts such as flexible infrastructure, automatic vehicle guidance, modular and intelligent roads. This programme started in

1997 and is managed by Rijkswaterstaat which acts as a process manager. The programme is little more than an exploration project, with a focus on how the existing infrastructure may be better utilized. In terms of alternative technology options, light rail (tram-like trains) is viewed as a solution for the Randstad area and some urban areas that are heavily congested. In Eindhoven a guided bus system is being constructed, and it is to be introduced more widely, together with light rail systems. Both the rolling material and the system concept come from abroad.

The existing policy in the field of new transport technologies (for example, hybrid or electric cars) is very fragmented, being scattered over various stimulation programmes of NOVEM, and is very opportunistic. Experiments are carried out more or less ad hoc. They are not linked to a vision of sustainability or to transition programmes. There is a strong opportunistic element in them. There have been a lot of experiments with electric vehicles because of the concerns about air pollution. The experiments are not taking place in a systematic trajectory of experimentation and learning. Very often, the outcome of the experiments is that the particular technology is not yet ready for the (existing) market, which means the (temporal) end of the experiment. Learning *across* experiments (where learning experiences from one project are used as input in a new project) rarely takes place.

There was an interesting initiative in the mid-1990s called '*Innovatie in Inland Transporttechnologie*' (innovations in domestic transport technology) which explored new concepts for transport such as intermodal chain mobility, underground city freight transport, fast ships, dynamic traffic management systems, dynamic transit information systems, modular vehicles, automatic vehicle guidance and tele-activities, but this never led to concrete programmes and pathway policies pursued as part of a wider transition agenda.

Another diagnosis is that the Dutch transport technology policy is mainly focused on congestion and not so much on environmental issues. With regard to the environment, SO_2 and NO_x problems in particular have been addressed. The CO_2 issue has barely been dealt with. The Dutch government and Dutch transport authorities regard CO_2 issues more as belonging to the European policy arena. This does not mean that it considers CO_2 unimportant; within the EU it is quite active in establishing CO_2 standards.

In 2000, the transport minister, Mrs T. Netelenbos, fought hard to introduce road pricing in the Randstad metropolitan area but encountered great opposition from the ANWB (a very powerful organization with 3.6 million members representing the interests of car drivers) and the popular press. This demonstrates that it is hard to press for solutions in a top-down manner. A constituency has to be built for technological solutions, which can be done in two ways: through package deals (as the minister attempted to do through a large-scale mobility plan called '*Bereikbaarheidsoffensief Randstad*'

(Randstad accessibility offensive), and through the use of local experiments and consecutive 'upscaling' policies. A gradual introduction may alleviate worries and deal with opposition in a constructive way, by finding suitable designs. Road pricing met a lot of opposition in other countries, such as Norway and France, but once introduced people started to see the benefits and sense of it. Push policies can be detrimental; the pushing should be done cautiously, after a period of learning and should also contain an element of control so as to limit the side-effects, which are inexorably linked up with the use of new technology.

4 PATTERNS AND CHANGE MECHANISMS IN THE TRANSPORT AND MOBILITY DOMAIN

Authorities cannot plan for sustainable transport for the reasons enunciated, but there are many ways through which public authorities can influence transport developments – make it cleaner and safer, and reduce energy use and CO_2 emissions. Climate change benefits may be achieved through adaptation of the current trajectories and through a transformation, involving system innovation. Greener cars are an example of the first type of response (or greening of existing trajectories) while chain mobility and underground travel in vacuum tubes is an example of the second type of change. Both types of change have a role to play but so far policy has been oriented mainly towards system improvement, rather than system innovation. In our view there should be programmes aimed at system innovation involving radical change in the way in which we satisfy our mobility needs. It is not a matter of instruments (for example, subsidies or regulation). System innovation requires a different type of approach – of process management aimed at the *modulation* of dynamics and creating path dependencies in the right direction. For this there should be a sense of direction. A real problem here is that in the transport domain there is no vision of sustainability. One reason for this is that transport choices involve all kinds of trade-offs. No solution appears satisfactory. The challenge for policy makers is to manage the trade-offs in a way in that yields societal sustainability benefits without sacrificing user benefits. We could say that it is about achieving revolution through evolution, although a better way of putting it is that it is about achieving structural change in an evolutionary way. Short of a benign dictatorship, evolution appears the only feasible mechanism to achieve structural change. The multilevel model of sociotechnical change, expounded in Rip and Kemp (1998), and applied in Geels and Kemp (2000) and Geels (2001), offers some clues for working towards regime shifts: through the use of niche management, setting long-term goals, creating visions of sustainability, changing the mode of governance (for example through selective activation – the involvement of non-motorized actors in the policy process). The recommendations are very general and pertain to transition management in general and thus to any area that is in need of structural change. Transition policies for transport must be tuned to the special circumstances and take into account specific features of the transport and mobility domain. The following features of transport are important from a transition point of view.

The Importance of Cross-technical Linkages and Hybrid Forms

This aspect is illustrated in the historical transition from horse-based transportation to cars in the early twentieth century. The development of the internal combustion engine vehicle (ICEV) built upon the knowledge and experiences of the bicycle, gas-engine and coach transport regimes. In this early period, there were three competing propulsion technologies: steam, electric and petrol. Until 1911, electric vehicles were the dominating automobile technology, which was used for urban transport, and by the well-to-do, women and taxi companies. After the introduction of the electric starter in 1911, petrol-driven cars increasingly invaded these last niches of electric vehicles, and became the dominant transportation regime. This electric starter is an interesting example of positive *cross-technical influences*: the technological trajectory of petrol-driven cars was improved, because it borrowed an element (batteries and high-voltage ignition) from the trajectory of electric vehicles. This shows that the use of a new technology may require complementary technology which perhaps is not yet available or expensive to use. For example, the catalytic converter case showed that for the introduction of the three-way catalytic converter, unleaded petrol was needed for its optimal functioning in cars. The technology to unleaded petrol still had to be developed at that time. The EV case study showed the cross-fertilization between military technology knowledge and EV technology developments. In addition, the introduction of battery-powered electric vehicles requires the development of an infrastructure for charging batteries. Electric vehicles, therefore, do not fit well into the existing petrol-based transport regime.

In the 1980s and 1990s, EV technology did not replace the dominant ICEV propulsion technology, but *hybrid forms* emerged, combining the knowledge and competencies of the dominant ICEV transport regime with the potential new EV regime. Hybrid forms may be an important *transitional* element, which helps society to move to achieve a transition to a more sustainable system. The word 'transitional' does not just mean transitory. Hybrid forms may have a pathway function and act as a transition mechanism. In transport we do not have simple substitution processes, but complex interaction processes. This is illustrated by the ICEV, which benefited from the electric start engine and by the example of steam ships, which evolved out of sailing ships. The first steamboats were really sailing boats with auxiliary engines, to help out against bad winds. For a while the coal for steamships was transported through sailing ships (Geels 2002). Transitions in transport are thus complex, variegated processes, in which there is an *accumulation of niche developments*, besides regime and landscape developments. In transport many cross-technical influences and hybrid forms, which facilitate transition.

The Importance of External Developments for Transport Developments

The transport transitions just described – the development of the ICEV regime and the shift from sailing boats to steam ships – took place against wider developments that heavily shaped them. In past transitions external developments in the sociotechnical landscape (of changed prices, values, belief systems, politics and trade) opened up spaces for innovation and set overall directions for the transport regime. In the transition to individual motorized transport, two important landscape developments were the growing affluence of people and individualization. For policy makers, an important ideograph was the freedom to drive – that is, unconstrained automobility – and, later on, freedom from accidents and 'freedom' from pollution. The diffuse goal of the ideograph sustainability (narrowly understood as lower pollution) has been functioning as a *mobilizer* in both the catalytic converter and the alternative vehicles case studies.

A superficial look would suggest that climate protection is not really part of transport policy. For the US automobile manufacturers, however, the signing of the Kyoto Protocol and investment programmes of other car manufacturers in clean vehicles led to an important change with respect to cleaner vehicle concepts. In 1997 Ford made the decision to invest heavily in fuel-cell research. It also committed itself to the production of hybrid vehicles and in 1999 it acquired a majority interest in the Norwegian company PIVCO the producer of the Think, a small electric car with a plastic body.[11] Since Kyoto, GM has entered an alliance with Toyota to invest in a range of technologies. In addition, GM has formed partnerships with the oil company British Petroleum (BP), and Ford with BP and Mobil for research and development of new fuels and alternative vehicle concepts (van der Woerd et al. 2000).

Accordingly, strengthening of the ideograph climate change (sustainability) in the Kyoto Protocol led to renewed interest of automobile manufacturers in the niche developments of new fuels and alternative vehicles. Consequently, new possibilities have been created with impulse value for further change.

The Great Heterogeneity within Transport Systems

A third characteristic of transport is the great variety in transport systems and user needs, which allows for variety and multiple learning processes. In the Netherlands, the majority of trips below 5 km are made by bicycle, and about 300 000 cars run on LPG. Furthermore, car use is constrained in cities; many cities have car-restricted zones. In California geographical conditions make the Los Angeles area prone to smog, calling for tougher restrictions on car pollution than in other places. Apart from variety in travel behaviour and variety in geographical conditions there is also variety in terms of capability and politics. At the local level there may exist special capabilities, for example for producing light-weight electric vehicles, and green political coalitions. Historically the local conditions led to sometimes quite unique transport solutions or a unique mix of options. What we are saying is that local features may make possible developments that are (yet) impossible in other places. Variety is an important element of evolutionary change and could be exploited strategically for system innovation.

5 SUGGESTIONS FOR MODULATION

The current transport systems have given rise to a range of problems. Exact figures about the costs to society are hard to come by, but rough estimates are that congestion costs the EU 2 per cent of GDP every year, accidents another 1.5 per cent, air pollution (excluding global warming) 0.4 per cent and noise 0.2 per cent, in total 4.1 per cent.[12] At one level people have accepted these problems as a normal element of modern life but the problems also cause actors to search for solutions, both actors with an interest in the existing system and new actors. Governments have put in place policies to remedy the existing transport system but are careful not to create too great a cost. Only in exceptional circumstances, when the problem is viewed as unacceptable, will public authorities use their power to instigate radical change. As has been stated earlier and shown in the various transport cases, for governments a command-and-control approach is not feasible for dealing with problems of complex sociotechnical systems; the non-malleability of technology means that governments cannot call up desirable technologies by legislation. Incentives and constraints (including regulation) will have effects (in proportion to the level at which they are introduced) and have a role to play in transport policy, but governments cannot measure their content and timing. There is a dilemma of control, which was pointed out by Collingridge (1980) which is that governments have the greatest influence over technological choices when they know the least about the impacts and desirability of the technology; when the technology is fully developed and widely used, it is extremely difficult to control it, because of vested interests and high adjustment costs. This should not be taken to mean that technology is out of control, but rather that the dynamics of control do not always lead to acceptable outcomes (Rip and Kemp 1998).

A different type of approach is needed, which we have called modulation. Modulation policies are oriented towards dynamics, structures, strategic games and learning. This requires a new role for government: that of an alignment actor, a matchmaker and facilitator of change rather than a sponsor, a taxer and regulator (Rip and Kemp 1998 and Kemp 2000) and leads to a different set of policy recommendations. A modulation strategy does not mean that we should not use traditional policies of regulation and taxes that these should be more oriented towards long-term (transition) goals and system innovation. Within a modulation strategy policy instruments should be fine-tuned to the context in which they are applied and differentiated according to the stage of the transition process. In the predevelopment phase, policy should stimulate variation and societal discussions about sustainable transport. Once it has become clear what solutions and configurations are attractive, it should stimulate investment and the integration of new technologies in existing systems through public planning and system management while controlling for side-effects of new systems that (might) occur in the later phases. In general, there is a need for both generic and technology-specific policies (Kemp 2002 and Arentsen et al. 2000).

Examples of modulation policies have been described as 'transition management' in Rotmans et al. (2000 and 2001) and Kemp and Rotmans (2001). Below we shall offer a number of suggestions for modulation policies for achieving GHG emission reductions in transport – taking into account the specific features of transport developments and the transport system and the proclaimed need to differentiate policies to the stages of a transition.

Given the inertia in transport systems and uncertainty about what solutions are best from sustainability and user points of view, our first policy suggestion is to engage in the use of social experiments and create niches for promising technologies (*strategic niche management*). At the early phase of development, new technologies need protection especially in areas in which there is much lock-in (which is true for transport). Without protection (that is support) new technologies have difficulty coming into their own. Protection should be partial, temporary and phased out. In transport, governments have supported research in batteries and telematics. But apart from promoting research it is important for society to engage in the real use of new technologies. This fosters interactive learning and institutional adaptation, which is necessary for pushing the transition process forward. Government policy could be used towards this end. By using local opportunities afforded by special circumstances a transition path may be created in a bottom-up, non-disruptive manner. Of course, this raises the question of what technologies we should experiment with. The answer, given in the literature on strategic niche management (Kemp et al. 1998a and Hoogma et al. 2002) is that we should support especially pathway technologies – technologies that help to bridge the gap between the current regime and a new (sustainable) one, and thus help to escape lock-in.

This leads to our second policy suggestion, which is the *stimulation of pathway technologies*. This is well accepted in the transport technology policy. The *Perspectievennota Verkeer en Vervoer* (Perspectives note on traffic and transportation) in the Netherlands talks about '*sleuteltechnologieën voor systeemvernieuwing*' (breakthrough technologies for system changes) and mentions electronic vehicle identification, automatic vehicle control, interoperability and global positioning systems as key technologies for system innovation. To these we would like to add: electric propulsion and transport information, booking and reservation systems. Both electric propulsion and transport telematics have a great potential for development and for achieving environmental sustainability benefits, not in the short term but in the long term when they are part of an integrated mobility system. They are supported by public policies and there has been investment in these technologies by industry but there is a gap between research and diffusion. Electric propulsion is often dismissed because people think it is about batteries, which it is not. Batteries are one way of providing electricity and not a very attractive one if the electricity is generated in power plants. Fuel cells appear more attractive as are hybrid, partial electric propulsion systems. It is important to explore all kinds of electric propulsion systems because they may create varying benefits and because society should not bet on one horse. The fuel cell vehicle is viewed as a magical environmental solution because it does not emit any pollutants but it may be less safe. We have to ensure that the hydrogen produced or stored in the vehicle will not explode. Specific technology programmes often meets criticism from people who say that the government cannot pick winners. The government programme concerning batteries is often criticized because it did not lead to the widespread use of batteries; but the support given to batteries has led to innovations in electric power systems, to lightweight construction and aerodynamic designs that will be widely applied. Nevertheless, it did not lead to the development of a cheap, long range battery and we should now probably focus attention on hybrid vehicles and fuel cell vehicles. The need for stimulation should be continuously assessed, and policies should be flexible.

To increase the chance that a transition will occur and ensure that the path chosen is the best one, we should explore different paths and the possibilities for cross-linkages and cumulation. This leads to our third suggestion, which is to *focus on routes of niche accumulation* that may lead to transport regime changes. There cannot be transition without a transition path. In transport, there is not one path but many possible paths of which it is impossible to tell which one is best. There is a need to identify possible paths and explore these. By initiating a little bit of irreversibility in the right direction a path or trail may be created. To create a good trail or to identify one, we should evaluate present transport regimes and the possibility of shifting them in desirable directions, and identify opportunities to influence niche branching and niche piling. Active stimulation of the development of *hybrid forms* as interludes between the old and new regimes, such as the hybrid car, could facilitate transitions to a new transportation regime. For example, the on-board reformers in cars that extract hydrogen from petrol could be regarded as *gateway technology* (Unruh 2001) enabling new fuel cell components to function within the existing transport regime.

Furthermore, consideration should be given to interrelations between developments. Cross-technical influences may give a development momentum. A clear example in the transport case is that the ICE powered car development had been improved by borrowing components and knowledge from the EV development trajectory. A recent example is the use of new materials for electric cars and telematics. In addition, the optimization of battery-powered electric vehicles towards energy efficiency will have favourable effects on the performance of hybrids and fuel cell vehicles, as will the modular components and lightweight construction principles developed for battery-powered vehicles. Even petrol and diesel cars may benefit from this.

Stimulation of battery-powered vehicles may lead to a 'favourable detour' to fuel cell vehicles and perhaps an energy regime based on renewable sources with hydrogen as the power source.[13] This approach also increases the chances of realizing very economical hybrids (such as 'hypercars'), which potentially have biofuels as an energy source (Hoogma et al. 2002). The focus should be on experimenting with a wide range of niche technologies, which in the long term could serve as stepping-stones for a new transportation regime. The experiments should be more than demonstration projects. They should be set up in such a way that both suppliers and users learn about new possibilities. Basic assumptions and existing expectation should be tested through second-order learning. For example, car manufacturers should be stimulated to rethink their assumptions about what a car should do whereas users should be stimulated to rethink their mobility needs and how to satisfy these. This is necessary for processes of co-evolution to occur.

This leads to our fourth suggestion, which is to *modulate 'promise-requirement' cycles of perceptions and expectations*. New technologies have been characterized as 'hopeful monstrosities' (Mokyr 1990): they hold promise but are still poorly developed in terms of user requirements. The requirements themselves may not be clear yet or in a flux. This calls for the need to stimulate promise-requirement cycles and the attendant resource-mobilization activities, so as to build a forceful agenda (for development work in the technological niches) on which general interests appear in addition to (short-term) actors' interests (Rip and Schot 1999). We have already seen that car manufacturers' strategies are based on certain expectations of user preferences and the way in which technologies fit these. Expectations and views on this may diverge, as in the case of the catalytic converter. The Japanese industry had high expectations of this new technology, while the American car industry perceived it as an imposed technology, which they frustrated for a while.

The fact that the EV developments largely took place outside the rather conservative car industry led to the exploration of a wider range of new ideas and concepts in relation to transport and mobility. The most innovative electric vehicles were developed outside the automobile industry (Hoogma et al. 2002).

Promise-requirement cycles may give rise to new markets, opening up for wider change. Examples are the use of the electric vehicle as a second vehicle in households, the use of toll roads, and the success of car sharing on which city authorities could draw in their transport policies.

A fifth suggestion is to *develop programmes for system innovation* such as integrated mobility and mobility management. Opportunities for system innovation producing sustainability benefits should be explored and exploited. An example of system innovation in transport is integrated mobility or chain mobility (in Dutch: *ketenmobiliteit*): the multiple uses of aligned transport services, which makes a positive contribution to all dimensions of sustainability. It offers benefits in the form of reduced congestion and leads to fewer emissions and accidents through a more selective use of cars and trucks. At present there is a gap between individualized and collective transport but various innovations may help to bridge the gap, for example:

- *Individual forms of collective transport* that make public transport more flexible and more directly tied to the transport needs of the consumer. This concerns innovations such as 'dial-a-bus' or collective taxi experiments (such as the Dutch '*trein-taxi*'), using information technology for route planning and vehicle tracking.
- *Collective use of private means of transport*, such as car sharing, bicycle sharing, ride-sharing (for example, car-pooling), a market for sharing long-distance trips, or voluntary schemes for transporting disabled or elderly people. These options could be enhanced through improved

information systems for accessing the vehicles, tracking their location and fleet management.

- *Transit information systems and mobility information services* that tell people how they can combine different modes of transport. These information schemes have to be complemented by Park and Ride facilities. The existence of such information services may help different transport companies integrate their services better and optimize the overall transport system.

These innovations will help car drivers to move away from the one-vehicle-for-all-purposes paradigm. There really should be a comprehensive programme for integrated mobility because it is attractive and requires a collaborative effort in terms of infrastructure (transfer places), reorganization of the sector (the creation of mobility agencies and cooperation between transport companies), technology (information and ticketing systems) and the setting of standards of interoperability. It is strange that the *Perspectievennota Verkeer en Vervoer* and the new national plan for traffic and transport (NVVP) only mention chain management when talking about public transport, and do not make it a central issue for the whole of the transport sector. This is an area in which the Netherlands can achieve a great deal on its own because it is not dependent on foreign efforts and there are no competitive disadvantages involved. Even the truck sector stands to benefit from it because intermodal travel leads to a more efficient use of existing infrastructure.

A second guidance image is that of *mobility management*, the control of traffic streams from a sustainability point of view, as a new guiding principle. This principle has already been officially adopted but is still poorly implemented. Local authorities have an important role to play here, as they know better how to accommodate the different needs. Sustainability has a local and global dimension. Sustainability is about finding locally suitable solutions. Mobility management can also be pursued by regular companies who may offer their employees informational incentives and economic incentives to use more sustainable modes of transport. This is already occurring in the Netherlands.

Finally, transition management should be seen as an *integrative framework* for achieving greater coherence in policy action. The suggestions made above should be pursued as part of an overall transition endeavour, not as isolated actions. They are best undertaken as part of a *transition programme* with development rounds in which progress is assessed and goals and instruments are evaluated and adjusted and a transition agenda is formulated. Ideas about transition management are worked out in a project for the National Milieubeleids Plan 4 (NMP-4) (the Fourth Dutch National Environmental

Policy Plan) and described in Rotmans et al. (2000). Transport would be a suitable case for transition management because of the uncertainty, complexity and inertia that exist in this domain, which make it hard to deal adequately with issues through the use of single instruments.

Three key elements of transition management are:

- the establishment of a transition goal (involving a basket of goals), based on visions of sustainability;
- the use of societal experiments with technological options or product-service systems that fit the sustainability vision; and
- the use of development rounds in which policies and transition goals are reassessed and redefined.

Transition management involves the use of a wide range of policies, with their timing gauged to the particular circumstances of transition phases and external developments. It does not offer a step model to get to state Y through steps $X_1, \ldots X_N$. Some policy interventions are linked to stages, such as the exploration of many solutions in the predevelopment stage, and policies towards system integration in the take-off stage; others are recurrent, such as the periodic reassessment of goals, visions and policies. Some interventions should be continued throughout the entire transition process, such as the internalization of external costs, the support of science and technology research for sustainable transport, environmental research and policy research for example. Figure 14.1 gives an overview of the type and timing of the range of policies involved in managing the transition.

Transition management differs from a planning and implementation approach. It does not operate on the basis of a blueprint (*greenprint*) but on the basis of a set of goals and quality images. The goals are not fixed and the policies to further the goals are constantly assessed and periodically adjusted in development rounds. This creates some flexibility but maintains a sense of direction. Through its focus on long-term ambitions and its attention to dynamics it aims to overcome the conflict between long-term ambition and short-term concerns. Learning, maintaining variety and institutional change are important policy aims. Control policies are part of transition management. Transition management does not exclude the use of control policies, such as the use of standards and emissions trading or trading systems of 'mobility miles', but uses these as part of a transition endeavour, in tandem with other policies and using windows of opportunity. It offers an *integrative framework* for policy deliberation and the choice of instruments and individual and collective action. Managing the transition to sustainable mobility requires changes in the mode of governance. Outsiders should be involved in the policy process and there should be a commitment to change and clear stakes. The

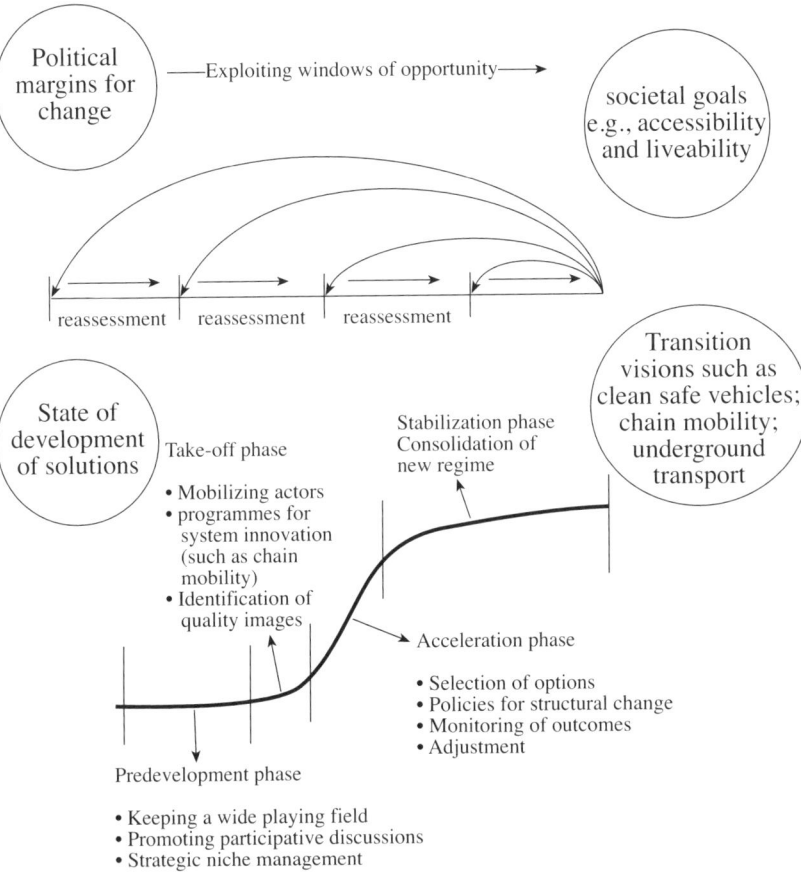

Figure 14.1 *Types of policies involved in managing the transition to sustainable transport*

transition endeavour should be institutionalized. Even when all this is done, there is no guarantee of success. What it does is to increase the chance for an actual transition towards sustainable transport to occur.[14]

When we propose transition management and modulation policies as a solution, we do not underestimate the difficulties of making transport climate friendly and sustainable. It is hard to deal with the problems of transport in a forward-looking manner. On a niche level something can be done, for example, by setting up various experiments, but that is no guarantee for wider success. Many different actors are involved in many places and their influence on the outcome is limited. At the same time transport is a very hierarchical system in which there is a lot of planning and control. All the different types of action are taken with reason. To paraphrase Shakespeare, there is method in madness. The 'method' consists of existing structures and ways of doing things that act as *de facto* forms of governance. The modulation approach seeks to exploit the method in the madness by intervening in the dynamic games. It is about the public use of private interests and modulation of ongoing dynamics into socially desirable directions. This chapter has outlined ways of doing this and offered some practical suggestions for policy, such as the creation of a large-scale programme for chain mobility. Policy for climate protection is not only about the use of instruments but also about patterns of interaction, the roles of private and public decision makers and intervention approaches and the guiding principles for policy. The current corporatist arrangement serving motorized interests is a barrier to achieving sustainable mobility and should be widened to include other non-motorized and social interests and be based on new guiding principles such as advanced public transport (in Dutch: *hoogwaardig openbaar vervoer*), underground transport, climate-friendly transport and chain mobility.

NOTES

1. Based on a chapter written for the MATRIC project for the NOP research programme on climate change. We thank Boelie Elzen, Aad Correljé and Michael Faure for comments made on earlier versions of this chapter.
2. Based on Banister et al. (2000, pp. 159-61).
3. Ibid. Technology, too, can be seen as an instrument (which is how Banister et al. look at it).
4. Energy saving in transport is an important policy goal of EU policy. The Netherlands relies on EU policies here.
5. Governance is thus a much broader concept than government. Governance refers to the actor configurations operating in certain technological domains and government can be part of these configurations but not necessarily.
6. In transport in the Netherlands the four most important policy principles are
   - facilitating (auto) mobility;
   - accessibility;
   - variabilisation of costs (shifting transport costs from fixed to variable costs);
   - users should pay (for infrastructure use);
   - external costs should be internalized;
   - markets should be liberalized.

   The last three principles are combined into the EU fair and efficient pricing concept in the Communication to the Commission (Kinnock 1995), which also includes transparency of transport costs and non-discrimination across transport modes and nations as additional principles.
7. A 'niche' is the functionally or geographically constricted domain of technology use, development and experimentation. A 'technological regime' is the rule set or grammar embedded in a complex of engineering practices, production process technologies, product characteristics, skills and procedures, ways of handling relevant artifacts and persons, ways of defining problems (Rip and Kemp 1998, pp. 340). The rule set consists of the things to which actors adhere to: technical standards, practices (ways of doing things), roles and shared assumptions. Because the actors adhere to them, willingly or unwillingly, they can be

viewed as rules. The 'sociotechnical landscape' is a landscape in the literal sense, something around us that we can travel through; and in a metaphorical sense, something that we are part of, that sustains us.
8. Corporatism is one of the four ideal types of arrangements identified by Eising and Kohler-Koch (1999). The other three types are: statism, pluralism and network arrangements.
9. On a normal working day between 350,000 and 500,000 cars experience congestion on the Dutch highways where the number of traffic jams doubled between 1990 and 1997 (NVPP 2000). In cities the problem of congestion has grown from bad to worse.
10. According to one author (Wallace 1995) the impact vehicle of GM triggered the Californian mandate.
11. The development of the Think automobile by PIVCO is described in Schwartz and Maruo (1998) and Hoogma et al. (2002).
12. The estimates are based on various studies and are reported in the Communication to the commission (Kinnock 1995). Ninety per cent of the external costs of transport come from road transport which gives rise to an external cost of €250 billion, with cars being responsible for a total of €164 billion.
13. The use of batteries may also lead to a rebound effect, for example, when people use battery-assisted bicycles instead of normal bicycles. This shows that the impact of a technology will depend on how it is used. Impacts are co-produced (Rip and Kemp, 1998). This calls for policies that limit the side-effects, either through regulation or pollution taxes. External costs should be internalized.
14. The idea of transition management is described in Rotmans et al. (2000 and 2001) and Kemp and Rotmans (2001). Limitations of space prevent us from describing it in more detail here. We just want to say that transition management aims for long-term change in a stepwise and flexible way. It is based on a philosophy of 'learning by doing' and 'doing by learning' (in Dutch: 'al-lerende-doen' and 'al-doende-leren'). Complexity, uncertainty and interdependence are not shunned but highlighted and made an explicit consideration for policy. When engaging in transition management, climate goals are pursued as part of a larger sustainability programme for which private and public resources and support is mobilised (and legitimized) through the political process. In our view, this will be a better strategy than taxing energy consumption or asking people to drive less for the sake of climate protection. Increasing fuel taxes will be a no-go strategy anyhow, as shown by the fuel price protests in the year 2000.

REFERENCES

Arentsen, Maarten, René Kemp and Esther Luiten (2002), 'Technological change and innovation for climate protection: the governance challenge', in M.T.J. Kok, W.J.V. Vermeulen, A.P.C. Faaij and D. de Jager (eds), *Global Warming and Social Innovation. The Challenge of a Climate-neutral Society*, London: Earthscan, pp. 59–82.

Banister, David, Dominic Stead, Peter Steen, Jonas Akerman, Karl Dreborg, Peter Nijkamp and Ruggero Schleichter-Tappeser (2000), *European Transport Policy and Sustainable Mobility*, London and New York: Spon Press.

Collingridge, D. (1980), *The Social Control of Technology*, London: Pinter.

Eising, R. and B. Kohler-Koch (1999), 'Introduction', in: B. Kohler-Koch and R. Eising (eds), *The Transformation of Governance in the European Union*, London: Routledge.

Elzen, B., R. Hoogma and J. Schot (1996), 'Mobiliteit met Toekomst: Naar een vraaggericht technologiebeleid' (Mobility with a future: towards a demand-oriented technology policy): report to the Ministry of Traffic and Transport (in Dutch).

Feitelson, Eran, Ilan Salomon and Galit Cohen (2001), 'From policy measures to policy packages: a spatially, temporally and institutionally differentiated approach', in Eran Feitelson and Erik T. Verhoef (eds), *Transport and the Environment. In Search of Sustainable Solutions,* Cheltenham, UK and Northampton, MA, USA: Edward Elgar, pp. 34-53.

Geels, F. (2001), 'Technological transitions as evolutionary reconfiguration process: a multi-level perspective and a case study', Paper presented at ECIS conference 'The Future of Innovation Studies', Eindhoven, 20-23 September, forthcoming in *Research Policy.*

Geels, Frank (2002), *Understanding the Dynamics of Technological Transitions. A Co-evolutionary and Socio-technical Analysis,* Enschede: Twente University Press.

Geels, Frank and René Kemp (2000), Transities vanuit sociotechnisch perspectief (Sociotechnical transitions), for a study called 'Transities en Transitiemanagement' (Transition and transition management) at ICIS and MERIT for the Dutch NMP-4, November, UT, Enschede and MERIT, Maastricht.

Hoogma, R. (2000), *Exploiting Technological Niches. Strategies for Experimental Introduction of Electric Vehicles,* Enschede: Twente University Press.

Hoogma, Remco, René Kemp, Johan Schot and Bernhard Truffer (2002), *Experimenting for Sustainable Transport Futures. The Approach of Strategic Niche Management,* London and New York: EF&N Spon.

Kemp, René (2000), 'Technology and environmental policy: innovation effects of past policies and suggestions for improvement', OECD proceedings *Innovation and the Environment,* Paris: OECD, pp. 35-61.

Kemp, René, (2002), 'An integrated policy for the environment', in B. Patries (ed.), *Innovation Policy and Sustainable Development. Can Public Incentives make a Difference?,* Brussels: IWT.

Kemp, René and Jan Rotmans (2001), 'The management of the co-evolution of technological, environmental and social systems', Paper for conference 'Towards Environmental Innovation Systems', Eibsee, 27-29 September.

Kemp, René, Johan Schot and Remco Hoogma (1998a), 'Regime shifts to sustainability through processes of niche formation: the approach of strategic niche management', *Technology Analysis and Strategic Management,* **10** (2), 175-95.

Kemp, René, and Benoît Simon (2001), 'Electric vehicles. A socio-technical scenario study', in Eran Feitelson and Erik T. Verhoef (eds), *Transport and the Environment. In Search of Sustainable Solutions,* Chelenham, UK and Northampton, MA, US: Edward Elgar, pp. 103-35.

Kemp, René, Bernhard Truffer and Sylvia Harms (1998b), 'Strategic niche management as a tool for transition to a sustainable transportation system', Paper presented at the conference 'Policy Instruments for Incorporating Social Costs in the Transportation System', New York, 1-4 October. Published in K. Rennings, O. Hohmeier, and R.L. Ottinger (eds), *Social Costs and Sustainable Mobility - Strategies and Experiences in Europe and the United States',* Heidelberg, New York: Physica Verlag (Springer), 167-187.

Kinnock, N. (1995), 'Towards fair and efficient pricing in transport', Communication by Mr Neil Kinnock to the Commission, COM(95) 691.

Kirsch, David (2000), *The Electric Vehicle and the Burden of History,* New Brunswick, NJ: Rutgers University Press.

Mokyr, Joel (1990), *The Lever of Riches: Technological Creativity and Economic Progress,* New York: Oxford University Press.

Mom, Gijs (1997), *Geschiedenis van de auto van morgen: Cultuur en techniek van de elektrische auto* (History of the car of tomorrow: culture and technology of the

electric car), Thesis (in Dutch), University of Nijmegen, Deventer: Kluwer Bedrijfsinformatie BV.
Nester, W. (1997), *American Industrial Policy*, New York: St Martin's Press.
Nouwen, P.A. (1995), *Ontsnapping uit de 20ste eeuwse vervoersdilemma's in Verkeerschaos en vervoershonger – Perspectief op Mobiliteit* (Escape from the 20th century transportation dilemmas in traffic chaos and transportation needs – perspectives on mobility), The Hague: SMO.
NVPP (2000), *National Verkeers- en Verveorsplan: van A naar Beter* (National Traffic and Transportation Plan: from A to better), Ministry of VWS, Netherlands.
Rip, Arie and René Kemp (1998), 'Technological change', in Steve Rayner and Liz Malone (eds), *Human Choice and Climate Change, Vol. 2: Resources and Technology*, Washington, DC: Batelle Press, pp. 327–99.
Rip, Arie, and John Schot (1999), 'Anticipating on contexualization: loci for influencing the dynamics of technological developments', in D. Sauer and C. Lang (eds), *Paradoxien der Innovation: Perspectiven social vissenschaftlicher Innovationsforschung*, Frankfurt and New York: Campus Verlaf, pp. 129–46.
Rotmans, Jan, René Kemp and Marjolein van Asselt (2001) 'More evolution than revolution. Transition management in public policy', *Foresight*, **3** (1), 15–31.
Rotmans, Jan, René Kemp, Marjolein van Asselt, Frank Geels, Geert Verbong and Kirsten Molendijk (2000), 'Transities en Transitiemanagement. De casus van een emissiearme energievoorziening'. (Transitions and transition management. The case of a low emission energy system), Final Report for the study 'Transities en Transitiemanagement' for the Dutch NMP-4, ICIS and MERIT, Maastricht, October.
Sabatier, Paul A. and Hank C. Jenkins-Smith (1993), *Policy Change and Learning. An Advocacy Coalition Approach*, Boulder, CO: West View Press.
Schwartz, Birgitta and Kanehira Maruo (1998), 'An outsider initiative in the emerging EV market. The PIVCO adventures in Norway and California', final report of case study for the project 'Strategic Niche Management as a Tool for Transition to a Sustainable Transportation System', Göteborg: Göteborg University.
Unruh, G.C. (2000), 'Understanding carbon lock-in', *Energy Policy*, **28**, 817–30.
Unruh, G.C. (2001), 'Escaping carbon lock-in', Working paper. Instituto ae Empresa, Madrid.
van der Woerd, K.F., C.M. de Wit, A. Kolk, D.L. Levy, P. Vellinga and E. Behlyarova (2000), *Diverging Business Strategies towards Climate Change. A USA-Europe Comparison for Four Sectors of Industry*, Dutch National Research Programme on Global Air Pollution and Climate Change, Report No.: 410 200 052, Institute for Environmental Studies, Amsterdam.
Verhoef, E. (1994), *The Economics of Regulating Road Transport*, Aldershot: Edward Elgar.
Wallace, David (1995), *Environmental Policy and Industrial Innovation. Strategies in Europe, the US and Japan*, London: Earthscan.
Weber, Matthias, and Andreas Dorda (1999), *Strategic Niche Management: A Tool for the Market Introduction of New Transport Concepts and Technologies*, IPTS Report, Institute for Prospective Technological Studies Seville, February, pp. 20–27.
Weber, M., R. Hoogma, B. Lane and J. Schot (1999), *Experimenting with Sustainable Transport Innovations. A Workbook for Strategic Niche Management*, Seville/Enschede: University of Twente, The Netherlands, January.
World Commission of Environment and Development (WCED) (1987), *Our Common Future* (Brundtland Report), Oxford: Oxford University Press.

15. Institutional change in Europe and the implications for climate control measures

Graham Bennett

1 INTRODUCTION: INSTITUTIONAL CHANGE IN A CHANGING INSTITUTIONAL CONTEXT

It is a curious fact that most current debates on the feasibility of securing institutional change in Europe fail to appreciate that the past 50 years have seen two separate and unprecedented institutional revolutions on the continent. The process that started with the establishment of the European Coal and Steel Community in 1952 and progressed through the European Economic Community and the European Community to today's European Union has transformed the economic and political architecture of Western Europe. Then, nearly four decades later, the events of 1989 triggered a complete reordering of the political, economic and social institutions of 20 states in a region extending from the Baltic Sea to the Bering Strait.

Institutional change of such character, scale, magnitude and impact is unique. Moreover, both revolutions are far from spent – witness the current processes of EU and NATO (North Atlantic Treaty Organization) enlargement, the recent EU Intergovernmental Conference on institutional reform and the huge transition challenges still facing the countries of Central and Eastern Europe. The capacity for institutional change in Europe cannot therefore be questioned, nor the existence of powerful cultural, political, economic, social and technological forces that are driving further change. Unlike the US, where climate policy will develop within a remarkably stable institutional environment, the long-term prospects for climate control actions in Europe are certain to be profoundly influenced by the course of institutional change.

An analysis of institutional change in Europe shows that time and again it is powerful social, economic and political driving forces that determine the course and timing of institutional change rather than the operations of the institutions themselves; indeed, by their very nature, institutions have a vested interest in stabilizing institutional patterns. In other words, given a political or socio-economic critical mass, certain developments take place whether or not formal institutional competences or requirements exist; conversely, in the absence or decay of this critical mass, other developments will not take place, even where these may be legally required. The past 50 years have produced examples enough. For instance, the original Treaty of Rome laid down the primary task of the European Economic Community as the establishment of a common market, a task that was only seriously taken up some 30 years after the Treaty came into effect when economic recession stimulated the member states to embark on the '1992' project. Again, in the period up to 1987, broad social and political consensus on the need to mitigate increasingly conspicuous environmental problems persuaded the EU to adopt some 200 legal measures concerning the environment, despite the fact that the original Treaty of Rome provided no explicit legal basis for the Community to regard the environment as a legitimate object of Community action. Or again, today, 14 years after Article 130r(2) of the Single European Act required that 'environmental protection requirements shall be a component of the Community's other policies', integration remains a major challenge for EU policy making. Yet again, the more recent 'velvet revolutions' in the Central and Eastern European (CEE) countries were driven by broad social dissatisfaction with the prevailing political regimes and institutions which, although apparently strongly embedded in the legal, economic and social fabric of all the countries in the region, had failed to generate sufficient legitimacy to be able to resist the upswelling of popular pressure for fundamental change.

The key issue is therefore how the driving forces behind the continuing institutional revolution in Europe will shape the boundary conditions that largely determine the future course and substance of climate control actions. Consequently, this chapter argues from the premise that any assessment of the longer-term institutional context for climate control actions in Europe should focus on the forces that drive institutional change rather than on institution-specific analyses. Over a period of several decades, the most important institutional changes will be the consequence of responses to external needs and pressures rather than to short-term internal dynamics. Reforming voting procedures in the EU Council of Ministers will not significantly influence the course of European climate policy; a crisis concerning the legitimacy of EU governance that leads to a paradigm shift in the democratic accountability of EU institutions will.

2   DEMYTHOLOGIZING EUROPEAN MYTHOLOGY

Before discussing these driving forces, it is worth pausing to reflect on a number of commonly held but largely erroneous perceptions on the character, operation and consequences of certain key European institutional processes. A first misconception is that the complex institutional architecture of the EU operates to obstruct any attempt to develop and adopt effective measures to resolve a 'global-commons' problem such as climate change. It is true that the predominantly intergovernmental character of EU decision making has difficulty in transcending the national interests of the member states; but it is also true that the emergence of an EU institutional culture and the wealth of Community procedures and competences that have evolved over several decades – and which apply to most aspects of environmental policy making – enable the EU to function on the international stage with a far greater degree of unity, confidence and purpose than other broad-based intergovernmental organizations. The agreement to phase out the production of chlorofluorocarbons (CFCs) through the 1987 Montreal Protocol is a prominent example of how the EU can successfully shape global environmental action. Although in this case the original pressure for international action came from the US government, it was concerted action by the EU in support of production limits as the most effective instrument for reducing CFC emissions that persuaded first US environmental organizations, and then US industry and government, of the merits of such an approach in preference to the initial proposal by the US government for controls on CFC use.

It is also widely assumed that Community policies and legislation are invariably uniform with respect to the policy objectives and the formal obligations that are imposed on the member states. Thus, EU environmental directives lay down common protection levels, environmental standards and implementing instruments that apply without exception to all member states. Bound by this straitjacket of uniformity, the more ambitious member states and the policy-making institutions of the EU are constantly forced to accept Community measures that represent the 'lowest common denominator' of the 15 national policies of the member states. But in fact, an analysis of the *acquis communautaire* – and especially EU environmental law – shows there to be not only a remarkable degree of explicit or implicit differentiation in the level of obligations imposed on the member states but also considerable scope for each member state to adapt implementing measures to its own circumstances and requirements. See, for example, the various EU water pollution directives, the Habitats Directive and the EU position on implementing the Kyoto Protocol. This degree of flexibility, in combination with the persuasive pressures that negotiating processes always impose on small minorities, ensures that policy outcomes which represent the lowest common denominator are the exception rather than the rule in the EU.

A tendency in most analyses of policy making in the Community arena is to focus solely on the role of the most prominent EU institutions: the Council, the

Parliament, the Commission, the Court. However, the institutional environment that shapes the way in which the myriad actors in this arena behave is far more complex and subtle. In reality these actors are influenced by an enormous number of overlapping institutional footprints: at the international level alone, actors in the EU operate within frameworks developed through, among others, the Council of Europe, the Commonwealth of Independent States, Benelux, the Organization for Economic Cooperation and Development, NATO, the United Nations, the World Trade Organization and a range of specific multilateral agreements, in addition to the multitude of cooperative arrangements and interests in the economic sectors – aviation, energy, chemicals, agriculture, banking and so on. Europe is an arena characterized by an enormous variety of interlinked institutional constituencies in which the EU, although politically the most prominent and integrated, is increasingly interacting with and becoming dependent on the actions of an interwoven lattice of institutions.

From a Western European perspective, it is all too easy to confuse the widespread desire in the CEE countries to enjoy the economic fruits of EU membership with a wish to discard eastern for western culture: citizens in Central and Eastern Europe are Western Europeans in all but name. The perspective from the east is far removed from such simple notions. To be sure, the candidate member states are well aware that they are being offered a binary choice: EU membership or not? A negotiated settlement on the future blueprint of the Union is not on offer, at most an agreement on where derogations from the *acquis* will apply and for how long. But there exists throughout the region a strong cultural preference for strong political institutions as a means to secure social stability and welfare. And in the long term, with the prospect that within a generation CEE countries will comprise about half the total number of EU member states, this cultural factor could prove to be of enormous importance for the future institutional development of an enlarged EU.

A final misconception which in the context of long-term institutional change in Europe merits correction concerns the effect of EU policies. Much has been written on the so-called 'implementation deficit' of EU legislation, and particularly environmental directives which in the recent history of the Union have comprised about 40 per cent of all infringement proceedings against the member states. Implementation shortcomings have indeed caused serious problems for EU environmental policy: even after more than 20 years, the majority of member states have still failed to ensure full compliance with the Birds Directive, to give just one example. Yet recent years have seen a major improvement in the implementation record of EU environmental directives. Serious enforcement efforts by the European Commission, judgments by the Court of Justice and the provision introduced by the Treaty of Maastricht whereby financial sanctions can be imposed on member states which fail to comply with a ruling of the Court have progressively had their effect. The present-day reality is that compliance with EU legislation is in general relatively good. More significantly, EU legislation is a far more effective policy instrument than other kinds of international agreement, where regimes with the capability to impose effective legal enforcement and, if necessary, persuasive sanctions are generally in their infancy. Moreover – again a point that is poorly appreciated – EU environmental measures are monitored far more rigorously than the provisions of multilateral environmental agreements, with a wide range of Community procedures and agencies ensuring that progress in implementing environmental measures and their effect on the environment is relatively well monitored and reported.

3 THE DRIVING FORCES

Which driving forces might determine the course of institutional change in Europe, especially with respect to climate change actions? I suggest that four forces may prove to be particularly influential in this respect during the coming decades: globalization, EU enlargement, climate change research, and changes in social values and individual perceptions.

Globalization – the process through which the markets for products, services and investments and the operational sphere of companies become increasingly international in character – is currently responsible for a major shift in the balance of power and influence from government to market actors. Specifically, the capability of transnational companies and investors to take actions that have an impact on the environment is outpacing the capacity of governmental institutions to manage the processes that cause those impacts. An 'institutional deficit' is evolving as national governments become increasingly constrained in their opportunities to impose environmental controls on companies. These constraints are both legal in character – as more and more competences and environmental policy measures become responsibilities of the EU rather than the member states themselves – and economic, as companies become increasingly capable of shifting production operations away from countries where operating costs are relatively high and expansion opportunities are limited, both of which are adversely affected by strict environmental controls.

These developments are not being matched by a proportionate strengthening of international government institutions. However, companies are becoming increasingly aware of the need to strengthen their relationships with their consumers. Internationally renowned brand names can be worth billions of dollars: the potential economic damage to transnational companies from behaviour that is widely perceived to be socially or environmentally irresponsible is enormous – witness the experience of Shell with the Brent Spar oil platform example – particularly given the fact that the investment required to attract a new customer is on average four to five times that required to keep the custom of an existing consumer.

The second key driving force on institutional change in Europe is EU enlargement. A Union with, shortly, 25 member states, in the medium term with perhaps 27 and in the longer term with possibly 35 will impose substantially different needs and pressures on the structure and working of EU institutions. These changing needs and pressures are already the subject of serious attention and analysis – as demonstrated by the focus of the recent EU Intergovernmental Conference on institutional reform in Nice. The enlargement will inevitably have three important consequences for EU institutions: first, it will increase even further the already substantial degree of diversity within the EU – cultural traditions and perceptions, political regimes, socio-economic profiles, institutional structures and processes, environmental conditions – with a concomitant decline in Community cohesion; second, the greater number of actors will complicate even further Community decision-making procedures and the allocation of competences; and, third, the Union will face even greater challenges in ensuring that EU measures are appropriately, consistently and promptly implemented across a greater number and a more diverse family of member states.

A third driving force of enormous potential impact is improved scientific understanding of the greenhouse effect – 'potential' because it remains to be seen what results further research on climate change will produce. However, it is almost certain that improved atmospheric models will further clarify the relation between emissions of greenhouse gases and climate change. It is also likely that in the foreseeable future further studies will confirm to an acceptably high level of confidence the current hypotheses on climate change. Associated advances may make it feasible to construct more detailed and confident prognoses of global and regional climate changes and to demonstrate a credible causal link between climate change and specific natural disasters. The impact of such developments on public, political and corporate perceptions would be enormous. One conceivable consequence would be the formation of a vocal and active constituency comprising the existing and potential victims of climate change and popular opinion that accepts the reality and seriousness of the threat. Such a coalition might be sufficient to drive a process that could lead to a global climate regime of far greater substance and enforceability than Kyoto. The '30% Club' and its role in promoting acid-emission control measures is in this respect an instructive precedent.

The fourth driving force that is likely to have a substantial impact on

European institutions is the pressure of popular opinion that will follow from changing values and perceptions. The values of European societies are evolving rapidly in response to economic and communication developments: traditional forms of social organization – families, communities, religious groupings, trade unions – are being superseded by common-interest networks which are more specialized, more extensive, more informal, more flexible, more transitory and more consumer-oriented in character. The traditional structures through which groups have secured representation and promoted their interests in societal and political processes are degrading and being replaced by a new evolution in civil society in which network interests are being organized and promoted through more direct, focused and flexible means, such as television and the Internet. With a greater proportion and volume of goods and services being provided through competitive markets, citizens are also becoming increasingly aware of their power as consumers. This development can be linked to a growing demand for more responsive democratic institutions which do not necessarily follow the ground rules of traditional government organizations, not least at the European level. In the longer term, these developments will ensure that the way in which societal values impact on those public and private decision-making processes that shape actions which have consequences for environmental quality will operate very differently from current mechanisms.

4 INSTITUTIONAL IMPACTS

The impacts of these driving forces on institutions in Europe will be profound. The precise impacts, however, cannot be predicted with any degree of accuracy in the medium-to-long term: one of the most important lessons of recent European history is how events that are unexpected and to a large extent unpredictable can transform apparently highly resilient political and socio-economic institutions. (Note that this is not to say that such 'surprises' cannot to a certain degree be anticipated.) As a consequence, any perspective on institutional change that measures its time horizon in decades cannot pretend to foresee the details of future institutional structures, competences and procedures. But although details can take on decisive importance in particular cases, it is the broad topography of the future institutional landscape of Europe and the main opportunities and obstacles that this implies for climate control actions that is important.

As a powerful and, by definition, global driving force, globalization will have far-reaching impacts at all institutional levels, across virtually all institutional sectors and on the relation between government and business. It will drive the process of harmonizing the economic policies and legislation of the main trading blocs – the EU, NAFTA/FTAA (North American Free Trade Agreement/Free Trade Area of the Americas) and parallel initiatives in Asia and Africa – and thereby reduce the scope for autonomous EU policy on many environmental issues. But this development is also likely to feed countervailing needs for, first, more explicit and more elaborate international rules on the scope for local, national or regional differentiation with respect to trade regulations and instruments where this is necessary in the interests of environmental protection and, second, more effective international enforcement regimes. The current proposal to establish a World Environment Organization is a reflection of such a perceived need. To be sure, business will be reluctant to accept a significant degree of regulatory differentiation, but in a parallel development companies themselves will appreciate the advantages of launching initiatives that demonstrate a high level of social and environmental responsibility, thereby strengthening consumer trust in particular brand names and, in the words of an IBM executive, securing a 'societal residents permit'.

For Europe itself, the greatest institutional impacts during the next two decades will in all probability follow from the EU enlargement process. The greater diversity, institutional complexity and implementation challenges that are the inevitable consequences of enlargement will drive EU policy making away from the traditional practice of negotiating highly specific and detailed regulations and directives; instead a far greater emphasis will be placed on framework measures that lay down basic rules and targets for a particular policy object but which allow the member states a greater degree of discretion in how the objectives are achieved and which instruments are applied for that purpose. This in turn infers a shift towards longer-term policy making and a need to develop policy frameworks, mechanisms and instruments that can prove to be effective in formulating and securing long-term goals and objectives. In other words, the achievement of clearly defined and enforceable ends – to an increasing extent directed at the medium-to-long term – will become a more prominent component of EU policy making rather than the precise means by which these ends are to be achieved. For environmental policy, this implies a shift to the formulation of locally or regionally appropriate environmental and ecological quality and performance targets rather than detailed emission or technological standards. Where feasible, groups of member states may establish particular forms of flexible cooperation, for example with regard to the use of certain economic instruments. As in the case of globalization, an important consequence of EU enlargement will therefore be a greater acceptance of more explicit rules concerning differentiated measures that are locally or regionally appropriate.

The impact of scientific research on institutional change is subject to greater uncertainties, in the main because it cannot be predicted how further research will alter our understanding of the greenhouse effect and the role of atmospheric emissions in the process. However, it is probable, if further research were to confirm the more pessimistic viewpoints – and almost certain if a number of conspicuous natural disasters were to be attributed to climate change – that the impact on public, corporate and political perceptions will be sufficient to drive changes to those institutions and mechanisms that are associated with the development of climate policies and control measures. That implies a strengthening of international mechanisms for dealing with global-commons problems or the creation of a dedicated and substantive global climate regime. Consumer pressure will also be such as to force business to demonstrate its environmental responsibility through initiatives that reduce the climate impact of branded products through innovations in product design and manufacture.

Perhaps the most interesting and potentially the most volatile driving force for institutional change in Europe is public perception. The unprecedented rate at which individual values, social organization and demographic patterns across Europe are changing suggests that the longer-term impacts on institutions could be profound. Some of the greatest impacts could result from the increasing need by individuals, groups and organizations – and, through continuing developments in information and communications technology, their expanding capability – to exert direct and multi-focused pressure on public and private institutions on matters of concern. The way in which two aspects of this process will operate are of special interest and importance. First, an area in which potential tension will arise is the question of how these developments will interact with the cultural preference in the CEE countries for strong political institutions, particularly if and when these countries make up a substantial proportion of the EU member states. It should not be forgotten that the long road to EU institutional integration and reform will carry two-way traffic: impulses for institutional change will come from the new as well as the old member states. Second, it is instructive in considering the potential for EU institutional change to bear in mind the institutional revolution that is taking place in the countries of Central and Eastern Europe and the forces that triggered this revolution: the EU as a political entity, and certainly its institutions as a means of formulating and implementing socially relevant measures, have never captured and enthused the public imagination for any prolonged period of time. Indeed, today they fail to command widespread popular respect and support. This makes the institutional constructs of the EU particularly vulnerable to a capricious event that could catalyse public opinion in the same way that the relaxation by the Hungarian government of controls along the Austrian border for citizens of the German Democratic Republic and the subsequent fall of the Berlin Wall inspired popular imagination in 1989.

The potential consequences of a crisis in the democratic accountability of EU governance could be enormous – and the eventual outcome in terms of institutional reform is certainly unpredictable.

A final observation can be made concerning the significance of these institutional impacts for the future of climate control actions. Most of the projected consequences offer new and interesting opportunities for securing mitigating measures. For example, policy harmonization between the major trading blocs will require new global institutional mechanisms – possibly including an enforcement regime – which could be exploited for environmental purposes, particularly if new rules are agreed defining the scope for differentiation in the interests of environmental protection; companies will better appreciate the advantages of taking voluntary environmental initiatives; EU policy making will feature a more prominent long-term dimension, will focus more on the definition of enforceable environmental targets and will offer greater scope for flexibility by the member states in the choice of measures appropriate to achieve those targets; credible scientific evidence confirming and clarifying greenhouse processes, if forthcoming, would lead to greatly increased pressure on government and business to take effective action and to develop appropriate institutional arrangements; and changing public perceptions on democratic accountability and the legitimacy of EU institutions could force radical changes in the architecture of European governance.

To be sure, these are foreseeable opportunities that are likely to be created by the institutional impact of driving forces that show evidence of persisting into the medium term at the very least. In that sense they represent a surprise-free scenario. But Europe's future will not be surprise free – the continent has for centuries proved to be a remarkably complex and dynamic entity. In this respect, at least, the future holds no surprises, for the process of radical institutional change across the continent will certainly continue well into the 21st century; indeed, it is difficult from the perspective of 2001 to foresee a time when this process will stabilize or to predict the form into which Europe's institutional architecture will eventually evolve. That this process receives so little popular attention can only be accounted for by two factors: that Europeans have become accustomed to institutional change and, particularly in the EU, that citizens are not actively involved in shaping the process, which remains essentially an intergovernmental matter. But, as noted above, it would be naive to assume that both of these conditions will persist indefinitely. When either or both cease to apply, the institutional outcome will be unpredictable and possibly, as so often in the past, chaotic. The importance of developing response strategies as a means of exploiting events that may not be predictable but can at least be anticipated cannot be underemphasized.

NOTE

* The author is grateful to Tim O'Riordan, Magnus Andersson and Willemijn Tuinstra for their comments on a draft of the chapter. The preparation of an earlier version of the paper was funded by the Dutch national research programme on global air pollution and climate change (NOP) as a contribution to the COOL project (Climate OptiOns for the Long term).

Index

accounting system for emissions 299
Adaptation fund 35
adaptation tax 34–5
additionality 177
administrative costs 54
Agreement on Subsidies and Countervailing Measures (SCM) Agreement 84, 85–7, 99
Agreement on Technical Barriers to Trade (TBT) 87
  Article 2 96
  principles 89
agricultural policies 204
agriculture, Netherlands 302
agroforestry case study, Mexico 189–90
air pollution 315
ALARA principle 9, 58, 139, 140–143, 154
allowance factors 286–7
alternative design for national permit trading 53–5
Alternative Energy Requirement, Ireland 10
Annex I countries 25, 128, 249
  estimated reduction requirements 29–30
Arentsen, M. 329
Armington assumption 255
assigned amount 78
assigned amount units (AAUs) 25, 33, 38
ATIS (Advanced Travel Information System) 322
auctioning of permits 49, 50, 117, 150
  compared with grandfathering 110
Austria
  feed-in tariffs 215–16
  grants for promotion of electricity from RES 220
  'green electricity' 212
  revenues from RES promotion 218
automobile industry, response to climate change issues 318–19

Bandsma, J. 177, 182
Banister, D. 313
Basel Convention on Transboundary Movements of Hazardous Waste 75
BAT (best available techniques) principle 9, 58, 131–2, 133, 139, 140–143
  and European environmental law 142–3
battery powered vehicles 331
Begg, K.G. 250, 263
Belgium, grants for promotion of electricity from RES 220
Benitez, P. 188
Berlin Mandate 234
biomass 197, 198, 202, 293, 303
Birds Directive 343
Bode, J.W. 236
Bohm, P. 46, 48, 53, 59, 111
Bollen, J. 32, 250, 252, 258, 260
Bonn Agreement on climate change 147, 225
Boom, J.T. 50, 57, 62
Bosi, M. 177, 182
Boyer, J.G. 28
Brundtland Report 19
burden sharing 280
  and moral ambiguity 281
Bush, G.W. Jnr 70
business as usual (BaU) 250, 256

cap-and-trade schemes 5, 23, 46–55, 56, 60
Capros, P. 108
caps 297
  lowering 153
car industry, response to climate change issues 318–19
car manufacturers' R&D strategies 317–18

carbon leakages 283–4
Carraro, C. 248, 252
Castells, N. 273
catalytic converter technology 322
CDM (clean development mechanism) 4, 12, 25, 36–7, 249, 250–251, 282
 baselines 270–271
 case studies 181–90
 criteria for certification 172
 definition 172
 economic effects 266–70
 Egyptian case study 185–8
 formulation 262–3
 and less developed countries 26
 macroeconomic indicators 268
 Netherlands 9–10, 171–94
 simulation studies 180
 tax on 35
Central and Eastern Europe (CEE) 341
 EU membership 343
certified emissions reductions (CERs) 25, 34, 37, 262
chlorofluorocarbons (CFCs) 3342
CICERO (Centre for International Climate and Environmental Research Oslo) 14, 285
cigarette case, Thailand 92–3
Cini, M. 118, 119, 123
Clean Development Fund 34
clean development mechanism see CDM
'clean' technology 135
Climate Action Network (CAN) 241
climate change
 technological options 293
 and trade 12–13
 and transport 18–20
climate change negotiation and initiatives 70–73
climate change policy, and institutional change 20–22
Climate OptiOns for the Long-term (COOL) project 16, 292–3, 294, 301, 305–8
climate policies 16
climate-friendly subsidies 84–7
CO_2 303
 removal and storage 305
 setting price for 300
 see also emissions: emissions trading
CO_2 emissions 262

CO_2 emissions trading, Netherlands 134–43
CO_2 reduction, Netherlands 128–46
collective transport systems 332
Collingridge, D. 329
COM 108, 109, 113, 116, 119, 120, 121, 123, 124
commitment thresholds 286–7
competition 47
competitive distortion concept 109–12
compliance and enforcement provisions, in emissions trading schemes 153–4
compliance rules 38
Conference of the Parties (COP) 6, 22, 34, 68, 80, 81–2, 114, 171, 233, 249, 315
Convention on International Trade in Endangered Species (CITES) 74–5
'cool air' 34
COP/MOP 80, 82
cost effectiveness 283–4
cost-even mechanism 151
Cox, R.W. 231
CPB (Centraal Planbureau, Netherlands) 171
Cramton, P. 49, 53
credit trading 46, 55–9
 and international permit trading 59–63
crediting periods 36–7
cross-technical linkages, transport 326–7
Cupta, J. 236

de Groot, H.C.F. 252
de Jong, B.H.J. 190
de Leeuw, G.J. 181
deep purse theory 110
demand management instruments, and electricity 211–17
den Elzen, M. 280
Denmark
 feed-in tariffs 216
 'green electricity' 212, 214
 permit trading 120–123
 RES policy 219
 RES promotion 218
developing countries 279
Dijkstra, B.R. 49, 57, 62
distributive justice, in international climate change negotiations 280–82
Dolman, A.J. 232

Dorda, A. 314
downstream approach to national permit trading 50–51
draft emission trading guidelines 82
Duijse, P. van 53
Dutilh, C. 240
dynamic efficiency 283–4

Ebert, U. 57
EC Treaty 112–14, 124, 133
ECN (Energy Research Centre of the Netherlands) 14, 29, 179, 180, 285
Economic Modelling Forum 30
Economies in Transition (EIT) 251
Ecuador, waste management case study 188–9
efficiency 303
and equity 283
Egypt
fuel switch case study 185–8
baseline approach 185, 187
emission reductions and costs 187–8
electric propulsion 330
electric vehicles 320–321, 326–7, 332
electrical equipment, case study on application of international trade law 90–96
electricity
cost 207
demand management instruments 211–17
from RES 208–11, 222–3
instruments of promotion 208–11
'side effects' 225
supply management instruments 217–21
electricity markets, liberalization 205–8, 224
Ellerman, A. 46
Ellis, J. 177, 182
Elzen, B. 315
emission reduction units (ERUs) 25, 39
emission reductions 78, 81
cost 178–80
international market 26–8
supply and demand 27
emissions
international exchange 22
per capita allowances 14

'emissions ceiling' 135
emissions trading 4, 25, 135–6, 249, 295–6
alternative design options 5, 45–67
competition distortion and state aid 6–7
identification of parties 297–8
and international trade law 6
legal aspects 138
legal considerations 149–54
legal feasibility 147–70
national rules 83
prerequisites 296–7
emissions trading schemes
compatibility with existing environmental statutes 154
compliance and enforcement provisions 153–4
structure and method 148–9
emissions trading systems, national rules on 97–9
Endangered Species Act, US 95
energy, from renewable sources 195–229
energy efficiency 128, 218
Energy Journal 260
Environment News Service 35
environmental law 138
environmental law system, Netherlands 139
environmental non-governmental organizations (ENGOs) 11, 235, 237, 239, 241, 242, 243
environmental permits, Netherlands 139
Environmental Protection Agency (EPA), US 55, 70
equality, development from grandfathering 285–6
equality of rights 280–281
equity
and competitive distribution 111–12
and efficiency 283
equity issues, for global climate policy regimes 290
ERUPT (Emissions Reduction Procurement Tender), Netherlands 173, 174, 176, 178
EU 4, 6, 11–12, 23
Ad Hoc Group on climate 236, 237
burden sharing agreements 128, 233, 236–8

CDM 267
Central and Eastern Europe (CEE)'s
    membership 343
  and climate change 342
  climate change policy 71–2
  cross border trading systems 108
  energy consumption and production
    1980–1999 196
  energy imports 197
  enlargement 340, 347
    and European institutional change
      345
  environmental law 342
  greenhouse gas reduction targets 238
  IEM Directive 206, 208, 210, 216
  'implementation deficit' 343
  institutional impacts 21–2
  permit allocation 108
    and competitive distortions 109–12
    political analysis 118–24
  R&D transport programmes 322
  renewable energy, in inland energy
    consumption 199
  renewable energy production 200, 202
  renewable energy sources (RES)
    195–229
    statistical overview 195–203
  renewables in total energy balance
    198
  state aid, and permit allocation
    112–18
  'Towards a European Consensus'
    workshop 233–4
EU Troika 232
Europe
  environmental law, BAT principle
    142–3
  institutional change 340–350
    and globalization 344–5
    and the greenhouse effect 345
    and popular opinion 346
European Climate Change Programme
  (ECCP) 71, 119, 148, 162
European Commission 7, 52, 108, 113,
  116
European Commission Green paper 136
European Convention of Human Rights
  (ECHR) 151
European Court of Justice 113, 114–15
European Environmental Council 234

European Parliament (EP) 124
European Union see EU
EU–Netherlands Kyoto protocol
  negotiation 230–247
EVA (Austrian Energy Agency) 210
existing rights, withdrawal 151

feed-in tariffs 214–17
  Austria 215–16
  Denmark 216
  Germany 215
  Portugal 216–17
  Spain 216
Feitelson, E. 313
Festa, D.H. 50, 53
Figueres, C. 32
Finger, M. 243
Finland, grants for promotion of
  electricity form RES 219–20
fiscal instruments 217–18
  and RES 205
Fischer, C. 53
fluorocarbons 156
Folketinget 120
fossil fuel sales 48–9
fossil fuels 181
France
  asbestos products 89
  public service 208
free riding 300
Freedman, R. 124
Friends of the Earth 235, 241, 315
fuel cell electric vehicles (FCEVs) 318
'fungibility' 27

G77 11, 241, 242
game strategies 248
Gasoline dispute, US 92, 93, 94
gateway technology 331
GATT 74, 75, 77, 78
  Article III 91–2, 98
  Article XI 91–2, 97
  Article XX 91, 97
    environmental exemptions 92–5
    principles 88–9
GATT/WTO law, compatibility with
  unilateral national measures 84–99
Geels, F. 314, 325
general exemptions 80
general technical standards 150–151

Germany
  feed-in tariffs 215
  RES promotion 218–19
Glaverbel case 115
global burden sharing 279–91
  rules 279
global climate policy regimes, equity
    issues 290
Global Commons Institute 285
Global Environment Facility 35
global environmental change regimes
    248–75
Global Trade Analysis Project 251, 252
globalization 21
  and European institutional change
    344–5, 346–7
Goulder, L.H. 110, 115
governance arrangements, in transport
    314–23
government
  role in emissions trading 297–8
  and societal demand for climate
    policies 299–300
grandfathering 6–7, 14, 15, 49, 50, 51,
    53, 62, 98, 108–27, 138, 150
  compared with auctioning 110
  development to equality 285–6
  disadvantages of 110
  fairness, and the polluter pays
    principle 162–3
  and new entrants 281
  as state aid 114–16
'green cars' 317–18
'green electricity' 211, 224
  Austria 212
  certificates 213–14
  Denmark 212
  Ireland 212–13
  Italy 211
  Netherlands 211–12
  UK 212
greenhouse effect 148
  scientific understanding and European
    institutional change 345, 347–8
greenhouse gas emission trading within
    the European Union, Green paper
    108
greenhouse gases (GHGs) 25
  emission reduction 78
  reduction targets, EU 238

Greenpeace 235
Groeneberg, H. 284
Grubb, M. 31, 33, 34, 36, 231, 252
GTAP model 12, 13
  economic impacts 260
  effects without CDM 257–62
GTAP-CDM model 251–2, 252–6
GTAP-E model 252, 253
  constant elasticity of substitution
    (CES) 253, 254
  production of energy in 255
Gupta, J. 34, 231, 284

Haas, E.B. 249
Hahn, R.W. 46, 59, 62
Haites, E. 29, 32, 37, 38, 39, 120
Hargrave, T. 47, 48, 49, 50, 51, 52, 53,
    118
Hertel, T.W. 252, 253
Higgott, R. 243
Hildebrand, D. 123
Hill, J.N. 30
Hischemöller, M. 293
historic claims 281
historical contribution to the problem
    280
Hoogma, R. 314, 315, 320, 322, 331
'hot air' 32–4
  restrictions 33–4
  trading 33
Hourcade, J.-C. 31, 33, 34, 252
housing sector, Netherlands 301, 302
Huber, M. 233
hybrid approach 15–16
hybrid forms, in transport 326–7, 331
hybrid schemes, national permit trading
    51–2
hydropower 197, 204, 205

imperfect competition 112
imports
  and subsidies 85
  of trading permits 32
'industrial rationalization' 158–9, 160
industry and energy sector, Netherlands
    301, 302
initial reduction targets 281–2
institutes 240
institutional change
  and climate change policy 20–22

Europe 340–350
and scientific research 347–8
institutions 240
Intergovernmental Panel on Climate Change see IPCC
internal combustion engine (ICE) vehicles 317
international competitiveness 261
international coordination 23
international credit trading 61–3
international emissions trading 22–3, 45, 46, 250
International Energy Agency (IEA) 32, 251
international environmental agreements 248
International Human Dimensions Programme of Global Environmental Change (IHDP), Institutional Dimensions of Global Environmental Change conference 4
international market, emission reductions 26–8
international multilateral negotiation, in leadership models 231–2
international permit trading 59–61
and credit trading 59–63
international regime building 230–32
international tradable environmental permits 155–6
international trade 88–9
international trade conflicts 70, 80
international trade law 69
  case study on electrical equipment 90–96
  and emissions trading 6
  and MEAs 74–8
  and national rules on emission trading systems 97–9
  principles 87–8
intervention approaches, transport 313
IPCC 4, 36, 71, 303, 318
  Third Assessment Report 302
IPCC Directive, Integrated Pollution Prevention and Control 131, 133, 139, 142
Ireland
  Alternative Energy Requirement 10
  'green electricity' 212–13

Italy, 'green electricity' 211, 213

Jansen, J. 285
Janssen, J. 62
Jenkins-Smith, H.C. 313
Jensen, J. 110, 250
Jepma, C.J. 47, 48, 49, 50, 51, 52, 53, 111, 113, 171
JIN (Joint Implementation Network) 173, 178, 182
JIRC (Joint Implementation Registration Centre) 178
Johannesburg World Summit on Sustainable Development (WSSD) 73
joint implementation (JI) 4, 25, 37, 46, 147, 249

Kamerstuk 56
Kemp, R. 314, 316, 322, 325, 329
Kerr, S. 49, 53
Kirsch, D. 321
Klaassen, G. 50
Koutstaal, P.R. 49, 51, 53, 112
Krasner, S.D. 249
Kremers, H. 251
Kyoto Conference 241–2
Kyoto mechanisms
  design aspects 30–39
  quantitative evidence on role of 28–30
Kyoto Protocol 3
  instruments 4–5
  negotiation process 232–42
  provisions 78–84
  ratification 73

land use, land use change and forestry (LULUCF) 36
Larsson, K. 71
leadership models, in international multilateral negotiation 231–2
leadership theory 230, 231
Lefevere, J. 111
Leontief form 253
less developed countries, and CDM 26
level playing field approach 111–12
liability provisions 38
like products concept 89–90
Lindblom, C.E. 58

long-range transboundary air pollution
(LRTAP) 248–9
Luxembourg Environment Council
conclusions 71
Lyon, R.M. 110

MacCracken, C.N. 33
McGowan, L. 118, 119, 123
Manne, A.S. 33
marginal costs 45
marginal costs of abatement 25
Marrakesh Accords 3
Massachsetts Institute of Technology
(MIT) 32
MEAs *see* multilateral environmental
agreements
Meeting of Parties (MOP) 34
Metz, B. 280
Mexico, agroforestry case study 189–90
Meyers, S. 177, 182
Michaelowa, A. 31, 177, 183
'middle power' countries 231
mixed systems, of national permit
trading 52–3
mobility management 333
modes of governance, transport 313
modulation, in transport 328–36
modulation approach 316–17
Mokyr, J. 331
Mom, G. 321
Montreal Protocol 8, 76, 155–6, 159,
160, 191, 342
moral ambiguity, and burden sharing
281
most-favoured nation (MFN) principle
88, 89
multi-sector convergence approach
(MSC) 13–16, 279–91, 285–9
numerical illustration 287–9
persuasive function 289–90
multilateral environmental agreements
(MEAs) 69
and international trade law 74–8
and WTO law 74
multilateral negotiations 10–12

national climate change measures 74–84
National Dialogue Integration Workshop
17–18
National Energy Policy, US 71

national permit trading
alternative design 53–5
with a cap 46–55
credit trading 55–9
criteria for evaluation 47–8
design of schemes 47–55
downstream approach 50–51
hybrid schemes 51–2
mixed systems 52–3
upstream approach 48–9
national policies and programmes 79–80
national rules, on emission trading
systems 97–9
national treatment principle (NT) 88
NATO 340
negotiation process, Kyoto Protocol
232–42
negotiations, horizontal versus vertical
284
Nentjes, A. 46, 50, 53, 54, 59, 62, 109,
110, 111
Nester, W. 319
Netherlands 11–12
Activities Implemented Jointly
programme (AIJ) 174
pilot phase 176
agriculture 302
benchmarking agreement on energy
efficiency 132–4
CDM 9–10, 171–94
tender procedure 173, 176
under Ministry of Development
Cooperation 175
Climate OptiOns for the Long-term
(COOL) project 16, 292–3, 294,
301, 305–8
climate policies 173–6
Climate Policy Implementation Plan
72
CO_2 emissions trading 58, 134–43
CO_2 reduction 128–46
contribution to EU burden sharing
agreement 128, 236–8
cost of emission reductions 178–80
Council of State 141–2, 154
domestic measures 129
'DOMILO' 239
'DOMILO PLUS' 239, 242
emissions reduction 173
energy transition 292–311

environmental covenant 243
environmental law system 139
Environmental Management Act 132, 133, 138, 139, 141, 154
environmental permits 139
ERUPT (Emissions Reduction Procurement Tender cross) 173, 174, 176, 178
evaluation of energy saving plans 131
government conditions for flexible instruments 173
'green electricity' 211–12, 213
greenhouse gas reduction targets 238
housing sector 301, 302
implementation of Kyoto Protocol 8–9
industry and energy sector 301, 302
institutional basis 243
integrated assessment of environmental impacts 139–40
JI 176
multiannual agreements on energy saving 131–2
National Climate Policy Implementation Plan 129
National Dialogue 301
national environmental law and the ALARA principle 141–2
national environmental policy plan 131
National Environmental Policy Plan (NEPP) 239
NO_2 emissions trading scheme 56, 140, 151
NOVEM 131, 132, 210, 323
polder model of consultation and accommodation 299, 301
presidency of the European Council 234–6
quality of emission reduction 176–7
total energy principle 130
tradable NOx emission reductions 136–8
transport 18, 315–16
transport associations 315
transport technology policy 323–5
voluntary environmental agreements 130–34
VROM (Ministry of Housing, Spatial Planning and the Environment) 173, 174, 175, 177, 178, 235, 239

Netherlands–EU Kyoto Protocol negotiation 230–47
New International Economic Order 232
NGOs (non-governmental organizations) 152–3, 293
Nijkamp, P. 273
non-compliance 38
non-ratification of Kyoto Protocol, impact 39–41
Nordhaus, W.D. 28
Nouwen, P.A. 324

OECD 171, 177, 182
Official Journal of the European Communities (OJ) 109, 112, 114, 115, 117
Olson, M. 49
opportunity costs 111
ozone-depleting substances (ODSs) 8
  administrative interventions with trades 164–5
  compliance and enforcement provisions 165–6
  criteria for transfer of emission rights 158–9
  European permit market 155–62, 167
  allocation of rights 157–8
  compliance and enforcement provisions 161–2
  trades within member states 159–60
  transfers between member states 160
  transfers with third parties 160–161
  phasing out 163–4
  trading 148–9

Parry, I.W.H. 250
parts of assigned amounts (PAAs), trading in 82–3
pathway technologies 330
permit allocation
  and competitive distortions, EU 109–12
  perceptions of political decisions 118–24
  political analysis 118–24
  political support in favour of harmonizing 123–4
  and state aid 112–18
    exemptions 114–18

and state aid criteria 112–14
permit trading 46
  Denmark 120–123
  UK 120–123
Philippines
  emission reductions and costs 182–5
  renewable energy case study 181–5
  baseline 181, 182, 184–5
Phylipsen, G.J.M. 236, 284
physical planning, for RES 204
policies and measures (PAMs) 74
policy belief systems 313
polluter pays principle 124, 280, 295–6
  and grandfathering 162–3
popular opinion, and European institutional change 346
Portugal, feed-in tariffs 216–17
predatory pricing theory 110
Princen, T. 243
prisoner's dilemma 274
private emissions trading 108
private law, and tradable rights 149–50
private parties 298, 300
private party credit trading 63
private party trading 45–6, 60
product standards 87
programmes and measures (PAMs) 80, 81
'promise-requirement' cycles 331–2
public perception 22
  and European institutional change 348–9
public R&D programmes 321–3
public service obligation (PSO) 208

quantified emissions limitation and reduction commitment (QELRCs) 25, 68, 78, 232, 238, 244

R&D 303
  public programmes 321–3
  and transport 317–18
Rasmussen, T.N. 110, 250
ratification of the Kyoto Protocol 73
Reconsidering Energy Policy group 302
regional organization 230, 242–4
regulation 294–5
regulatory frameworks 70
regulatory standards 79

regulatory tools 142
relative standards 55, 57
renewable energy 195–229
  Philippines 181–5
renewable energy sources (RES) 10
  electricity from 205–8, 208–11, 222–3
  EU 195–229
  statistical overview 195–203
  expansion of electricity generation capacity 219
  and fiscal instruments 205
  instruments of promotion 208–11
  level playing field for promotion 218–19
  and physical planning 204
  policies for 203–11
  raising revenues from promotion of 218
renewable obligations 211–13
research and development see R&D
research and technical development (RTD), grants 204
Richels, R.G. 33
Rietveld, P. 53, 54, 62
Ringius, L. 233, 234, 237, 280
Ringius, R. 284
Rio Declaration 152
Rip, A. 314, 316, 325, 329
Rose, A. 280
Rotmans, J.R. 299, 329, 334
Russia 45, 73
Rutten, G.J. 133

Sabatier, P.A. 313
Scarbrough, E. 119
Schmalensee, R. 50
scientific research, and institutional change 347–8
SCM Agreement see Agreement on Subsidies and Countervailing Measures
sectoral allocations 286
security issues 319
seller versus buyer liability 60
Senter International 171, 173, 176
shared responsibility 16–17, 298–301
Shrimp-Turtle dispute 9, 94
Sijm, J.P.M. 179, 285
Simon, B. 322
Single European Act 341

sinks 280
  inclusion of 36
Sjostedt, G. 242
societal demand for climate policies, and government 299–300
Spain
  feed-in tariffs 216
  grants for promotion of electricity from RES 220
Special Climate Change Fund 35
stakeholder assumptions 301–7
state aid, and permit allocation, EU 112–18
state aid concept 113
Stavins, R.N. 46, 50, 59, 62
Stewart, R. 172, 177, 178
strategic niche management 314, 329–30
subsidies 84–7
  actionable 85–6
  definition 84–5
  legality of national programmes 85
  non-actionable 86
supplementarity 173
supplementarity requirements 31–2
  approaches to setting 31
  negative impacts 32
supply management instruments, electricity 217–21
sustainable development 173, 177, 178
sustainable transport 335
Svendsen, G.T. 49, 57
Sweden 71
  grants for promotion of electricity from RES 220
  system innovation programmes 332

taxes 52, 87
TBT see Agreement on Technical Barriers to Trade
technical change 314
technological options, and climate change 293
Thailand, cigarette case 92–3
third parties (non-governmental organizations) 152–3
'30% Club' 345
Tietenberg, T.H. 50
tradable NO_x emission reductions, Netherlands 136–8
tradable permit schemes 46

tradable permits, imports of 32
tradable permits market, start up 150–151
tradable right, legal qualification of 149–50
trade, and climate change 12–13
trade disputes 79, 80
transaction/institutional costs 38–9
transboundary policy 248
transfers, administrative criteria for intervention 151–2
transition management 329, 333–6
transport 312–39
  alternative modes 319
  battery powered vehicles 331
  car industry 317–19
  catalytic converter technology 322
  and climate change 18–20
  collective 332
  cross-technical linkages 326–7
  electric vehicles 320–321, 326–7, 330, 332
  fuel cell electric vehicles (FCEVs) 318
  governance arrangements in 314–23
  hybrid forms 326–7
  instruments to reduce emissions 312–13
  intervention approaches 313
  modes of governance 313
  modulation in 328–36
  Netherlands 18, 315–16
  patterns and change mechanisms 325–6
  and R&D 317–18
  sustainable 335
  US 319
transport development, and external developments 327–8
transport regimes, governance arrangements in 319–20
transport systems, heterogeneity 328
transport telematics 330
Treaty of Aarhus 152
Treaty of Rome 21, 341
Triptych approach 11–12, 14, 236–7, 243, 284–5
Truong, T.P. 251, 252, 254, 255, 257, 258, 260
Tsigas, M. 252, 253

Tuna/Dolphin case, US 74, 93–4
UK
  electricity from RES 209, 210
  'green electricity' 212, 214
  lean burn engine technology 322–3
  permit trading 120–123
  revenues from RES promotion 218
Ulph, A. 62
UN (United Nations) 173
UNCTAD 50, 58
Underdal, A. 230, 231
UNEP 34
UNFCCC (United Nations Framework Convention on Climate Change) 3, 11, 35, 36, 60, 61, 68, 78, 80, 81, 128, 171, 173, 177, 233, 243, 249, 279
unilateral leadership 230
unilateral national measures, compatibility with GATT/WTO law 84–99
Unruh, G.C. 319, 331
upstream approach, national permit trading 48–9
US 250, 256
  California zero-emissions mandate 320–321
  Clean Air Act 70, 150
  climate change policy 70–71, 72
  emissions trading for air pollution 148
  Endangered Species Act 95
  Environmental Protection Agency (EPA) 55, 70
  Gasoline dispute 92, 93, 94
  in GATP model 256–7
  impact of non-ratification of Kyoto Protocol 39–41, 271–3
  and Kyoto Protocol negotiations 3
  National Energy Policy 71
  road building 319
  transport 319
  Tuna/Dolphin case 74, 90

van der Gaast, W.P. 111, 177

Van der Laan, R. 109, 111
van der Linden, N.H. 178, 180
van der Woerd, K.F. 319, 327
Van Deth, J.W. 119
van Harmelen, A.K. 284
van Ierland, E.C. 177, 181
Verhoef, E. 313
Vienna Convention 8, 77, 155, 156
voluntary environmental agreements, Netherlands 130–134
Vrolijk, C. 36
VROM (Ministry of Housing, Spatial Planning and the Environment), Netherlands 173, 174, 175, 177, 178, 235, 239

waste management case study, Ecuador 188–9
WCED (World Commission of Environment and Development) 319
Weber, M. 314
Welch, W.P. 113, 117
Weyant, J.P. 30
Williams, R.C. III 250
Woerdman, E. 49, 109, 111, 118, 177
World Bank 180
World Trade Organization (WTO) 6, 69, 77, 80–81, 89
  Appellate Body 90
  Committee on Trade and Environment (CTE) 76–7
  Dispute Settlement Body 85
World Trade Organization (WTO) law
  compatibility with unilateral national measures 84–99
  and MEAs 74
WorldScan model 252
Worsley, R. 124
WWF (World Wide Fund for Nature) 235, 241

Yamin, F. 32, 60, 62, 111
Young, O.R. 230

Zhang, X.Z. 29, 33, 46, 53, 59, 60, 250